Light Alloys

To
Andrea, Sally and David

Light Alloys

From Traditional Alloys to Nanocrystals

Fourth edition

I. J. Polmear

Professor Emeritus and formerly Foundation Chairman,
Department of Materials Engineering, Monash University,
Melbourne, Australia

ELSEVIER

AMSTERDAM • BOSTON • HEIDELBERG • LONDON
NEW YORK • OXFORD PARIS • SAN DIEGO
SAN FRANCISCO • SINGAPORE • SYDNEY • TOKYO
Butterworth-Heinemann is an imprint of Elsevier

Butterworth-Heinemann is an imprint of Elsevier
Linacre House, Jordan Hill, Oxford OX2 8DP
30 Corporate Drive, Burlington, MA 01803

First published 1981
Second edition 1989
Reprinted 1993
Third edition 1995
Reprinted 1995
Transferred to digital printing 2003
Fourth edition 2006

British Library Cataloguing in Publication Data
A catalogue record for this book is available from the British Library

Library of Congress Control Number: 2005932508

ISBN 0 7506 6371 5

For information on all Elsevier Butterworth-Heinemann publications
visit our website at http://books.elsevier.com

Typeset by Integra Software Services Pvt. Ltd, Pondicherry, India
www.integra-india.com
Printed in United Kingdom

CONTENTS

PREFACE TO THE FIRST EDITION

The fact that the light metals aluminium, magnesium and titanium have traditionally been associated with the aerospace industries has tended to obscure their growing importance as general engineering materials. For example, aluminium is now the second most widely used metal and production during the next two decades is predicted to expand at a rate greater than that for all other structural metals. Titanium, which has a unique combination of properties that have made its alloys vital for gas turbine engines, is now finding many applications in aircraft structures and in the chemical industry.

Light alloys have never been the subject of a single book. Moreover, although the general metallurgy of each class of light alloys has been covered in individual texts, the most recent published in English appeared some time ago—aluminium alloys in 1970, magnesium alloys in 1966 and titanium alloys in 1956. Many new developments have occurred in the intervening periods and important new applications are planned, particularly in transportation. Thus it is hoped that the appearance of this first text is timely.

In preparing the book I have sought to cover the essential features of the metallurgy of the light alloys. Extraction of each metal is considered briefly in Chapter one, after which the casting characteristics, alloying behaviour, heat treatment, properties, fabrication and major applications are discussed in more detail. I have briefly reviewed the physical metallurgy of aluminium alloys in Chapter two although the general principles also apply to the other metals. Particular attention has been devoted to microstructure/property relationships and the role of individual alloying elements, which provides the central theme. Special features of light alloys and their place in general engineering are highlighted although it will be appreciated that it has not been possible to pursue more than a few topics in depth.

The book has been written primarily for students of metallurgy and engineering although I believe it will also serve as a useful guide to both producers and users of light alloys. For this reason, books and articles for further reading are listed at the end of each chapter and are augmented by the references included with many of the figures and tables.

The book was commenced when I was on sabbatical leave at the Joint Department of Metallurgy at the University of Manchester Institute of Science and Technology and University of Manchester, so that thanks are due to Professor K. M. Entwistle and Professor E. Smith for the generous facilities placed at my disposal. I am also indebted for assistance given by the Aluminium Development Council of Australia and to many associates who have provided me with advice and information. In this regard, I wish particularly to mention the late Dr E. Emley, formerly of The British Aluminium Company Ltd; Dr C. Hammond, The University of Leeds; Dr M. Jacobs; TI Research Laboratories; Dr D. Driver, Rolls-Royce Ltd; Dr J. King and Mr W. Unsworth, Magnesium Elektron Ltd; Mr R. Duncan, IMI Titanium; Dr D. Stratford, University of Birmingham; Dr C. Bennett, Comalco Australia Ltd; and my colleague Dr B. Parker, Monash University. Acknowledgement is also made to publishers, societies and individuals who have provided figures and diagrams which they have permitted to be reproduced in their original or modified form.

Finally I must express my special gratitude to my secretary Miss P. O'Leary and to Mrs J. Colclough of the University of Manchester who typed the manuscript and many drafts, as well as to Julie Fraser and Robert Alexander of Monash University who carefully produced most of the photographs and diagrams.

IJP
Melbourne
1980

PREFACE TO THE SECOND EDITION

In this second edition, the overall format has been retained although some new sections have been included. For the most part, the revision takes the form of additional material that has arisen through the development of new compositions, processing methods and applications of light alloys during the last eight years.

Most changes have occurred with aluminium alloys which, because of their widespread use and ease of handling, are often used to model new processes. Faced with increasing competition from fibre-reinforced plastics, the aluminium industry has developed a new range of light-weight alloys containing lithium. These alloys are discussed in detail because they are expected to be important materials of construction for the next generation of passenger aircraft. More attention is given to the powder metallurgy route for fabricating components made from aluminium and titanium alloys. Treatment of this topic includes an account of techniques of rapid solidification processing which are enabling new ranges of alloys to be produced having properties that are not attainable by conventional ingot metallurgy. Metal-matrix composites based on aluminium are also finding commercial applications because of the unique properties they offer and similar magnesium alloys are being developed. New methods of processing range from methods such as squeeze casting through to advances in superplastic forming.

In preparing this new edition, I have again paid particular attention to microstructure/property relationships and to the special features of light alloys that lead to their widespread industrial use. In addition to an expanded text, the number of figures has been increased by some 40% and the lists of books and articles for further reading have been extended. Once more, the book is directed primarily at undergraduate and postgraduate students although I believe it will serve as a useful guide to producers and users of light alloys.

I am again indebted for assistance given by colleagues and associates who have provided me with information. Acknowledgement is also made to publishers, societies and individuals who have provided photographs and diagrams which they have permitted to be produced in their original or modified form.

Finally I wish to express my gratitude to Mesdames J. Carrucan, C. Marich and V. Palmer, who typed the manuscript, as well as to Julie Fraser, Alan Colenso and Robert Alexander of Monash University who carefully produced most of the photographs and diagrams.

I. J. Polmear
Melbourne 1988

PREFACE TO THE THIRD EDITION

The central theme of the first two editions was microstructure/property relationships in which special attention was given to the roles of the various alloying elements present in light alloys. This general theme has been maintained in the third edition although some significant changes have been made to the format and content.

As before, much of this revision involves the inclusion of new material which, in this case, has arisen from developments during the seven years since the second edition was published. The most notable change in format has been to group together, into a new chapter, information on what have been called new materials and processing methods. Examples are metal matrix and other composites, structural intermetallic compounds, nanophase and amorphous alloys. Interest in these and other novel light alloys has increased considerably during the last decade because of the unceasing demands for improvements in the properties of engineering materials. Since light alloys have been at the forefront of many of these developments, the opportunity has been taken to review this area which has been the focus of so much recent research in materials science.

Another feature of the third edition is the greater attention given to applications of light alloys and their place in engineering. More case studies have been included, such as the use of light alloys in aircraft and motor cars. Economic factors associated with materials selection are also discussed in more detail. Moreover, since the light metals are often placed at a competitive disadvantage because of the high costs associated with their extraction from minerals, more attention has been given to these processes. This has led to a considerable increase in the size of the first chapter. Joining processes are described in more detail and, once again, service performance of light alloys is discussed with particular regard to mechanical behaviour and corrosion resistance.

As a result of these various changes, the text has been expanded and the number of figures has been increased by a further 20%. Lists of books and articles for further reading have been updated. While the book continues to be directed primarily at senior undergraduate and postgraduate students, I believe it will again serve as a useful guide to the producers and users of light alloys.

I am again indebted for assistance given by colleagues and associates who have provided me with information and helpful discussions. General acknowledgement is made to publishers, societies and individuals who have responded to requests for photographs and diagrams that have been reproduced in their original or modified form. Finally I wish to express my gratitude to my wife Margaret for her constant encouragement, to Carol Marich and Pam Hermansen who typed the manuscript and to Julie Fraser and Robert Alexander who once again carefully produced so many of the photographs and diagrams.

I. J. Polmear
Melbourne 1995

PREFACE TO FOURTH EDITION

Since the third edition of Light Alloys appeared in 1995, developments with new alloys and processes have continued at an escalating rate. Competition between different materials, metallic and non-metallic, has increased as producers seek both to defend their traditional markets and to penetrate the markets of others. New compositions of aluminium, magnesium and titanium alloy have been formulated and increasing attention has been given to the development of novel and more economical processing methods. Because of their ease of handling, aluminium alloys in particular have been used as experimental models for many of the changes. Recently, potential automotive applications have led to a resurgence of interest in cast and wrought magnesium alloys.

The central theme of earlier editions was microstructure/property relationships and particular attention was given to the roles of the various alloying elements present in light alloys. This general theme has been maintained in the fourth edition although further significant changes have been made to format and content. Special consideration has again been given to the physical metallurgy of aluminium alloys and many of the general principles also apply to magnesium and titanium alloys. The description of changes occurring during the processing of the major class of non-heat treatable aluminium alloys has been extended. Although a century has now elapsed since the discovery of age hardening by Alfred Wilm, new observations are still being made as the latest experimental techniques reveal more details of the actual atomic processes involved. As examples, more information is now available about the role of solute and vacancy clusters during the early stages of ageing, as well as other phenomena such as secondary hardening. Some success has been achieved with the modelling of precipitation processes. Precipitation hardening was hailed as the first nanotechnology and now it is possible to develop fine-scale microstructures in a much wider range of alloys through the use of novel processing methods.

Some new topics in this fourth edition are strip and slab casting, creep forming, joining technologies such as friction stir and laser welding, metallic foams, quasicrystals, and the production of nanophase materials. Economic factors associated with the production and selection of light metals and alloys are considered in more detail and information on recycling has been included. Sections dealing with the commercial applications of light alloys and their general place in engineering have also been expanded. This applies particularly to transportation as aluminium alloys face increasing competition from fibre-reinforced polymers for aircraft structures, and where higher fuel costs make both aluminium and magnesium alloys more attractive for reducing the weight of motor vehicles.

Because of these and other developments, the text has been updated and expanded by about 20%. A further 50 figures and several new tables have been added. References to original sources of information are shown with most figures and tables but are not included in the general text. Relevant articles and books for further reading have been revised and are listed at the end of each chapter. As originally intended, the book is directed primarily at senior undergraduate and postgraduate students, but it is also believed it will to serve as a useful general guide to producers and users of light alloys.

I am again indebted for the assistance given by colleagues and associates who have provided me with information and advice. In this regard, special mention should be made of J. Griffiths, E. Grosjean J. Jorstad, R. Lumley, J-F Nie, R. R. Sanders, H. Shercliff, J. Taylor and the Australian Aluminium Council. General acknowledgment is again made to publishers, societies and individuals who have responded to requests for photographs and diagrams. Facilities provided by the Department of Materials Engineering at Monash University have been much appreciated and, as with all other editions, many of the figures have been skilfully reproduced in their original or modified form by Julie Fraser. Finally I wish to express my gratitude to my wife Margaret for her constant support.

I. J. Polmear
Melbourne, 2005

I

THE LIGHT METALS

1.1 GENERAL INTRODUCTION

The term 'light metals' has traditionally been given to both aluminium and magnesium because they are frequently used to reduce the weight of components and structures. On this basis, titanium also qualifies and beryllium should be included although it is little used and will not be considered in detail in this book. These four metals have relative densities ranging from 1.7 (magnesium) to 4.5 (titanium) which compare with 7.9 and 8.9 for the older structural metals, iron and copper, and 22.6 for osmium, the heaviest of all metals. Ten other elements that are classified as metals are lighter than titanium but, with the exception of boron in the form of strong fibres contained in a suitable matrix, none is used as a base material for structural purposes. The alkali metals lithium, potassium, sodium, rubidium and caesium, and the alkaline earth metals calcium and strontium are too reactive, whereas yttrium and scandium are comparatively rare.

1.1.1 Characteristics of light metals and alloys

The property of lightness translates directly to material property enhancement for many products since by far the greatest weight reduction is achieved by a decrease in density (Fig. 1.1). This is an obvious reason why light metals have been associated with transportation, notably aerospace, which provided great stimulus to the development of light alloys during the last 50 years. Strength: weight ratios have also been a dominant consideration and the central positions of the light alloys based on aluminium, magnesium and titanium with respect both to other engineering alloys, and to all materials are represented in an Ashby diagram in Fig. 1.2. The advantages of decreased density become even more important in engineering design when parameters such as stiffness and resistance to buckling are involved. For example, the stiffness of a simple rectangular beam is directly proportional to the product of the elastic modulus and the cube of the thickness. The significance of this relationship is illustrated by the nomograph

I

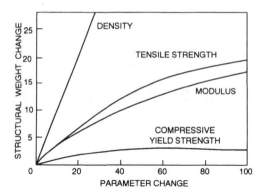

Fig. 1.1 Effect of property improvement on structural weight (courtesy of Lockheed Corporation).

shown in Fig. 1.3 which allows the weights of similar beams of different metals and alloys to be estimated for equal values of stiffness. An iron (or steel) beam weighing 10 kg will have the same stiffness as beams of equal width and length weighing 7 kg in titanium, 4.9 kg in aluminium, 3.8 kg in magnesium, and only 2.2 kg in beryllium. The Mg–Li alloy is included because it is the lightest (relative density 1.35) structural alloy that is available commercially. Comparative stiffness for equal weights of a similar beam increase in the ratios 1:2.9:8.2:18.9 for steel, titanium, aluminium, and magnesium respectively.

Concern with aspects of weight saving should not obscure the fact that light metals possess other properties of considerable technological importance, e.g. the high corrosion resistance and high electrical and thermal conductivities of aluminium, the machinability of magnesium, and extreme corrosion resistance of titanium. Comparisons of some physical properties are made in Table 1.1.

Beryllium was discovered by Vauquelin in France in 1798 as the oxide in the mineral beryl (beryllium aluminium silicate), and in emerald. It was first isolated independently by Wöhler and Bussy in 1828 who reduced the chloride with potassium. Beryl has traditionally been a by-product of emerald mining and was until recently the major source of beryllium metal. Currently more beryllium is extracted from the closely associated mineral bertrandite (beryllium silicate hydroxide).

Beryllium has some remarkable properties (Table 1.1). Its stiffness, as measured by specific elastic modulus, is nearly an order of magnitude greater than that for the other light metals, or for the commonly used metals iron, copper and nickel. This has led to its use in gyroscopes and in inertial guidance systems. It has a relatively high melting point, and its capture cross-section (i.e. permeability) for neutrons is lower than for any other metal. These properties have stimulated much interest by the aerospace and nuclear industries. For example, a design study specifying beryllium as the major structural material for a supersonic transport aircraft has indicated possible weight savings of up to 50% for

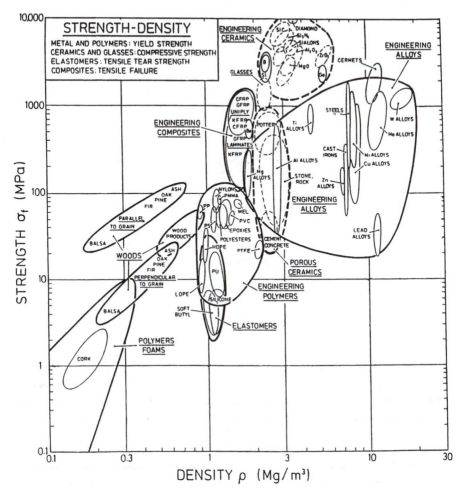

Fig. 1.2 Strength: density relationships for light alloys and other engineering materials. Note that yield strength in used as the measure of strength for metals and polymers, compressive strength for ceramics, tear strength for elastomers and tensile strength for composites (courtesy M. F. Ashby).

components for which it could be used. However, its structural uses have been confined largely to components for spacecraft, and for applications such as satellite antenna booms. In nuclear engineering it has had potential for use as a fuel element can in power reactors. Another unique property of beryllium is its high specific heat which is approximately twice that of aluminium and magnesium, and four times that of titanium. This inherent capacity to absorb heat, when combined with its low density, led to the selection of beryllium as the basis for the re-entry heat shield of the Mercury capsule used for the first manned spacecraft developed in the United States. In a more general application, it has served as

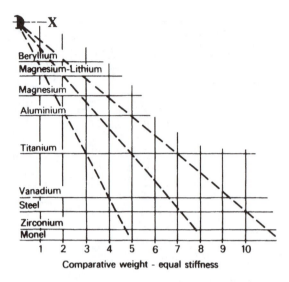

Fig. I.3 Nomograph allowing the comparative weights of different metals or alloys to be compared for equal levels of stiffness. These values can be obtained from the intercepts which lines drawn from point X make with lines representing the different metals or alloys (courtesy Brooks and Perkins Inc.).

a heat sink when inserted in the centre of composite disc brakes used in the landing gear of a large military transport aircraft. Beryllium also shows outstanding optical reflectivity, particularly in the infrared, which have led to its combat use in target acquisition systems, as well as in space telescopes.

Despite much research in several countries, wider use has not been made of beryllium because it is costly to mine and extract, it has an inherently low ductility at ambient temperatures, and the fact that the powdered oxide is extremely toxic to some people. The problem of low ductility arises because of the dimensions of the close-packed hexagonal crystal structure of beryllium. The c/a ratio of the unit cell is 1.567 which is the lowest and most removed of all metals from the ideal value of 1.633. One result of this is a high degree of anisotropy between mechanical properties in the a and c crystallographic directions. At room temperature, slip is limited and only possible on the basal plane, which also happens to be the plane along which cleavage occurs. Furthermore, there has also been little opportunity to improve properties by alloying because the small size of the beryllium atom severely restricts its solubility for other elements. One exception is the eutectic composition Be-38Al in which some useful ductility has been achieved. This alloy was developed by the Lockheed Aircraft Company and became known as Lockalloy. Because beryllium and aluminium have little mutual solid solubility in each other, the alloy is essentially a composite material with a microstructure comprising stiff beryllium particles in a softer aluminium matrix. Light weight (specific gravity 2.09) extrusions and sheet have found limited aerospace applications.

Table 1.1 Some physical properties of pure metals (from Lide, D.R. (Ed), *Handbook of Chemistry & Physics*, 72nd edn, CRC Press, Boca Raton, USA, 1991–92; *Metals Handbook*, Volume 2, 10th edn, ASM International, Metals Park, Ohio, USA, 1990)

Property	Unit	Al	Mg	Ti	Be	Fe	Cu
Atomic number	–	13	12	22	4	26	29
Relative atomic mass (C = 12.000)	–	26.982	24.305	47.90	9.012	55.847	63.546
Crystal structure		fcc	cph	cph	cph	bcc	fcc
a	nm	0.4041	0.3203	0.2950	0.2286	0.2866	0.3615
c	nm	–	0.5199	0.4653	0.3583	–	–
Melting point	°C	660	650	1678	1289	1535	1083
Boiling point	°C	2520	1090	3289	2472	2862	2563
Relative density (d)	–	2.70	1.74	4.51	1.85	7.87	8.96
Elastic modulus (E)	GPa	70	45	120	295	211	130
Specific modulus (E/d)	–	26	26	26	160	27	14
Mean specific heat 0–100 °C	$J\,kg^{-1}\,K^{-1}$	917	1038	528	2052	456	386
Thermal conductivity 20–100 °C	$W\,m^{-1}\,K^{-1}$	238	156	26	194	78	397
Coefficient of thermal expansion 0–100 °C	$10^{-6}\,K^{-1}$	23.5	26.0	8.9	12.0	12.1	17.0
Electrical resistivity at 20 °C	$\mu\,ohm\,cm^{-1}$	2.67	4.2	54	3.3	10.1	1.69

Note: Conversion factors for SI and Imperial units are given in the Appendix.

Beryllium is now prepared mainly by powder metallurgy methods. Metal extracted from the minerals beryl or bertrandite is vacuum melted and then either cast into small ingots, machined into chips and impact ground, or directly inert gas atomised to produce powders. The powders are usually consolidated by hot isostatic pressing (HIP) and the resulting billets have properties that are more isotropic than are obtained with cast ingots. Tensile properties depend on the levels of retained BeO (usually 1 to 2%) and impurities (iron, aluminium and silicon) and ductilities usually range from 3 to 5%. The billets can then be hot worked by forging, rolling to sheet, or extruding to produce bar or tube. Lockalloy (now also known as AlBeMet™ 162) is now also manufactured by inert gas atomisation of molten pre-alloyed mixtures and the resulting powders are consolidated and hot worked as described above.

1.1.2 Relative abundance

The estimated crustal abundance of the major chemical elements is given in Table 1.2 which shows that the light metals aluminium, magnesium, and titanium are first, third and fourth in order of occurrence of the structural metals. It can also be seen that the traditional metals copper, lead, and zinc are each present in amounts less than 0.10%. Estimates are also available for the occurrence of metals in the ocean which is the major commercial source of magnesium. Sea water contains 0.13% of this metal so that 1.3 million tonnes are present in each km^3, which is approximately equivalent to three times the annual world consumption of magnesium in 2004. Overall, the reserves of the light metals are adequate to cope with anticipated demands for some centuries to come. The extent to which they will be used would seem to be controlled mainly by their future costs relative

Table 1.2 Crustal abundance of major chemical elements (from Stanner, R. J. L., *American Scientist*, **64**, 258, 1976)

Element	% by weight
Oxygen	45.2
Silicon	27.2
Aluminium	8.0
Iron	5.8
Calcium	5.06
Magnesium	2.77
Sodium	2.32
Potassium	1.68
Titanium	0.86
Hydrogen	0.14
Manganese	0.10
Phosphorus	0.10
Total	99.23

to competing materials such as steel and plastics, as well as the availability of electrical energy that is needed for their extraction from minerals.

1.1.3 Trends in production and applications

Trends in the production of various metals and plastics are shown in Fig. 1.4 and it is clear that the light metals are very much materials of the twentieth century. Between 1900 and 1950, the annual world production of aluminium increased 250 times from around 6000 tonnes to 1.5 million tonnes. A further eightfold increase took place during the next quarter century when aluminium surpassed copper as the second most used metal.

During this period the annual rate of increase in aluminium production averaged 9.2%. Since the late 1970s the demand for most basic materials has fluctuated and overall annual increases have been much less (Fig. 1.4). These trends reflect world economic cycles, the emergence of China and the Russian Federation as major trading nations, and the greater attention being given to recycling. World

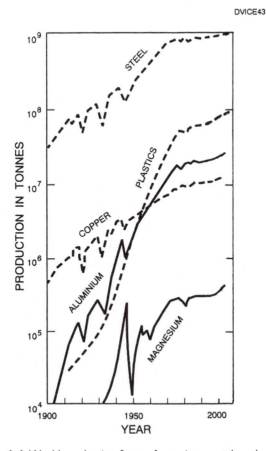

DVICE43

Fig. 1.4 World production figures for various metals and plastics.

production of primary (new) aluminium grew by an annual average of around 4% during the two decades 1980 to 2000 had reached an estimated 29.2 million tonnes in 2004. During this period, the production of secondary (recycled) aluminium is estimated to have risen from 3.8 to 8.4 million tonnes in 2004. Production of iron and steel, which still amounts to about 95% of all metal consumed, and which remained effectively static at around 700 million tonnes for a decade between 1980 and 1990, exceeded 1 billion tonnes in 2004. Here it is interesting to record that the cumulative total of aluminium that has ever been produced (estimated as being around 720 million tonnes up to 2004) is less than one year's supply of iron! It may also be noted that the production of commodity plastics and other polymeric materials, which enjoyed spectacular growth during the period 1950 to 1980, had reached an estimated 185 million tonnes by 2004.

Two political events have had a major impact on the production and pricing of the light metals in recent years. The first was the change from the former Soviet Union to the Commonwealth of Independent States which was followed by the release of large quantities of metals for sale in Western nations, often at discounted prices. More significant has been the rapid transition of China over the last 20 years from a largely agrarian society to an increasingly industrialised economy. China now leads the world in the production of aluminium with Russia placed second followed by Canada, the United States, Australia and Brazil (Table 1.3). Further major expansion is expected in China where the per capita consumption of aluminium is still low being only 3.4 kg in 2002 compared with 29.3 kg in the United States and 27.9 kg in Japan. China has also become the world's largest producer of magnesium and is rapidly increasing its capacity to produce titanium sponge.

Metal prices change from year to year and depend on factors that influence supply and demand. For example, the release of large quantities of aluminium by the C.I.S. that was mentioned above, combined with the effects of the world depression in the early 1990s, caused the price of aluminium to fall from an average of $US 1675 per tonne in 1990 to under $US1100 in 1993.

Table 1.3 World production of primary aluminium for the years 1996, 2000 and 2004 (from International Aluminium Institute, London)

	1996	2000	2004
Africa	1 015	1 178	1 711
Asia	1 624	2 221	2 735
China	?	2 794	6 589
Europe Western	3 192	3 801	4 295
Europe Eastern (includes Russian Fed.)	3 185	3 689	4 138
Latin America	2 107	2 167	2 356
North America	5 860	6 041	5 110
Oceania	1 656	2 094	2 246

Thousands of metric tonnes

One consequence of this was the closure of some Western smelters, and cut-backs in others, even though smelters in the C.I.S. were generally regarded as being less efficient. During the decade 1995 to 2004 prices on the London Metals Exchange have ranged from $US 1325 per tonne in 1999 to $US 1806 in 1995, with an average over the whole period of $US 1506.

During the decade 1983–1993, the price of magnesium was relatively constant and, on average, was around twice that of aluminium on the basis of weight. However, the volatility in the price of aluminium has meant that the magnesium:aluminium price ratio has at times ranged from 2.5 to just below 1.5, the level at which magnesium becomes competitive on a volumetric basis. Direct cost comparisons of aluminium with steel are also difficult to make because of density differences, although it may be noted that aluminium alloy sheet is normally three to four times more expensive than mild steel sheet of the same weight. This price differential is reduced with products made from recycled metal, or those for which volume or area are prime considerations.

In most countries, aluminium is used in five major areas: building and construction; containers and packaging; transportation; electrical conductors; machinery and equipment. Patterns vary widely from country to country depending on levels of industrialisation and economic wealth. As examples, Fig. 1.5 compares the consumption of aluminium in 2002 for the current two major users of aluminium, the United States and China. In the United States, transportation (31.6%) had become the major user followed by packaging (20.7%) whereas the reverse was true a decade ago. In China, the market is heavily weighted to building and construction (33%) whereas packaging only accounts for 8%. Transportation consumed 24% of the aluminium used in

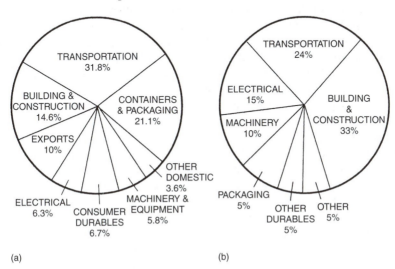

(a)   (b)

Fig. 1.5 Outlets for the consumption of aluminium in (a) United States and (b) China in the year 2002. (from Hunt, W.H., *JOM*, **56**, No. 9, 21, 2004).

China and it is interesting to note that, whereas in the United States in 2003 there were around 800 automobiles for each 1000 people, this figure was less than 20 in China. The largest potential for further growth in the use of aluminium does appear to reside in the automobile which, in the United States has risen from an average of 8 kg per vehicle in 1971 to 90 kg in 1994 and 130 kg in 2004. In Europe, the amount of aluminium in each vehicle was about 100 kg in 2000 and is predicted to increase to 150 kg in 2005 and 200 kg by 2015.

As shown in Fig. 1.4, global production of magnesium was relatively constant at about 250 000 tonnes per annum for the more than two decades from around 1970 until the late 1990s. During this period, some three-quarters of magnesium metal was consumed as alloying additions to aluminium (~55%), as a desulphurising agent for steels (~15%), and to produce nodular cast iron (~6%). Less than 20% was actually used to produce magnesium alloys, mostly as die castings for the aerospace and general transport industries. However, during the decade 1994 to 2004 the increase in the production of magnesium alloy die castings averaged 16% per annum and the amount used for this purpose was estimated to have risen to 152 000 tonnes in 2004. This was approximately one-third of all magnesium produced. Global demand for magnesium, excluding China and Russia, rose by 8% from 2003 to 2004 to reach 410 000 tonnes and is expected to exceed 500 000 tonnes by 2010. To date, wrought magnesium alloys have accounted for relatively little use of this metal because the hexagonal crystal structure makes them less amenable to hot or cold working.

Titanium was not produced in quantity until the late 1940s when its relatively low density and high melting point (1678 °C) made it uniquely attractive as a potential replacement for aluminium for the skin and structure of high-speed aircraft subjected to aerodynamic heating. Liberal military funding was provided in the decade 1947–57 and one of the major metallurgical investigations of all time was made of titanium and its alloys. It is estimated that $400 million was spent in the United States during this period and one firm examined more than 3000 alloys. One disappointing result was that titanium alloys showed relatively poor creep properties bearing in mind their very high melting points. This factor, together with a sudden change in emphasis from manned aircraft to guided weapons, led to a slump in interest in titanium in 1957–58. Since then, selection of titanium alloys for engineering uses has been made on the more rational bases of cost-effectiveness and the uniqueness of certain properties. The high specific strength of titanium alloys when compared with other light alloys, steels, and nickel alloys is apparent in Fig. 1.6. The fact that this advantage is maintained to around 500 °C has led to the universal acceptance of certain titanium alloys for critical gas turbine components. Titanium alloys have also found increasing for critical structural components in aircraft including forged undercarriages, engine mountings and high strength fasteners. Traditionally, aerospace applications have accounted for as much as 75% of titanium mill products, but this level has now fallen below 50%. Much of the remainder is finding increasing applications in the chemical

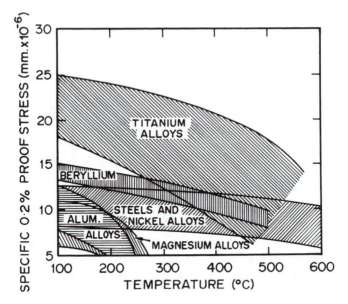

Fig. 1.6 Relationship of specific 0.2% proof stress (ratio of proof stress to relative density) with temperature for light alloys, steels and nickel alloys.

and power generation industries where the outstanding corrosion resistance of titanium is the key factor.

World production of titanium as sponge (Section 1.4) peaked at around 180 000 tonnes in the mid-1980s when the Cold War had a dominant influence, and had halved by the year 2002. At that time, worldwide production of titanium and titanium alloys mill products was 57 000 tonnes, and is predicted to reach 80 000 tonnes in 2010. In 2002, The USA (28%), CIS (26%) and Japan (25%) supplied nearly 80% of all the world's titanium mill products. China (7%) was then the fifth largest producer and plans are being implemented to double its output from 7 000 tonnes in 2003 to 14 000 by the year 2010.

1.1.4 Recycling

Due to the realisation that the supply of minerals is finite, much attention is now being paid to recycling as a means of saving metals, as well as reducing both the amount of energy and the output of greenhouse gases involved in their extraction. Based on known reserves, Fig. 1.7 shows one estimate of the dramatic effect that recycling rate may have on the remaining supply of several metals. With aluminium, for example, it has been predicted that a recycling rate of 50% would allow the reserves to last for about 320 years, whereas a rate of 80% would sustain these reserves for more than 800 years. In the United states, Japan and much of Europe, some restrictions have been introduced that require

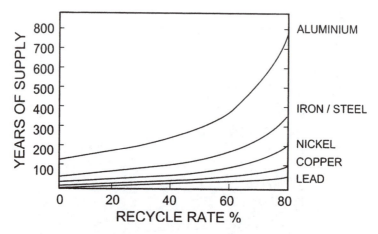

Fig. 1.7 Estimated relationship between remaining years of supply of some metals and recycling rates based on known reserves of minerals (courtesy of J. Rankin and T. Norgate, Commonwealth Scientific and Industrial Research Organization, Minerals Division, Melbourne).

materials used for consumer products to be recycled back into the same products. From 2006, for example, the European Union will require the fraction of metals recovered for re-use from old motor vehicles must be increased to at least 85% of the average vehicle weight.

The incentive for recycling the light metals is particularly strong because their initial costs of production are relatively high. For aluminium, remelting of scrap requires only about 5% of the energy needed to extract the same weight of primary metal from its ore bauxite. Currently the ratio of secondary (scrap) to primary aluminium is around 30% which is well below that recovered from steel or copper. However, as shown in Table 1.4, the relatively long life span of some aluminium-containing products can limit the supply of used metal.

Table 1.4 Global estimates of service lives and recycling rates for products made from aluminium (courtesy International Aluminium Institute, London)

Major End Markets	Average Product Life Years	Average Recycling Rate Per Cent
Building and construction	25–50	80–85
Transportation – cars	10–15	90–95
Transportation – aerospace	15–25	90–95
Transportation – marine	15–40	40–90
Transportation – trucks, buses, rail	15–30	50–90
Engineering – machinery	10–30	30–90
Engineering – electrical	10–50	40–80
Packaging – cans	0.1–1.0	30–90
Packaging – foil	0.1–1.0	20–90

The light alloys in general do present a special problem because they cannot be refined, i.e. alloying elements cannot be extracted or removed. Consequently, unless alloys for particular products are segregated and confined to a closed circuit, remelting tends to downgrade them. With aluminium alloys, the remelted general scrap is used mainly for foundry castings which, in turn, are limited in the amount they can absorb. Thus there is a need to accommodate more secondary aluminium alloys into cast billets that are used to produce wrought materials. In this regard, aluminium provides by far the highest value of any recyclable packaging material and, as an example, the efficient collection of all-aluminium alloy beverage cans to produce canstock is essential for the competitive success of this product (Section 3.6.5). Worldwide, four out of every five beverage cans are made from aluminium alloys and the current recycling rate averages 55%. This figure now exceeds 70% in the United States and is as high as 90% in Sweden and Switzerland.

A schematic that seeks to represent the complex interactions involved in the production, use, and recycling of aluminium is shown in Fig. 1.8. The automotive industry is now the second largest provider of aluminium alloy scrap although sorting from other materials after discarded vehicles have been shredded into small pieces presents technical challenges. Currently flotation methods are commonly used. In the European Union in 2002, some 90% of automotive aluminium alloys were recovered which compares with 80% from discarded building products and electrical appliances, and only 40% from packaging. The aluminium industry in the United States has a vision of 100% recycling of aluminium alloys by the year 2020. Such an ambitious target will require major technological advances, notably the development of a low-cost process for metal purification so that the recycled scrap can become a source of primary aluminium.

In 2004, approximately half the world's output of magnesium was used as an alloying addition to aluminium and much of this is recycled along with these alloys. Otherwise, the recycling of magnesium alloy scrap is less advanced than for aluminium because of magnesium's high reactivity with oxygen and nitrogen, and problems with contamination by metals such as copper and nickel that adversely affect corrosion resistance. Clean and sorted magnesium alloy scrap arising from sprues, risers and other discards from casting processes may be recycled within individual foundries. Other scrap is usually remelted under a refining flux that consists of a mixture of alkali and alkaline earth metal chlorides and fluorides. To maximize metal recovery, and reduce toxic gases, it is desirable to remove surface coatings such as paint, lacquer and oil which adds to the cost of recycling. Because of magnesium's high vapour pressure and relatively low boiling point (1090 °C) distillation may offer a promising alternative solution to the problem of contamination from scrap. It may also provide the opportunity to produce pure magnesium metal from alloys which is much less feasible with aluminium or titanium (boiling points 2520 °C and 3289 °C respectively).

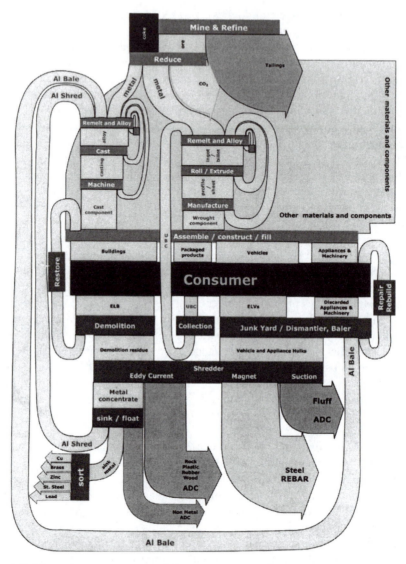

Fig. 1.8 Schematic representation of the interactions in aluminium production, uses and recycling (courtesy A. Gesing).

Less attention has been directed to the recycling of titanium alloys despite the fact that, on average, only 0.4 kg of each 1.3 kg of titanium metal sponge ends up in the finished product. High quality titanium alloy scrap can be remelted within a closed circuit to make ingots or slabs. Other scrap, including sponge, can be used to produce ferro-titanium for adding to speciality steels.

1.2 PRODUCTION OF ALUMINIUM

Although aluminium is now the second most used metal, it is a comparative newcomer among the common metals because of the difficulty in extracting it from its ores. Unlike iron, for example, it combines so strongly with oxygen that it cannot be reduced with carbon. An impure form of aluminium was first isolated in 1809 in England by Sir Humphry Davy which he produced from alum, its bisulphate salt. He called this new metal "aluminum", which is the name still used in the United States, whereas it is now known as "aluminium" in Europe and most other countries.

The first commercial preparation of aluminium occurred in France in 1855 when H. Sainte-Claire Deville reduced aluminium chloride with sodium. As is so often the case, the potential military applications of this new metal led to government support because Napoleon the Third foresaw its use in lightweight body armour. During the period 1855–59, the price of aluminium per kg fell from over $US500 to $US40 but all the Emperor is reported to have received were some decorative military helmets, an aluminium dinner set and some aluminium toys for the children of the Imperial Court. The aluminium produced by Sainte-Claire Deville's process was less than 95% pure and it proved to be more expensive than gold at that time.

Independent discoveries in 1886 by Hall in the United States and Héroult in France led to the development of an economic method for the electrolytic extraction of relatively high-purity aluminium which remains the basis for production today. By 1888 the price had fallen to less than $US4 per kg and in recent times it has varied between $US1 and $US2 per kg.

Aluminium is extracted from bauxite which was discovered by the French chemist P. Berthier, and named after the town of Les Baux in Provence, Southern France, where the ore was first mined. Bauxite is the end product of millions of years of surface weathering of aluminium silicates (e.g. feldspars) and clay minerals, usually in tropical locations. The principal aluminium-bearing minerals in bauxite exist as several forms of hydrated aluminium oxide, notably gibbsite ($Al_2O_3 \cdot 3H_2O$), which is also known as hydragillite or trihydrate, and boehmite ($Al_2O_3 \cdot H_2O$) also known as monohydrate. The actual chemical composition varies with location and the geology of each deposit. Bauxite that is mined simply by open cut methods and the largest known reserves exist in Australia which, in 2004, produced 38% (55.6 million tonnes) of the world's supply followed by Guyana 10.6%, Jamaica 9.25%, Brazil 9.0% and China 8.6%.

Typically ore bodies contain 30–60% hydrated Al_2O_3 together with impurities comprising iron oxides, silica and titania. High-grade bauxites with low silica contents are expected to be depleted in 20–30 years time. When it becomes necessary to use grades with higher silica contents, the ore will first need to be processed to remove this impurity which will introduce a cost penalty. One method that has been developed in Russia is called thermochemical alkaline conditioning. This involves roasting the bauxite at high temperatures to convert the two most important silica-bearing minerals, kaolinite and quartz, into products that can be dissolved in caustic soda.

Immense amounts of aluminium are also present in clays, shales and other minerals and the amphoteric nature of aluminium provides the opportunity to use acid as well as alkaline processes for its recovery. As one example, some attention has been given to acid extraction of alumina from kaolinite which is widely distributed as a clay mineral and is a major constituent of the ash in coal. The minerals, nepheline, $Na_3K(AlSiO_4)_4$, and alunite, $KAl_3(SO_4)_2(OH)_6$, are processed commercially in the Commonwealth of Independent States in plants located in regions remote from sources of bauxite. However, alumina obtained from these and other alternative sources is 1.5–2.5 times more costly than that produced from the Bayer process which is described below.

1.2.1 Bayer process for alumina recovery

The Bayer process was developed and patented by Karl Josef Bayer in Austria in 1888 and essentially involves digesting crushed bauxite in strong sodium hydroxide solutions at temperatures up to 240 °C. Most of the alumina is dissolved leaving an insoluble residue known as 'red mud' which mainly comprises iron oxides and silica and is removed by filtration. The particular concentration of sodium hydroxide as well as the temperature and pressure of the operation are optimized according to the nature of the bauxite ore, notably the respective proportions of the different forms of alumina (α, β, or γ). This first stage of the Bayer process can be expressed by the equation:

$$Al_2O_3 \cdot xH_2O + 2NaOH \rightarrow 2NaAlO_2 + (x + 1)H_2O.$$

Subsequently, in the second stage, conditions are adjusted so that the reaction is reversed. This is referred to as the decomposition stage:

$$2NaAlO_2 + 2H_2O \rightarrow 2NaOH + Al_2O_3 \cdot 3H_2O.$$

The reverse reaction is achieved by cooling the liquor and seeding with crystals of the trihydrate, $Al_2O_3 \cdot 3H_2O$, to promote precipitation of this compound as fine particles rather than in a gelatinous form. Decomposition is commonly carried out at around 50 °C in slowly stirred vessels and may require up to 30 h to complete. The trihydrate is removed and washed, with the sodium hydroxide liquor being recycled back to the digestors.

Alumina is then produced by calcining the trihydrate in rotary kilns or, more recently, fluidized beds. Calcination occurs in two stages with most of the water of crystallization being removed in the temperature range 400–600 °C. This produces alumina in the more chemically active γ-form which further heating to temperatures as high as 1200 °C converts partly or completely to relatively inert α-alumina. Each form has different physical characteristics and individual aluminium smelters may specify differing mixtures of α- and γ-alumina. Typically the Bayer process produces smelter grade alumina in the range 99.3–99.7% Al_2O_3.

In 2004, the total capacity of alumina refineries worldwide was 65 million tonnes per annum. Approximately one third of the world's aluminia was produced in Australia which had three largest refineries ranging in size from 3.25 to 3.75 million tonnes, one of which was being expanded to a capacity of 4 million tonnes per annum. In most countries, aluminia is costs for US$180–200 per tonne to produce.

1.2.2 Production of aluminium by the Hall–Héroult Process

Alumina has a high melting point (2040 °C) and is a poor conductor of electricity. The key to the successful production of aluminium lies in dissolving the oxide in molten cryolite (Na_3AlF_6) and a typical electrolyte contains 80–90% of this compound and 2–8% of alumina, together with additives such as AlF_3 and CaF_2. Cryolite was first obtained from relatively inaccessible sources in Greenland but is now made synthetically.

An electrolytic reduction cell (known as a pot) consists essentially of baked carbon anodes that are consumed and require regular replacement, the molten cryolite–alumina electrolyte, a pool of liquid aluminium, a carbon-lined container to hold the metal and electrolyte, and a gas collection system to prevent fumes from the cell escaping into the atmosphere (Fig. 1.9). There are also alumina feeders that are activated intermittently under some form of automatic control. A typical modern cell is operated at around 950 °C and takes up to 500 kA at an anode current density around 0.7 A cm^{-2}. The anode and cathode

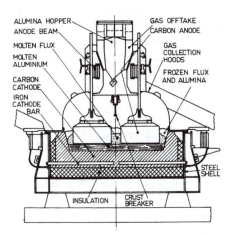

Fig. 1.9 Hall–Héroult electrolytic cell for producing aluminium (Courtesy Australian Aluminium Council).

Fig. 1.10. Potline of electrolytic cells for producing aluminium (courtesy Comalco Ltd).

are separated by 4–5 cm and there is a voltage drop of approximately 4.5 V across each cell. The cell is operated so that the carbon side-linings are protected with a layer of frozen cryolite and the upper surface of the bath is covered with a crust of alumina. The molten aluminium is siphoned out regularly to be cast into ingots and alumina is replenished as required. The largest and most productive cells operate at a current efficiency of around 95% and have a daily output of aluminium close to 4000 kg. Typically, some 300 cells are connected in series to make up a potline (Fig. 1.10).

The exact mechanism for the electrolytic reaction in a cell remains uncertain but it is probable that the current-carrying ions are Na^+, AlF_4^-, AlF_6^{3-} and one or more ternary complex ions such as $AlOF_3^{2-}$. At the cathode it is considered that the flouroaluminate anions are discharged via a charge transfer at the cathode interface to produce aluminium metal and F^- ions while, at the anode, the oxoflouro-aluminate ions dissociate to liberate oxygen which forms CO_2. The overall reaction can be written simply as follows:

$$2Al_2O_3 + 3C \rightarrow 4Al + 3CO_2.$$

Figure 1.11 shows a flow diagram for the raw materials needed to produce 1 tonne of aluminium. Commonly some 3.5–4 tonnes of bauxite are needed from which 2 tonnes of alumina are extracted that, in turn, yield 1 tonne of aluminium. Significant quantities of other materials, such as 0.4 tonne of carbon, are also consumed. However, the most critical factor is the consumption of electricity which, despite continual refinements to the process, still required a world average of 15 700 kWh for each tonne of aluminium produced from

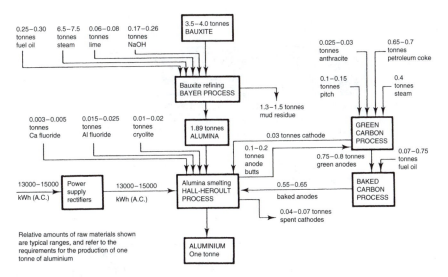

Fig. 1.11 Flow diagram for integrated production of aluminium from bauxite (courtesy Australian Aluminium Council).

alumina in the year 2004. The best performance reported for that year was an electricity consumption of 13 300 kWh per tonne. These values compare with 28 000 kWh per tonne needed shortly after the Hall-Héroult process was first commercialised late in the 19th century, and the theoretical requirement which is about 6 500 kWh. By 2004, it was estimated that the production of aluminium worldwide consumed 266 000 GWh of electricity, about half of which was generated by hydropower.

Of the total voltage drop of 4.5 V across a modern cell, only 1.2 V represents the decomposition potential or free energy of the reaction associated with the formation of molten aluminium at the cathode. The largest component of the voltage drop arises from the electrical resistance of the electrolyte in the space between the electrodes and this amounts to around 1.7 V, or 35–40% of the total. Efficiency can be increased if the anode–cathode distance is reduced and this aspect has been one focus of recent changes in cell design. A modification that shows promise is to coat the cathode with titanium diboride which has the property of being readily wetted by molten aluminium. This results in the formation of a thinner, more stable film of aluminium that can be drained away into a central sump if a sloped cathode is used (Fig. 1.12). Reductions in anode–cathode spacing from the normal 4–6 cm down to 1–2 cm have been claimed permitting a decrease of 1–1.5 V in cell voltage. Predictions on cell performance suggest that electrical consumption can be reduced to an average of 12 500 kWh per tonne of aluminium produced.

Fig. 1.13 summarizes one calculation of the total energy consumed during all stages in the production of the light metals, as well as that for copper, zinc

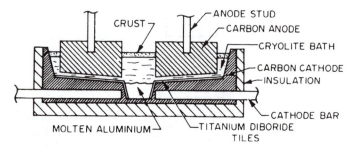

Fig. 1.12 Modified cell design for electrolytic reduction of aluminium (courtesy Kaiser Aluminium & Chemical Corporation).

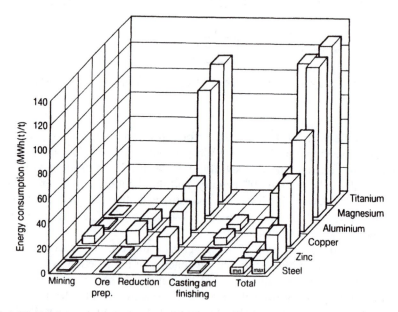

Fig. 1.13 Total energy consumption in megawatt hours (thermal) for each stage in the production of the light metals and copper, zinc and steel (from Yoshiki-Gravelsins, K.S. *et al.*, *J. Metals*, **45**, (5), 15, 1993).

and steel. To produce one tonne of primary aluminium from bauxite in the ground was estimated to require between 70 000 and 75 000 kWh (thermal), where the total energy required has been converted back to an equivalent amount of fossil fuel by assuming 1 kWh (electrical) = 3 kWh (thermal). This reduces to 30 000 kWh (thermal) per unit volume of aluminium, but is still much greater than the estimated 13 000–16 000 kWh (thermal) of energy required to produce one tonne of steel from iron ore in the ground to finished product.

1.2.3 Alternative methods for producing aluminium

Because of the large disparity between the theoretical and actual requirements for electrical energy to produce each tonne of aluminium, it is to be expected that alternative methods of production would have been investigated. One example has been a chloride-based smelting process developed by Alcoa which commenced operation in the United States in 1976 with an initial capacity of 13 500 tonnes of aluminium per year and the potential to achieve a 30% cost saving. This process also used alumina as a starting material which was combined with chlorine in a reactor to produce $AlCl_3$. This chloride served as an electrolyte in a closed cell to produce aluminium and chlorine, the latter being recycled back into the reactor. The process had the advantage of being continuous but the provision of materials of construction that could resist attack by chlorine over long periods of time proved to be difficult. This factor, together with improvements in efficiencies in conventional electrolysis, led to the process being discontinued in 1985.

Several companies have been investigating carbothermic methods for producing aluminium. One process involves mixing aluminium ore with coking coal to form briquettes, which are then reduced in stages in a type of blast furnace operating at temperatures ranging from 500 to 2100 °C. The molten metal product comprises aluminium combined with iron and silicon which is scrubbed and absorbed by a spray of molten lead at the bottom of the furnace. Since aluminium and lead are immiscible, the lighter aluminium rises to the surface where it can be skimmed off. Further purification of the aluminium is required. Although overall cost savings have been predicted, no commercially viable process has so far eventuated.

1.3 PRODUCTION OF MAGNESIUM

One of the novel explanations for the disappearance of the dinosaurs is the Chinese theory that this was due to a magnesium deficiency which had an adverse effect on the strength of egg shells thereby preventing reproduction. Today it is recognized that human beings require a daily intake of 300–400 mg of magnesium which means that some 600 000 tonnes should be ingested annually throughout the world! Until recently, this was more than double the amount of magnesium metal actually produced each year.

In 1808 Sir Humphry Davy established that magnesium oxide was the oxide of a newly recognised element. Magnesium metal was first isolated in 1828 by the Frenchman Antoine-Alexander Bussy who fused $MgCl_2$ with metallic potassium. The first production of magnesium by the electrolytic reduction of the chloride was accomplished by Michael Faraday in 1833. Commercial production commenced in Paris in the middle of the 19th century but had reached only some 10 tonnes per annum by the year 1890. Output increased to 3000 tonnes during the last year of the First World War, fell again afterwards and then rose to around 300 000 tonnes per annum late in the Second World War. In 2004 annual production exceeded 400 000 tonnes.

Magnesium compounds are found in abundance in solid mineral deposits in the earth's crust and in solution in the oceans and salt lakes. The most common surface minerals are the carbonates dolomite ($MgCO_3 \cdot CaCo_3$) and magnesite ($MgCO_3$). The mineral brucite ($MgO \cdot H_2O$) is somewhat rarer, as are the chlorides of which carnallite ($MgO \cdot KCl \cdot 6H_2O$) is one example. Concentrated aqueous solutions in the form of brine deposits occur at several places in the world including the Dead Sea in Israel and the Great Salt Lake in Utah, United States where a total of close to 80 000 tonnes is currently produced annually. Virtually unlimited reserves are present in the oceans which contain 0.13% magnesium and, until recently, sea water provided more than 80% of the world's supply of this metal which was extracted by the electrolytic reduction of $MgCl_2$. During the 1940's the largest sea water plant was constructed in the United States at Freeport, Texas, and production peaked at 120 000 tonnes in the 1970s. Output then declined during the next two decades and the plant was closed in 1998 following severe storm damage. At present the extraction of magnesium from sea water has become uncompetitive for reasons that are explained below.

Two processes have been developed to produce magnesium by the direct reduction of dolomite by ferrosilicon at high temperatures. One is the Pidgeon process originating from Canada in 1941 in which this reaction is carried out in the solid state. Until recently it was only economic in rare conditions where there was a natural site advantage. The other is the Magnétherm process that was developed in France and which operates at much higher temperatures so that the reaction mixture is liquid. This process is also producing magnesium at prices that are currently uncompetitive with metal produced in China by the Pidgeon process.

1.3.1 Electrolytic extraction of magnesium

Two types of electrolytic processes have been used to produce magnesium that differ in the degree of hydration of the $MgCl_2$ and in cell design. One was pioneered by IG Fabenindustrie in Germany in 1928 and was adopted later by the Norwegian company Norsk Hydro when it was a major European producer of this metal. Known as the IG process, it uses dry MgO derived from minerals in sea water which is briquetted with a reducing agent, e.g. powdered coal, and $MgCl_2$ solution. The briquettes are lightly calcined and then chlorinated at around 1100 °C to produce anhydrous $MgCl_2$, which is fed directly into the electrolytic cells operating at 740 °C. Other chloride compounds such as NaCl and $CaCl_2$ are added to improve electrical conductivity and to change the viscosity and density of the electrolyte. Each cell uses graphite anodes that are slowly consumed and cast steel cathodes that are suspended opposite one another. Typically the cells operate at 5–7 V and currents may now exceed 200 000 A. Magnesium is deposited on the cathodes as droplets which rise to the surface of the electrolyte, whereas chlorine is liberated at the anodes and is recycled to produce the initial $MgCl_2$ cell feedstock.

The second electrolytic process for extracting magnesium from sea water was developed by the Dow Chemical Company for use at the Freeport plant mentioned above. Magnesium in $MgCl_2$ was precipitated as the hydroxide by the addition of lime and then dissolved in HCl. This solution was then concentrated and dried although the process stopped short of complete dehydration of the $MgCl_2$ which was then available as the cell electrolyte. In contrast the IG-Norsk Hydro process, the cells required external heat with the steel cell box serving as the cathode. These cells operated at 6–7 V and a current of 90 000 A.

The energy consumed per kg of magnesium produced was around 12.5 kWh for the Norsk Hydro cell and 17.5 kWh for the Dow cell. However, each process required the additional consumption of approximately 15 kWh of energy per kg for preparation of $MgCl_2$ cell feed.

Two new processes for extracting the basic feedstock anhydrous $MgCl_2$ in a high-purity form for use in electrolytic cells have been developed in Canada and Australia, although neither has reached commercial production. In Canada, the Magnola process uses tailings from asbestos mines to take advantage of magnesium silicate contained in serpentine ore. The tailings are leached in strong HCl in a novel procedure to produce a solution of $MgCl_2$ which is purified by adjusting the pH. Ion exchange techniques are then used to generate concentrated, high-purity brine that is dehydrated for use in an electrolytic cell. The Australian process was developed to exploit magnesite ($MgCO_3$) mined from a huge surface deposit in Queensland that is estimated to contain 260 million tonnes of high-grade ore. $MgCl_2$ is leached from the magnesite with HCl and glycol is added to the solution, after which water is removed by distillation. Magnesium chloride hexammoniate is then formed by sparging with ammonia. Final calcining produces a relatively low cost, high-purity $MgCl_2$, and the solvent and ammonia are recycled.

1.3.2 Thermic processes

Production of magnesium by direct thermal reduction of calcined dolomite with ferrosilicon proceeds according to the simplified equation:

$$2CaO{\cdot}MgO + Si \rightarrow 2\,Mg + (CaO)_2SiO_2.$$

In the Pidgeon process, briquettes of the reactants are prepared and loaded in amounts of around 150 kg each into a number of tubular steel retorts that are typically 250–300 mm in diameter and 3 m long. The retorts are then evacuated to a pressure of below 0.1 torr and externally heated to a temperature in the range 1150–1200 °C, usually by burning coal. Magnesium forms as a vapour that condenses on removable water-cooled sleeves at the ends of the retorts that are located outside the furnace. Approximately 1.1 tonnes of ferrosilicon is consumed for each tonne of magnesium that is produced. Advantages of the Pidgeon process are the relatively low capital cost and the less stringent requirement that is placed on the purity of the raw materials. Major deficiencies are that it is a labour-intensive, batch process which only produces around 20 kg of magnesium from each retort, and requires a lengthy cycle time of around 8 h. The retorts must then be emptied, cleaned and recharged in conditions that are usually dusty and unpleasant. The Pidgeon process has been widely adopted in China where labour costs are low relative to Western nations, and there are readily available supplies of low cost ferrosilicon and anthracite. Several hundred plants of varying sizes have been constructed throughout China which, in 2004, supplied more that half the world's magnesium. Prices have ranged between $US 1200 and $US 1500 per tonne.

The Magnétherm process employs an electric arc furnace operating at around 1550 °C and with an internal pressure of 10–15 torr. Because the reaction takes place in the liquid phase, the time required for its completion is less than that needed for the Pidgeon process. The furnace may be charged continuously and discharged at regular intervals, and alumina or bauxite is added to the dolomite/ferrosilicon charge which keeps the reaction product, dicalcium silicate, molten so that it can be tapped as a slag. Magnesium is again produced as a vapour which is solidified in an external condenser. Batch sizes may be as high as 11 000 kg and plants have operated in France, Japan, the United States and in the former Yugoslavia.

Alternative thermic techniques have been proposed, although none is currently operating commercially. One idea was to use a plasma arc furnace in which pelletized MgO and coke are fed into a pre-melted MgO, CaO, Al_2O_3 slag. The high energy density of the plasma, and the fact that high temperatures (e.g. 1500 °C) are generated at the surface of the slag where the silicothermic reaction occurs, allows it to be sustained at normal atmospheric pressure. This advantage, combined with the efficient and near total silicon consumption were claimed to enhance economic competitiveness of the thermic route to magnesium production. However, this development occurred before use of the Pidgeon process was greatly expanded in China.

I.4 PRODUCTION OF TITANIUM

The existence of titanium was first recognized in 1791 by William McGregor, an English clergyman and amateur mineralogist, who detected the oxide of an unknown element in local ilmenite sand ($FeO \cdot TiO_2$). A similar observation was made in 1795 by a German chemist, Martin Klaproth, who examined the mineral rutile (TiO_2) and he named the element titanium after the mythological first sons of the earth, the Titans. An impure sample of titanium was first isolated in 1825 but it was not produced in any quantity until 1937 when Kroll, in Luxembourg, reacted $TiCl_4$ with molten magnesium under an atmosphere of argon. This opened the way to the industrial exploitation of titanium and the essential features of the process are as follows.

1. Briquette TiO_2 with coke and tar and chlorinate at $800\,°C$ to promote the reaction: $TiO_2 + 2Cl_2 + 2C \rightarrow TiCl_4 + 2CO$.
2. Purify $TiCl_4$ by fractional distillation.
3. Reduce $TiCl_4$ by molten magnesium or sodium under an argon atmosphere, one reaction being: $TiCl_4 + 2Mg \rightarrow Ti + 2MgCl_2$.

Titanium forms as an impure sponge around the walls of the reduction vessel and is removed periodically. The sponge produced by reacting with magnesium must be purified by leaching with dilute HCl and/or distilling off the surplus $MgCl_2$ and magnesium. The use of sodium has the advantages that leaching is more efficient and the titanium sponge is granular, making it easier to compact for the subsequent melting process.

The mineral rutile (TiO_2) is the most convenient source of titanium and is found mainly in beach sands along the eastern coast of Australia, India, Mexico and in estuaries in Sierra Leone. Most titanium metal is extracted from rutile although the much more plentiful, but more complex mineral ilmenite ($FeO\cdot TiO_2$) will become the major source in the future. Ilmenite is available in a number of countries with particularly large deposits being found in China and Russia. It is interesting to note that some 93% of the world's production of titanium minerals is processed into titanium oxide and amounts to more than 4.5 million tonnes annually. This compound is used as a white pigment in paint, in which it replaced white lead, as well as in papermaking, printing ink, ceramics and plastics. Titanium metal currently accounts for only some 3% of minerals production.

The controlling factor in refining titanium sponge is the metal's high reactivity with other elements, notably its affinity for oxygen, nitrogen, hydrogen, and carbon. As shown in Table 1.5, the solubility of these interstitial elements in titanium is greater by several orders of magnitude than in other commonly used metals. Since quite small amounts of these elements adversely affect the ductility and toughness of titanium, it is clearly impossible to melt in air or in a normal crucible because the metal will both absorb gases and react with any known oxide or carbide refractory. Accordingly, a radically new method of melting had to be devised leading to what is known as the consumable-electrode arc furnace (Fig. 1.14a).

Melting is carried out in a copper crucible cooled internally by circulating water or a liquid sodium-potassium eutectic. Heat is generated by a direct current arc that is struck between an electrode of titanium to be melted and a starting slug of this material contained in the crucible. An advantage of the liquid alloy coolant is that it does not react with titanium should the electric arc perforate the crucible, whereas water and steam can cause an explosion. The electrode is usually made from welded blocks of compressed titanium sponge into which alloying elements are incorporated in powder form (Fig. 1.14b). The entire

Table 1.5 Solubility at room temperature of O, N, C and H in titanium, iron and aluminium (from Morton, P.H., *The Contribution of Physical Metallurgy to Engineering Practice*, Rosenhain Centenary Conference, The Royal Society, 1976)

| Metal | Interstitial element | | | |
	Oxygen	Nitrogen	Carbon	Hydrogen
Titanium	14.5 wt%	~20 wt%	0.5 wt%	~100 ppm
Iron	~1 ppm	<5 ppm	100 ppm	<1 ppm
Aluminium	<1 ppm	<1 ppm	<1 ppm	<1 ppm

ppm = parts per million

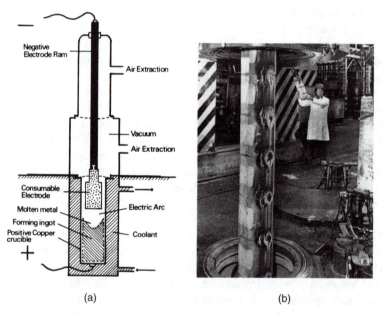

              (a)   (b)

Fig. 1.14 (a) Consumable-electrode arc furnace for melting and refining titanium; (b) consumable electrode made by welding together blocks of compressed titanium sponge (courtesy T.W. Farthing).

arrangement is encased in a vessel that can be evacuated and into which an inert gas such as argon is introduced. The capacity of the furnace is increased by using a retractable hearth and ingots are now commonly produced with diameters of 700–1200 mm and weights of 3–15 tonnes. Double and sometimes triple melting is carried out to improve the homogeneity of the ingots.

Because titanium alloys are used for critical rotating parts of gas turbine engines, special attention has been given to the elimination of inclusions in castings which may serve as sites for the initiation of fatigue cracks in components such as forged compressor discs. Sponge quality may be improved by resorting to vacuum distillation and more costly cold hearth melting furnaces have been tried which offer longer residence times for the molten alloy so that there is a greater opportunity for impurities to be dissolved or removed. In these furnaces, melting is carried out either in a high vacuum using electron beam guns as the heat source, or by a plasma torch operating in a controlled atmosphere.

Overall, the energy consumed in making pure titanium sponge is greater than that required for producing any other common metals in ingot form (Fig. 1.13). It is about 70% higher than that required for extracting an equal weight of aluminium and as much as 14 times the energy needed to produce steel. Production by the carbothermic or plasma reduction of titanium compounds has be shown to be feasible on a laboratory scale but other technical problems have so far prevented commercial exploitation. Similarly, extraction of titanium by electrolysis of fused salts has been investigated for many years with little success until a recent unexpected discovery was made in England. This has resulted in the new Fray, Farthing and Chen (FFC) process which is currently being evaluated for commercial production. This process involves using powder compacts of TiO_2 as the cathode and graphite as the anode in a bath of molten $CaCl_2$. Through careful selection of voltage, the favoured cathodic reaction is the ionization of O_2 rather than the deposition of Ca. The actual reactions are:

$$\text{Cathode: } TiO_2 + 4e^- = Ti + 2\ O^{2-}$$
$$\text{Anode: } C + x\ O^{2-} = CO_x + 2x\ e^-$$

A key factor in this process is that TiO_2, which is normally an insulator, becomes conducting once a small amount of O_2 is removed at the beginning of electrolysis. As the reaction proceeds, the porous oxide powder is converted to a high-purity, metallic sponge that can be melted and alloyed without the need of further refining, or used directly to make sintered powder products. Advantages of the FFC process over the traditional Kroll method are (i) all the electrolytic cell materials TiO_2, graphite and $CaCl_2$ are inexpensive, (ii) the titanium sponge is produced at a faster rate, and (iii) continuous production is possible. If the FFC process can be successfully scaled up, it is thought to have the potential to halve the cost of titanium which would significantly alter this metal's competitive position with respect, for example, to stainless steels.

FURTHER READING

Ashby, M.F., On the engineering properties of materials, *Acta Metall.*, **37**, 1273, 1989

Field III, F.R., Clark, J.P. and Ashby, M.F., Market drivers and process development in the 21st century, *MRS Bulletin*, September 2001, p. 716

Bever,, M.B., *Encyclopedia of Materials Science and Engineering*, Pergamon Press, Oxford, England, 1986

Altenpohl, D.G., *Aluminium: Technology, Applications and Environment*, The Aluminium Association Inc., Washington, D.C., USA and The Mineral, Metals & Materials Soc., Warrendale, Pa, USA, 1998

Altenpohl, D.G., *Materials in World Perspective*, Springer-Verlag, Berlin, 1980

Marder, J.M., Beryllium: alloying, thermomechanical processing, properties and applications, *Encyclopedia of Materials Science and Engineering*, Elsevier Ltd., Oxford, 506, 2001

Hogg, P.J., The role of materials in creating a sustainable economy, *Mater. Tech. & Advanced Performance Materials*, **19**, 70, 2004

Davis, J.R. Ed., Recycling technology, *ASM Specialty Handbook on Aluminium Alloys*, ASM International Materials Park, Ohio, USA, 47, 1993

Gesing, A., Assuring the continued recycling of light metals in end-of-life vehicles: a global perspective, *JOM*, **56**, 18, Aug., 2004

Stobart, P.D., (Ed.), *Centenary of the Hall and Heroult Processes 1886–1986*, International Primary Aluminium Institute, London, 1986

West, E.G., Aluminium – the first 100 years, *Metals and Materials*, **20**, 124, 1986

Grjotheim, K. and Kvante, H. (Eds), *Introduction to Aluminium Electrolysis*, 2nd Edn., Aluminium-Verlag, Düsseldorf, Germany, 1993

Welsh, B.J., Aluminium production paths in the new millennium, *JOM*, **51**, 24, May 1999

Hunt Jr., W.H., The China Factor: Aluminium Industry Impact, *JOM*, **56**, 21, Sept., 2004

Clow, B.B., History of primary magnesium since World War II, *Magnesium Technology 2002*, p. 3, Ed. H.I. Kaplan, The Minerals, Metals & Materials Society, Warrendale, Pa, USA, 2002

Emley, E.F., *Principles of Magnesium Technology*, Pergamon Press, London, 1966

Avedesian, M.M. and Baker, H., *ASM Specialty Handbook on Magnesium and Magnesium Alloys*, ASM International, Materials Park, Ohio, USA, 1999

Chen, G.Z., Fray, D.J. and Farthing, T.W., Direct electrochemical reduction of titanium oxide to titanium in molten calcium chloride, *Nature*, **407**, 361, 2000

Okura, Y., Titanium sponge technology, from Blenkinsop, P. A. *et al* (Eds.), *Titanium 95: Science and Technology, Proc. 8th World Conf. on Titanium*, The Institute of Metals, London, 1427, 1995

Farthing, T.W., The development of titanium alloys, *The Metallurgy of Light Alloys*, The Institute of Metallurgists, London, 9, 1983

Froes, F.H., Tenth world titanium conference, *Mat. Tech. & Adv. Perf. Mat.*, **19**, 109, 2004

2

PHYSICAL METALLURGY
OF ALUMINIUM ALLOYS

Although most metals will alloy with aluminium, comparatively few have sufficient solid solubility to serve as major alloying additions. Of the commonly used elements, only zinc, magnesium (both greater than 10 atomic %)†, copper and silicon have significant solubilities (Table 2.1). However, several other elements with solubilities below 1 atomic % confer important improvements to alloy properties. Examples are some of the transition metals, e.g. chromium, manganese and zirconium, which are used primarily to form compounds that control grain structure. With the exception of hydrogen, elemental gases have no detectable solubility in either liquid or solid aluminium. Apart from tin, which is sparingly soluble, maximum solid solubility in binary aluminium alloys occurs at eutectic and peritectic temperatures. Sections of typical eutectic and peritectic binary phase diagrams are shown in Figs 2.1 and 2.2.

High-purity aluminium in the annealed condition has very low yield strength (7–11 MPa). When it is desired to use annealed material, strength may be increased only by solid solution hardening. For this to be achieved, the solute must:

(i) have an appreciable solid solubility at the annealing temperature,
(ii) remain in solid solution after a slow cool, and
(iii) not be removed by reacting with other elements to form insoluble phases.

Figure 2.3 shows the increment in yield strength that occurs when selected solutes are added to high-purity aluminium. Some elements are shown in

---

†Unless stated otherwise, alloy compositions and additons are quoted in weight percentages.

Table 2.1 Solid solubility of elements in aluminium (from Van Horn, K.R. (Ed), *Aluminium*, Volume 1, American Society for Metals, Cleveland, Ohio,1967; Mondolfo, L.F., *Aluminium Alloys: Structure and Properties*, Butterworths, London, 1976)

Element	Temperature (°C)	Maximum solid solubility (wt%)	(at%)
Cadmium	649	0.4	0.09
Cobalt	657	<0.02	<0.01
Copper	548	5.65	2.40
Chromium	661	0.77	0.40
Germanium	424	7.2	2.7
Iron	655	0.05	0.025
Lithium	600	4.2	16.3
Magnesium	450	17.4	18.5
Manganese	658	1.82	0.90
Nickel	640	0.04	0.02
Silicon	577	1.65	1.59
Silver	566	55.6	23.8
Tin	228	~0.06	~0.01
Titanium	665	~1.3	~0.74
Vanadium	661	~0.4	~0.21
Zinc	443	82.8	66.4
Zirconium	660.5	0.28	0.08

Note:

(i) Maximum solid solubility occurs at eutectic temperatures for all elements except chromium, titanium, vanadium, zinc and zirconium for which it occurs at peritectic temperatures.

(ii) Solid solubility at 20 °C is estimated to be approximately 2 wt% for magnesium and zinc, 0.1–0.2 wt% for germanium, lithium and silver and below 0.1% for all other elements.

concentrations beyond their room temperature solubility but each alloy was processed to retain all the solute in solution. On an atomic basis, manganese and copper are the most effective strengtheners at 0.5% or less. However, manganese usually precipitates as the dispersoid Al_6Mn during ingot preheating (Section 3.1.4) and hot processing so that only 0.2–0.3% tends to remain in solution. Copper additions to the non-heat-treatable alloys are normally held to a maximum of 0.3% to avoid the possible formation of insoluble Al–Cu–Fe constituents. Magnesium is the most effective strengthener on a comparative weight basis because of its relatively high solid solubility, and annealed sheet and plate containing up to 6% of this element have yield strengths up to 175 MPa. It will also be noted that zinc, too, has a high solubility but causes little strengthening.

Aluminium alloys can be divided into two groups. One contains those alloys for which the mechanical properties are controlled by work hardening and

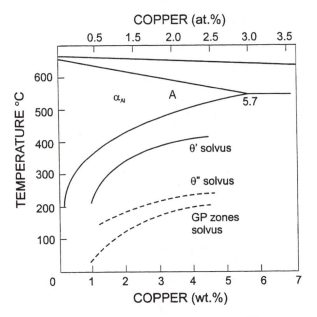

Fig. 2.1 Aluminium-rich corner of the Al–Cu eutectic diagram. The positions of the solvus lines for GP zones and the other metastable precipitates are also shown (from Murray, J.L., *Int. Met. Rev.*, **30**, 211, 1985).

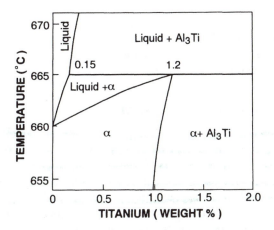

Fig. 2.2 Section of Al–Ti peritectic phase diagram.

annealing. Commercial purity aluminium and alloys based on the Al–Mg and Al–Mn systems are the common examples. The second group comprises the alloys such as Al–Cu–Mg, Al–Mg–Si, and Al–Zn–Mg–Cu that respond to age- or precipitation-hardening. In the next two sections, it is therefore desirable to

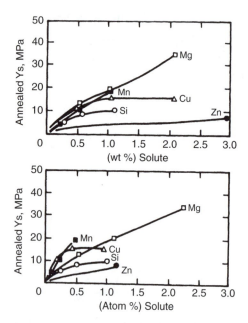

Fig. 2.3 Solid solution strengthening of high-purity binary aluminium alloys (from Sanders, R. E. *et al.*, *Proceedings of International Conference on Aluminium Alloys—Physical and Mechanical Properties*, Charlottesville, Virginia, Engineering Materials Advisory Services, Warley, U.K., 1941, 1986).

present brief reviews of the essential principles associated with these processes before considering specific alloy systems. These remarks are also relevant to most magnesium and titanium alloys.

2.1 WORK HARDENING AND ANNEALING

During deformation of metals and alloys, the dislocation content increases when dislocation generation and multiplication occur faster than annihilation can take place by dynamic recovery. Dislocation tangles, cells and subgrain walls are formed, grain shapes and internal structures change. All of these factors decrease mean free slip distance and give increased strength.

Fig. 2.4 is a schematic representation of the microstructure of a rolled aluminium alloy showing features that develop during deformation. Slip has occurred within the individual grains and grains have become elongated so that there is a large increase in total grain boundary area. A so-called deformation band or transition band (a) is shown within one grain that separates two internal regions that have developed two distinct orientations during the deformation process. Also there is a larger discontinuity known as a shear band (d) that tends to form in regions of high strain (true strain $\varepsilon > 1$). Such bands are

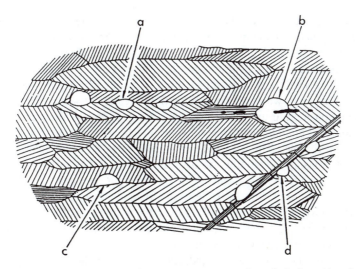

Fig. 2.4 Schematic representation of the microstructure of a rolled alloy. Also shown are possible nucleation sites for recrystallized grains (courtesy A. Oscarsson).

typically oriented at ~35° to the rolling plane and cut through existing grain structures. In sheet rolled to very high levels of strain, larger shear bands can develop that cross from one surface to another and provide paths along which failure may occur.

In multi-phase aluminium alloys containing coarse intermetallic particles and finer dispersoids, deformation becomes more inhomogeneous. As shown schematically in Fig. 2.5, substructures may develop in intensely deformed zones around

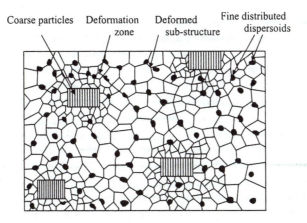

Fig. 2.5 Schematic representation of the substructure of a cold worked alloy containing coarse and fine intermetallic particles (from Nes E., *Proc. 1st Riso Int. Symp. on Metallurgy and Mater. Sci.*, Riso National Laboratory, Denmark, 1980, 36).

each of the coarse particles. The density of these deformed zones may become much higher in metal matrix composites that are reinforced with large volume fractions of ceramic particulates (Section 7.1.3). The finer dispersoids may serve to pin grain boundaries during thermomechanical processing of aluminium alloys (Section 2.5). As will be discussed later, both types of particles influence recrystallization and grain growth during annealing or hot working (Section 2.1.6).

Before considering the characteristics of work hardened aluminium alloys, it should be noted that elements in solid solution can influence deformation behaviour in several ways. These include enhancing rates at which dislocations can multiply, decreasing the mobility of dislocations so that they serve as more effective barriers to metal flow, and reducing rates of recovery during thermomechanical processing. Copper is the most effective element in this regard but, as mentioned above, additions to non-heat treatable aluminium alloys must normally be kept below 0.3%. Magnesium is less effective on an equi-atomic basis but has the greatest practical value because of its high solid solubility. As shown in Fig. 3.17, some heavily cold worked alloys based on the Al–Mg system may develop yield strengths exceeding 400 Mpa. Though having a high solid solubility, zinc has a negligible effect on the work hardening of aluminium alloys.

2.1.1 Strain-hardening characteristics

Strain-hardening occurs during most working and forming operations and is the main method for strengthening aluminium and those alloys which do not respond to heat treatment. For heat-treatable alloys, strain-hardening may supplement the strength developed by precipitation-hardening.

Tensile properties are the most affected and Fig. 2.6 shows work-hardening curves for 1100 aluminium and the alloys 3003 (Al–Mn)[1] and 5052 (Al–Mg), the latter two being representative of the main classes of non-heat-treatable alloys. Cold-working causes an initial rapid increase in yield strength, or proof stress, after which the increase is more gradual and roughly equals the change in tensile strength. These increases are obtained at the expense of ductility as measured by percentage elongation in a tensile test, and also reduced formability in operations such as bending and stretch forming. For this reason, strain-hardened tempers are not usually employed when high levels of ductility and formability are required. However, it should be noted that certain alloys, e.g. 3003, exhibit better drawing properties in the cold-worked rather than the annealed condition and this is an important factor in making thin-walled beverage cans (Section 3.6.5).

The work-hardening characteristics of heat-treatable alloys, in both the annealed and T4 tempers,[2] are similar to those described above. Cold-working prior to ageing some of these alloys may cause additional strengthening (T8 temper). In the fully hardened T6 temper, the increases in tensile properties by

---

[1]Alloy designations and compositions are shown in Tables 3.2 and 3.4
[2]Temper designations are described in Section 3.2

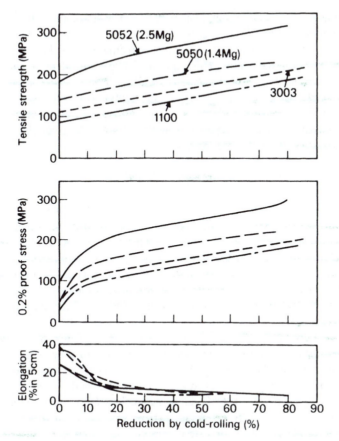

Fig. 2.6 Work-hardening curves for the alloys 1100 (99Al), 3003 (Al–1.2 Mn), 5050 (Al–1.4 Mg) and 5052 (Al–2.5 Mg) (from Anderson, W. A., in *Aluminium*, Vol. 1, K. Van Horn (Ed.), American Society for Metals, Cleveland, Ohio, 1967).

cold-working after ageing are comparitively small, except at very high strains, and are often limited by the poor workability of alloys in this condition. The principal use of this practice is for some extruded and drawn products such as wire, rod and tube which are cold-drawn after heat treatment to increase strength and improve surface finish. This applies particularly to products made from Al–Mg–Si alloys.

Work-hardening curves for annealed, recrystallized aluminium alloys, when plotted as a function of true stress and true strain, can be described by

$$\sigma = k\varepsilon^n$$

where σ is true stress, k is the stress at unit strain, ε is the true or logarithmic strain, which is defined as $\ln A_o/A_f$ where A_o and A_f are the initial and final cross-sectional areas of a sample, and n is the work-hardening exponent. As the initial strengths of the alloys increase, k also increases whereas, for values of k in the stress range 175–450 MPa, n decreases from 0.25 to 0.17.

Rates of strain-hardening can be calculated from the slopes of work-hardening curves. For non-heat-treatable alloys initially in the cold-worked or hot-worked condition, these rates are substantially below those of annealed material. For the cold-worked tempers, this difference is caused by the strain necessary to produce the temper and, if this initial strain equals ε_0, then the equation for strain-hardening becomes

$$\sigma = k(\varepsilon_0 + \varepsilon)^n$$

A similar situation exists for products initially in the hot-worked condition. The strain-hardening resulting from hot-working or forming is assumed to be equivalent to that achieved by a certain amount of cold-work. From a knowledge of the tensile properties of the hot-worked product, the amount of equivalent cold-work can be estimated by using the work-hardening curve for the annealed temper. By such procedures, it is usually possible to calculate work-hardening curves for hot-worked products that are in reasonable agreement with those for annealed products.

The work-hardening characteristics of aluminium alloys vary considerably with temperature. At cryogenic temperatures, strain-hardening is greater than at room temperature, as shown in Fig. 2.7 which compares the work-hardening characteristics of the alloy 1100 at room temperature and $-196\,°C$. The gain in strength by working at $-196\,°C$ can be as much as 40% although there is a significant reduction in ductility. At elevated temperatures, the work-hardening characteristics are influenced by both temperature and strain rate. Strain-hardening decreases progressively as the working temperature is raised until a temperature is reached above which no effective hardening occurs due to dynamic recovery and recrystallization. This behaviour is important in commercial hot-working processes and it is necessary to determine the strength/temperature/time relationships when deforming different alloys in order to optimize these operations.

2.1.2 Substructure hardening

Aluminium has a high stacking fault energy ($\sim$170 mJ m^{-2}) and, during deformation, a cellular substructure is formed within the grains, rather than twins or stacking faults. This cellular substructure causes strengthening which can be defined by a Hall-Petch type equation having the form

$$\sigma = \sigma_0 + k_1 d^{-m}$$

where σ is the yield strength, σ_0 is the frictional or Peierls stress, k_1 denotes the strength of the cell boundaries and m is an exponent that varies from 1 to 0.5.

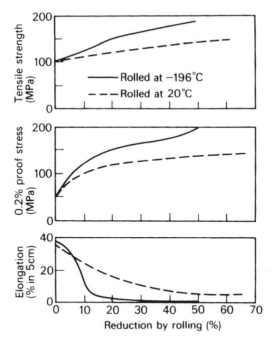

Fig. 2.7 Work-hardening curves for 1100–0 (99Al, annealed) sheet at room temperature and −196 °C (from Anderson, W. A., in *Aluminium*, Vol. I, K. Van Horn (Ed.), American Society for Metals, Cleveland, Ohio, 1967)

The substructure produced by working at relatively low temperatures is usually referred to as cells. These cells differ in orientation by only about 1° and have walls comprising tangled dislocations. On the other hand, deformation at higher temperatures produces 'subgrains' bounded by narrow, well-defined walls, and for which the misorientation is greater. The value of m changes from 1 to 0.5 if the alloys undergo the process of recovery which causes the substructure to change from cells to subgrains. The formation of subgrains is favoured if maximum substructure strengthening is desired.

2.1.3 Forming limit curves

The formability of sheet depends both on the work-hardening exponent n and the R-value which is a measure of the effect of metal thinning under axial strain. R is defined as the ratio of total width strain to total thickness strain. In simple terms, the significance of R is that, for values >1, a metal sheet characteristically resists thinning whereas there is a tendency for thinning to occur if $R < 1$. In deep drawing, for example, high R-values enable a deeper cup to be formed due to the greater resistance to thinning of the side walls. The determination of accurate values for n and R can prove difficult and the concept of the Forming

Limit Curve (FLC) was developed as a convenient tool to analyse the inhomogeneous strains occurring in sheet components. The curves identify the boundary limits between no failure during forming on the one hand, and either failure by necking or actual fracture on the other. They are obtained experimentally by covering a sheet specimen with a grid and then measuring major and minor (90° orientation) strains that develop during a particular forming operation.

Examples of FLC curves for a typical mild steel sheet, the naturally aged Al–Mg–Si–Cu alloy 6111, and for the Al–Mg–Mn alloy 5052 (in the annealed and cold rolled conditions) are shown in Fig. 2.8. For strain conditions that lie below each the respective curves, no failures are expected. For positive minor strains, the curve represents forming limits for stretching processes from uniaxial ($e_2 = 0$) to biaxial ($e_1 = e_2$) conditions. For negative minor strains, the curve represents forming limits for tension-compression strain conditions as found in the side walls of deep drawn components. As shown in the figure, steel sheet normally has forming characteristics that are superior to aluminium alloy sheet. This aspect is discussed further in Section 3.6.2. Also, it will be noted that alloy 5052 sheet displays a better capacity for forming in the annealed rather than the work hardened condition, although the reverse is true for the canstock alloy 3004 (Section 3.6.5).

2.1.4 Textures

Deformation of aluminium and its alloys proceeds by crystallographic slip that normally occurs on the {111} planes in the ⟨110⟩ directions. Large amounts of

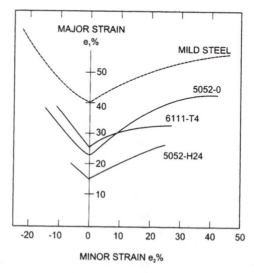

Fig. 2.8 Forming Limit Curves for a typical mild steel sheet, and for the aluminium alloys 6111–T4 and 5052 in the annealed (0) and cold rolled (H24) conditions (courtesy Australian Aluminium Council).

deformation at ambient temperatures lead to some strengthening through the development of textures, the nature of which depends in part on the mode of working. Aluminium wire, rod and bar usually have a 'fibre' texture in which the $\langle 110 \rangle$ direction is parallel to the axis of the product, with a random orientation of crystal directions perpendicular to the axis. In rolled sheet, the texture that is developed may be described as a tube of preferred orientations linking $(110) \langle \bar{1}12 \rangle$, $(112) \langle 11\bar{1} \rangle$ and $(123) \langle \bar{2}11 \rangle$.

The standard method used for determining textures is x-ray pole figures in which a stereographic projection is produced that shows the distribution of a particular crystallographic direction in an assembly of grains in a specimen. More information about the fine details of textures can now be achieved from patterns obtained using the technique of electron back scattered diffraction (ESBD). This technique involves the use of the scanning electron microscope in which the specimen is tilted to obtain the diffraction pattern. This technique has confirmed that nucleation of grains with a cube orientation texture is favoured at transition bands (Fig. 2.4). This characteristic is critical in the annealing stage in the rolling of canstock (Section 3.6.5).

When cold-worked aluminium or its alloys are recrystallized by annealing, new grains form with orientations that differ from those present in the cold-worked condition. Preferred orientation is much reduced but seldom eliminated and the annealing texture that remains has been extensively studied in rolled sheet. Some new grains form with a cube plane parallel to the surface and a cube edge aligned in the rolling direction, i.e. (100)[001] texture, and other strain-free grains develop in which the rolling texture is retained. The texture and final grain size in recrystallized products are determined by the amount of cold-work, annealing conditions (i.e. rate of heating, annealing temperature and time), composition and the size and distribution of intermetallic compounds which tend to restrict grain growth.

Textures developed by cold-working cause directionality in certain mechanical properties. Texture hardening causes moderate increases in both yield and tensile strength in the direction of working and it has been estimated that, with an ideal fibre texture, the strength in the fibre direction may be 20% higher than that for sheet with randomly oriented aggregates of grains. Forming characteristics of sheet may also be improved through an increase in the R-value for sheet which is the ratio of the strain in the width direction of a test piece to that in the thickness direction. A large R-value means there is a lack of deformation modes oriented to provide strain in the through-thickness direction. As a consequence, sheet will be more resistant to thinning during forming operations and this is desirable.

Preferred orientations in the plane of a sheet that are associated with textures may cause a problem known as earing. Earing describes the phenomenon of small undulations that may appear on the top of drawn cups (Fig. 2.9) and is wasteful of material because this uneven end of the cup must be trimmed off. Moreover, it may lead to production problems due to difficulties in ejecting

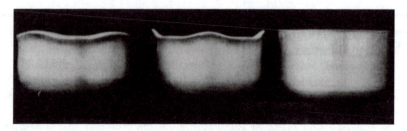

Fig. 2.9 Deep drawn aluminium cups showing 45° earing, 0–90° earing and no earing with respect to the rolling direction (from Anderson, W. A., in *Aluminium*, Vol. I, K. Van Horn (Ed.), American Society for Metals, Cleveland, Ohio, 1967).

products after a pressing operation. Four ears usually form because of non-uniform plastic deformation along the rim of deep drawn products. If the rolling texture is predominant, then ears will appear at 45° to the rolling direction of the sheet whereas, in the presence of the annealing (or cube) texture, they form in the direction of rolling and at right-angles to it. If there is a desirable balance between the two textures, there will be either eight small ears or none at all. Earing may be minimized by careful control of rolling and annealing schedules, e.g. sheet is sometimes cross-rolled so that textures are less clearly defined. This matter is considered further when discussing the production of canstock in Section 3.6.5.

Crystallographic textures should not be confused with mechanical fibring that occurs because of changes in grain shape, banding of small grains or the alignment of particles in worked alloys (Figs. 2.38 and 2.39). As mentioned in Sections 2.5.1 and 2.5.2 these effects also contribute to the anisotropy of mechanical properties and they are often more important than crystallographic texture.

2.1.5 Secondary work-hardening effects

Aluminium products formed by stretching, bending or drawing sometimes develop a roughened surface and this effect is known as 'orange peeling'. It is common to other materials and is caused by the presence of coarse grains at the surface.

A problem that occurs in a few aluminium alloys, but not in other non-ferrous materials, is the formation of stretcher strain markings, or Lüders lines, during the forming or stretching of sheet. These markings occur in one of two forms and may give rise to differences in surface topography of sheet during drawing and stretch-forming operations. One type occurs in annealed or heat-treated solid solution alloys, notably Al–Mg, and is produced when yielding takes place in some parts of a sheet but not in others. It is similar in origin to the well-known Lüders lines that may form during the deformation of certain sheet steels. The second type is associated with the Portevin–LeChatelier effect and it produces uneven or serrated yielding during a tensile test. Diagonal bands appear, oriented approximately 50° to the tension axis, which move up or down during stretching and terminate at the grips. This type of marking is rarely observed in commercial forming operations but may appear when strain-hardened sheet or plate is stretched to produce flatness.

Stretcher strain markings are generally undesirable because they cause uneven and roughened surfaces. They can be avoided by forming with strain-hardened rather than annealed sheet, providing the material has adequate ductility. The formation of these markings can also be avoided or minimized by forming or working at temperatures above 150 °C.

2.1.6 Annealing behaviour

Annealing at elevated temperatures allows partial or complete removal of the lattice distortions introduced by work hardening so that the pre-deformation properties can be progressively restored. Dislocation densities may be reduced from about 10^{12} in the severely cold worked condition to 10^{10} lines per cm^2 during recovery, and to between 10^7 and 10^8 lines per cm^2 after recrystallization. Strength properties decrease gradually during recovery and then more rapidly when recrystallization occurs, whereas there is an inverse increase in elongation. These changes are demonstrated in Fig. 2.10 for hard rolled, commercial purity aluminium sheet. Recrystallization does not commence until a definite temperature is reached which depends on alloy composition, annealing time, and the level of work hardening. The actual rate of recrystallization

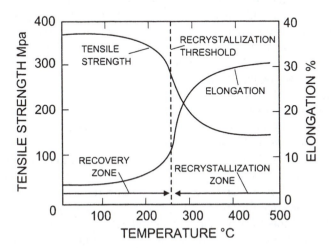

Fig. 2.10 Tensile strength and elongation plotted against annealing temperature for hard rolled, commercial purity aluminium sheet. The annealing time at each temperature is 5 minutes (from Altenpohl, D. *Aluminium Viewed from Within*, 1st ed., Aluminium-Verlag, Düsseldorf, Germany, 1982).

increases with increasing temperature; for hard rolled, pure aluminium sheet this process may take several hours to be complete at 280 °C, a few minutes at 380 °C and only seconds at 500 °C. Commercial aluminium alloys are normally annealed within the temperature range 300 to 420 °C. Further slight softening will occur if recrystallization is followed by grain growth.

Recovery in aluminium and its alloys occurs when most dislocations are either annihilated or re-arranged into walls leading to the formation of well defined subgrain boundaries within the individual grains. Some of these changes may already have occurred by dynamic recovery during deformation, particularly hot working. The division between recovery and classical recrystallization in which new strain-free grains are formed may be difficult to define in deformed aluminium alloys. This is because new grains may, in fact, evolve by a process of extended recovery in which the growth and coalescence of the pre-existing subgrains occurs until high angle grain boundaries are developed. Nevertheless, the appearance of the new grains is generally referred to as "recrystallization".

Four sites at which new grains may form have been identified in Fig. 2.4. as the (a) transition bands, (b) coarse intermetallic compounds, (c) previous grain boundaries and (d) shear bands. Of these locations, the first two have proved to be the most important for deformed aluminium alloys and the final texture that evolves during annealing is largely the result of competition between these two processes. A transition band develops when neighbouring volumes within a grain deform on different slip systems and rotate to adopt different orientations. Commonly they comprise a group of long, narrow cells or subgrains with a cumulative misorientation spreading from one side of the group to the other. Nucleation of grains with the cube orientation (100) [001] at transition bands is known to be favoured when aluminium and its alloys are annealed after cold rolling, and also seems likely to occur after warm or hot rolling.

The other main mode of forming new grains during annealing of most aluminium alloys involves their nucleation at coarse intermetallic compounds having sizes generally greater than about 1 μm. This is known as particle stimulated nucleation, or PSN, for which two conditions must be fulfilled in order for new grains to grow. One is that the deformation zone around the particles (Fig. 2.5) must have sufficient stored energy to facilitate rapid formation of the recrystallized grains. This involves rapid subgrain migration. The other is that the new grain must be sufficiently large and stable to continue to grow out into the surrounding matrix which has a lower stored energy.

Continued heating after recrystallization may promote grain growth and the driving force for this further change is the reduction in energy that is stored in the form of grain boundaries. The amount of stored energy is much less than that involved in stimulating primary recrystallization and grain boundary migration normally occurs at a much slower rate. Grain growth may be divided into two types which are defined as normal and abnormal. Normal grain growth involves the gradual elimination of small grains and the microstructure changes in rather a uniform way. During abnormal grain growth, a few grains grow excessively and consume the surrounding recrystallized grains so that the phenomenon is sometimes known as secondary recrystallization. As a result, a bimodal distribution of grain sizes develops. Where possible, coarse grain sizes are usually to be avoided except when there is a requirement for improved creep strength in an alloy. Tensile properties gradually decrease as grain growth occurs. Another problem is that coarse grained products formed by bending, stretching or drawing may develop roughened surfaces which is a phenomenon known as "orange peeling". This effect occurs because grains at a free surface are not constrained to deform like those in the interior and non-uniform deformation from grain to grain produces the orange peel appearance. Coarse surface grains that are shallow in depth result in less orange peeling than similar grains of greater thickness. Also hot worked products are less prone to this defect than are cold worked products.

The smaller dispersoid particles shown in Fig. 2.5 typically lie within the size range 0.05–0.5 μm and may serve to retard both recrystallization and grain growth by pinning the original grain boundaries in the deformed microstructure (Zener drag or pinning). They also delay recovery. As discussed in Section 2.5, and elsewhere, these dispersoids play several important roles in controlling the microstructure and properties of many aluminium alloys.

2.2 PRINCIPLES OF AGE HARDENING

It is now 100 years since the phenomenon of age or precipitation hardening was discovered by the German metallurgist, Alfred Wilm, who was working in Berlin where he was trying to develop a strong aluminium alloy to replace brass in ammunition. At that time, it was well known that steel could be hardened by quenching into water from a high temperature and Wilm was attempting to reproduce this behaviour in aluminium alloys. To his frustration, most of the aluminium alloys in fact became softer the faster they were quenched. It then happened that hardness measurements being made on some quenched Al–Cu alloy specimens were interrupted at the end of a week and, the following Monday, he was astonished to find that the hardness had increased considerably. What was not realised until some years later was that this increase in hardness was caused by precipitation of a fine dispersion of nanometre-sized particles. Also, he did not know that he had discovered what has since been described as the first nanotechnology!

2.2.1 Decomposition of supersaturated solid solutions

The basic requirement for an alloy to be amenable to age-hardening is a decrease in solid solubility of one or more of the alloying elements with decreasing temperature. Heat treatment normally involves the following stages:

1. Solution treatment at a relatively high temperature within the single-phase region, e.g. A in Fig. 2.1, to dissolve the alloying elements.
2. Rapid cooling or quenching, usually to room temperature, to obtain a super-saturated solid solution (SSSS) of these elements in aluminium.
3. Controlled decomposition of the SSSS to form a finely dispersed precipitate, usually by ageing for convenient times at one and sometimes two intermediate temperatures.

The complete decomposition of an SSSS is usually a complex process which may involve several stages. Typically, Guinier–Preston (GP) zones and an intermediate precipitate may be formed in addition to the equilibrium phase. In the Al–Cu system, which has been studied in much detail, four stages can be involved the crystal structures of which are depicted in Fig. 2.11. In addition, the new experimental technique of atom probe field ion microscopy (APFIM) has confirmed that small, disordered clusters of atoms (e.g. 20 to 50 in number) may precede the formation of GP zones in some alloys (see Fig. 2.20). These clusters may be retained on quenching, or form as ageing commences. In some alloys, it seems that they may contribute to early stages of age hardening.

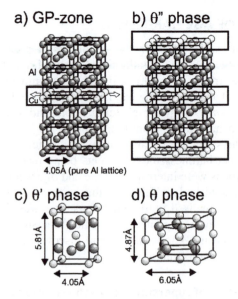

Fig. 2.11 Models showing the crystal structures of (a) GP zones, (b) θ'', (c) θ', and (d) θ (Al$_2$Cu) that may precipitate in aged binary Al–Cu alloys. Lighter balls represent copper atoms and darker balls represent aluminium atoms. (courtesy T. J. Bastow and S. Celottto).

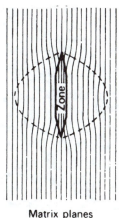

Matrix planes

Fig. 2.12 Representation of the distortion of matrix lattice planes near to the coherent GP zone (from Nicholson, R. B. *et al.*, *J. Inst. Metals*, **87**, 429, 1958–59).

GP zones are ordered, solute-rich groups of atoms that may be only one or two atom planes in thickness, e.g. Fig. 2.11 (a). They retain the structure of the matrix with which they are said to be coherent, although they can produce appreciable elastic strains in the surrounding matrix (Fig. 2.12). Diffusion associated with their formation involves the movement of atoms over relatively short distances and is assisted by vacant lattice sites that are also retained on quenching. GP zones are normally very finely dispersed and densities may be as high as 10^{17} to 10^{18} cm^{-3}. Depending on the particular alloy system, the rate of nucleation and the actual structure may be greatly influenced by the presence of the excess vacant lattice sites.

The intermediate precipitate is normally much larger in size than a GP zone and is only partly coherent with the lattice planes of the matrix. It has been generally accepted to have a definite composition and crystal structure both of which differ only slightly from those of the equilibrium precipitate. However, recent studies using atom probe field ion microscopy (APFIM) have revealed that the compositions may vary considerably. For example, the intermediate precipitate β', that precedes the equilibrium phase β (Mg_2Si) in aged Al–Mg–Si alloys (Table 2.3), may have Mg:Si ratios closer to 1:1 rather than the expected 2:1. This suggests that aluminium atoms may substitute for some magnesium atoms in the structure of this phase.

In some alloys, the intermediate precipitate may be nucleated from, or at, the sites of stable GP zones. In others this phase nucleates heterogeneously at lattice defects such as dislocations (Fig. 2.13). Formation of the final equilibrium precipitate involves complete loss of coherency with the parent lattice. It forms only at relatively high ageing temperatures and, because it is coarsely dispersed, little hardening results.

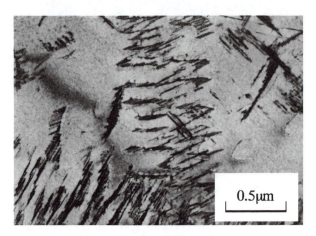

Fig. 2.13 Transmission electron micrograph showing the rods of the S-phase (Al₂CuMg) precipitated heterogeneously on dislocation lines. The alloy is Al–2.5Cu–1.5 Mg, aged 7 h at 200 °C (from Vietz, J. T. and Polmear, I. J., *J. Inst. Metals*, **94**, 410, 1966).

Most aluminium alloys that respond to ageing will undergo some hardening at ambient temperatures. This is called "natural ageing" and may continue almost indefinitely, although the rate of change becomes extremely slow after months or years. Ageing at a sufficiently elevated temperature ("artificial ageing") is characterised by different behaviour in which the hardness usually increases to a maximum and then decreases (e.g. Fig. 2.17). At one particular temperature, which varies with each alloy, the highest value of hardness will be recorded. Softening that occurs on prolonged artificial ageing is known as "overageing". In commercial heat treatment, an ageing treatment is usually selected that gives a desired response to hardening (strengthening) in a convenient period of time.

Maximum hardening in commercial alloys normally occurs when there is present a critical dispersion of GP zones, or an intermediate precipitate, or a combination of both. In some alloys, more than one intermediate precipitate may be formed. Some alloys are cold worked (e.g. by stretching 5%) after quenching and before ageing, which increases the dislocation density and provides more sites at which heterogeneous nucleation of intermediate precipitates may occur during ageing.

2.2.2 The GP zones solvus

An important concept is that of the GP zones solvus which may be shown as a metastable line in the equilibrium diagram (Fig. 2.1). It defines the upper temperature limit of stability of the GP zones for different compositions although its precise location can vary depending upon the concentration of excess vacancies. Solvus lines can also be determined for other metastable

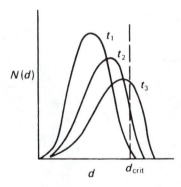

Fig. 2.14 Representation of the variation in GP zone size distribution with ageing time ($t_1 < t_2 < t_3$) (from Lorimer, G. W. and Nicholson, R. B., *The Mechanism of Phase Transformations in Crystalline Solids*, Institute of Metals, London, 1969).

precipitates. The distribution of GP zones sizes with ageing time is shown schematically in Fig. 2.14. There is strong experimental support for the model proposed by Lorimer and Nicholson whereby GP zones formed below the GP zones solvus temperature can act as nuclei for the next stage in the ageing process, usually the intermediate precipitate, providing they have reached a critical size (d_{crit} in Fig. 2.14). On the basis of this model, alloys have been classified into three types.

1. Alloys for which the quench-bath temperature and the ageing temperature are both above the GP zones solvus. Such alloys show little or no response to age-hardening due to the difficulty of nucleating a finely dispersed precipitate. An example is the Al–Mg system in which quenching results in a very high level of supersaturation, but where hardening is absent in compositions containing less than 5–6% magnesium.
2. Alloys in which both the quench-bath and ageing temperatures are below the GP zones solvus, e.g. some Al–Mg–Si alloys.
3. Alloys in which the GP zones solvus lies between the quench-bath temperature and the ageing temperature. This situation is applicable in most age-hardenable aluminium alloys. Advantage may be taken of the nucleation of an intermediate precipitate from pre-existing GP zones of sizes above d_{crit} using two-stage or duplex ageing treatments. These are now applied to some alloys to improve certain properties and this is discussed in more detail in Section 3.4.5. They are particularly relevant with respect to the problem of stress-corrosion cracking in high-strength aluminium alloys.

2.2.3 Precipitate-free zones at grain boundaries

All alloys in which precipitation occurs have zones adjacent to grain boundaries which are depleted of precipitate and Fig. 2.15a shows comparatively wide

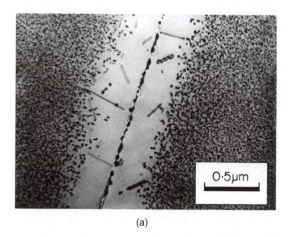

(a)

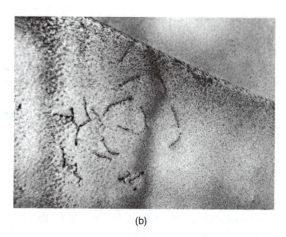

(b)

Fig. 2.15 (a) Wide PFZs in the alloy Al–4Zn–3Mg, aged 24 h at 150 °C; (b) effect of 0.3% silver on PFZ width and precipitate distribution in Al–4Zn–3Mg, aged 24 h at 150 °C (from Polmear, I. J., *J. Australian Inst. Met.* **17**, 1, 1972).

zones in an aged, high-purity Al–Zn–Mg alloy. These precipitate-free zones (PFZs) are formed for two reasons. First, there is a narrow (~50 nm) region either side of a grain boundary which is depleted of solute due to the ready diffusion of solute atoms into the boundary where relatively large particles of precipitate are subsequently formed. Second, the remainder of a PFZ arises because of a depletion of vacancies to levels below that needed to assist with nucleation of precipitates at the particular ageing temperature. It has been proposed that the distribution of vacancies near a grain boundary can take the form shown schematically in Fig. 2.16 (curve A) and that a critical concentration C_1 is needed before nucleation of the precipitate can occur at temperature T_1.

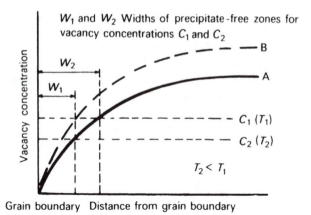

Fig. 2.16 Representation of profiles of vacancy concentration adjacent to a grain boundary in quenched alloys (from Taylor, J. L., *J. Inst. Metals*, **92**, 301, 1963–64).

The width of the PFZ can be altered by heat treatment conditions; the zones are narrower for higher solution treatment temperatures and faster quenching rates, both of which increase the excess vacancy content (e.g. curve B in Fig. 2.16), and for lower ageing temperatures. This latter effect has been attributed to a higher concentration of solute which means that smaller nuclei will be stable, thereby reducing the critical vacancy concentration required for nucleation to occur (C_2 in Fig. 2.16). However, the vacancy-depleted part of a PFZ may be absent in some alloys aged at temperatures below the GP zones solvus as GP zones can form homogeoneously without the need of vacancies.

2.2.4 Microalloying effects

In common with other nucleation and growth processes, precipitation reactions may be greatly influenced by the presence of minor amounts or traces of certain elements. These changes can arise for a number of reasons including:

1. Preferential interaction with vacancies which reduces the rate of nucleation of GP zones.
2. Raising the GP zones solvus which alters the temperature ranges over which phases are stable.
3. Stimulating nucleation of an existing precipitate by reducing the interfacial energy between precipitate and matrix.
4. Promoting formation of a different precipitate.
5. Providing heterogeneous sites at which existing or new precipitates may nucleate. These sites may be clusters of atoms or actual small particles.
6. Increasing supersaturation so that the precipitation process is stimulated.

   An early example of a microalloying effect was the role of minor additions of cadmium, indium or tin in changing the response of binary Al–Cu alloys to age-hardening. These elements reduce room temperature (natural) ageing because they react preferentially with vacancies and thereby retard GP zone formation (mechanism 1). On the other hand, both the rate and extent of hardening at elevated temperatures (artificial ageing) are enhanced (Fig. 2.17) because these trace elements promote precipitation of a finer and more uniform dispersion of the semi-coherent phase θ' (Al_2Cu) in preference to coherent θ'' (Table 2.3). It was first proposed that these elements are absorbed at the θ'/matrix interfaces, thereby lowering the interfacial energy required to nucleate θ' (mechanism 3). However, as shown in Fig. 2.18, recent observations have

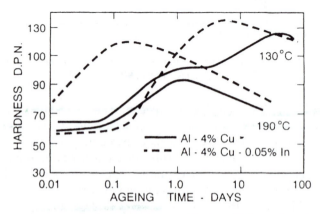

Fig. 2.17 Hardness–time curves for Al–4Cu and Al–4Cu–0.05In aged at 130 and 190 °C (after Hardy, H. K., *J. Inst. Metals*, **78**, 169, 1950–51).

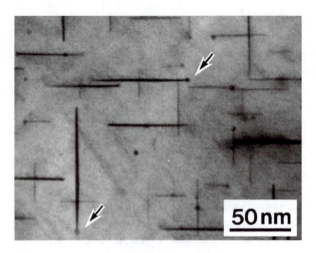

Fig. 2.18 Transmission electron micrograph showing θ' precipitates associated with small particles of tin (courtesy S. P. Ringer, K. Hono and T. Sakurai).

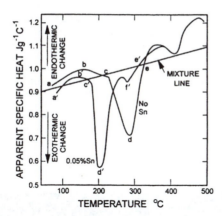

Fig. 2.19 Apparent specific heat-temperature curves showing thermal energy changes occurring during heating of as-quenched alloys Al–4Cu and Al–4Cu–0.05Sn at 2 °C a minute (from Polmear, I. J. and Hardy, H. K., *J. Inst. Metals*, **83**, 393, 1954–55).

indicated that θ' appears to be associated with small tin clusters or particles around 5 nm in diameter which suggests that heterogeneous nucleation has occurred at the sites (mechanism 5).

The potent effect of traces of tin in stimulating nucleation of θ' can also be demonstrated by observing changes in thermal energy that occur during heating in a differential scanning calorimeter. Fig. 2.19 compares the changes in the apparent specific heat during heating of the as-quenched alloys Al–4Cu and Al–4Cu–0.05Sn, and the areas cde and $c'd'e'$ between the mixture line (which assumes no phase changes) and the curves represent the respective heat evolutions associated with precipitation of θ'. Each value is approximately 17 J g^{-1} indicating that tin has not modified the volume fraction of θ' that has precipitated. What has changed is the rate of precipitation of θ' when tin is present. This effect is apparent in the tin-containing alloy in three ways - by the steep slope $c'd'$, the lower minimum value of the apparent specific heat, and the fact that it occurs at a lower temperature (200 °C) compared with 285 °C for Al–4Cu. This behaviour is characteristic of the way other microalloying additions affect precipitation in several aluminium alloys.

Another example of a microalloying effect is the role of small amounts of silver in modifying precipitation and promoting greater hardening in aluminium alloys that contain magnesium. Each system behaves differently. With Al–Zn–Mg alloys aged at elevated temperatures (e.g. Fig. 2.15b), silver stimulates the existing ageing process and this effect is attributed to an increase in the temperature range over which GP zones are stable (mechanism 2). In binary Al–Mg alloys, silver may induce precipitation in alloys in which normally it is absent (mechanism 6). Of particular interest is the Al–Cu–Mg system in which 0.1 at.% silver promotes formation of three new and quite different precipitates depending on the Cu:Mg ratio (Table 2.2).

Table 2.2 Details of precipitates formed by the addition of 0.4% (0.1 at.%) Ag to aged ternary Al–Cu–Mg alloys (Chopra H.D. *et al, Phil. Mag. Letters*, **71**, 319, 1995 and **73**, 351, 1996)

Cu:Mg Ratio	Precipitate	Crystal Structure	
High: e.g. Al–4Cu–0.3Mg	Ω	orthorhombic	a = 0.496 nm
			b = 0.859 nm
			c = 0.848 nm
Medium: e.g. Al–2.5Cu–1.5Mg	X′	c.p. hexagonal	a = 0.496 nm
			c = 1.375 nm
Low: e.g. Al–1.5Cu–4.0Mg	Z	cubic	a = 1.999 nm

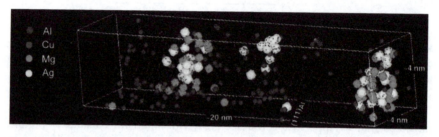

Fig. 2.20 Three dimensional elemental mapping of clusters in the alloy Al–4Cu–0.3Mg–0.4Ag aged 5 s at 180 °C. The silver and magnesium atoms are represented by large light and dark balls. The copper atoms are shown as small balls and the aluminium atoms are shown as dots. (courtesy M. Murayama and K. Hono).

Atom probe field ion microscopy has revealed that the effects of silver arise because clusters of silver and magnesium form within seconds after artificial ageing has commenced. Fig. 2.20 shows the appearance of these clusters that have formed after the alloy Al–4Cu–0.3Mg–0.4Ag has been aged for only 5 s at 180 °C. Copper atoms then diffuse to the clusters leading to nucleation of the precipitate Ω Al_2Cu (Fig.3.18) that grows along the $\{111\}_\alpha$ planes. As will be described in Section 3.4.1, Ω is relatively stable at elevated temperatures and its presence promote good creep resistance in aluminium alloys. Little is known of the properties of Al–Cu–Mg–Ag alloys hardened by the X′ or Z phases.

Minor additions (e.g. 0.2 wt.%) of scandium to aluminium alloys are also attracting interest despite the very high cost of extracting this metal from minerals containing rare earth elements, with which it is usually associated. Scandium combines with aluminium to form fine dispersions of coherent Al_3Sc particles that precipitate independently from any other phases that may be present in an alloy. Al_3Sc precipitates at relatively high temperatures (e.g. 350 °C), is resistant to coarsening, and in addition to causing dispersion strengthening, inhibits recrystallisation in wrought products at temperatures as high as 600 °C. Al_3Sc is an equilibrium phase with a cubic Ll_2 crystal structure and is isomorphous with a metastable form of the compound Al_3Zr that also inhibits recrystallisation. When added in the presence of zirconium, scandium combines to form particles of the stable compound $Al_3(Sc_x,Zr_y)$ that precipitate more rapidly and are more homogeneously dispersed than Al_3Zr.

Because extensive studies have already been made of the effects of major additions on the response of aluminium and other alloys to age hardening, it is to be expected that the role of microalloying elements will continue to receive attention. As indicated above, these minor elements can have important practical effects in changing ageing kinetics, microstructures and properties, some of which are discussed further when considering individual alloy systems.

2.2.5 Hardening mechanisms

Although early attempts to explain the hardening mechanisms in age-hardened alloys were limited by a lack of experimental data, two important concepts were postulated. One was that hardening, or the increased resistance of an alloy to deformation, was the result of interference to slip by particles precipitating on crystallographic planes. The other was that maximum hardening was associated with a critical particle size. Modern concepts of precipitation-hardening are essentially the consideration of these two ideas in relation to dislocation theory, since the strength of an age-hardened alloy is controlled by the interaction of moving dislocations with precipitates.

Obstacles to the motion of dislocations in age-hardened alloys are the internal strains around precipitates, notably GP zones, and the actual precipitates themselves. With respect to the former, it can be shown that maximum impedence to the dislocation motion, i.e. maximum hardening, is to be expected when the spacing between particles is equal to the limiting radius of curvature of moving dislocation lines, i.e. about 50 atomic spacings or 10 nm. At this stage the dominant precipitate in most alloys is coherent GP zones, and high-resolution transmission electron microscopy has revealed that these zones are, in fact, sheared by moving dislocations. Thus individual GP zones *per se* have only a small effect in impeding glide dislocations and the large increase in yield strength these zones may cause arises because of their high volume fraction.

Shearing of the zones increases the number of solute–solvent bonds across the slip planes in the manner depicted in Fig. 2.21 so that the process of clustering tends to be reversed. Additional work must be done by the applied stress in order for this to occur, the magnitude of which is controlled by factors such as relative atomic sizes of the atoms concerned and the difference in stacking-fault energy between matrix and precipitate. This so-called chemical hardening makes an additional contribution to the overall strengthening of the alloy.

Once GP zones are cut, dislocations continue to pass through the particles on the active slip planes and work-hardening is comparatively small. Deformation tends to become localized on only a few active slip planes so that some intense bands develop which allows dislocations to pile up at grain boundaries in the manner shown schematically in Fig. 2.22a. As will be discussed in Sections 2.5 and 3.4.6, the development of this type of microstructure may be

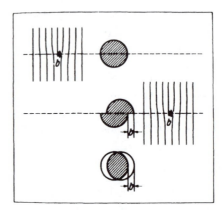

Fig. 2.21 Representation of the cutting of a fine particle, e.g. GP zone, by a moving dislocation (from Conserva, M. *et al.*, *Alumino E. Nuova Metallurgia*, **39**, 515, 1970).

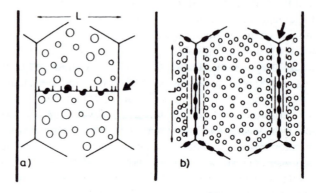

Fig. 2.22 (a) Shearing of fine precipitates leading to planar slip and dislocation pile-ups at grain boundaries; (b) Stress concentration at grain boundary triple points due to presence of precipitate-free zones (from Lütjering, G. and Gysler, A., *Aluminium Transformation, Technology and Applications*, American Society for Metals, Cleveland, Ohio, 171, 1980).

deleterious with respect to mechanical properties such as ductility, toughness, fatigue and stress corrosion.

If precipitate particles are large and widely spaced, they can be readily bypassed by moving dislocations which bow out between them and rejoin by a mechanism first proposed by Orowan (Fig. 2.23). Loops of dislocations are left around the particles. The yield strength of the alloy is low but the rate of work-hardening is high, and plastic deformation tends to be spread more uniformly throughout the grains. This is the situation with over-aged alloys and the typical age-hardening curve in which strength increases then decreases with ageing time has been associated with a transition from shearing (curve A) to bypassing

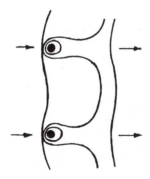

Fig. 2.23 Representation of a dislocation bypassing widely spaced particles.

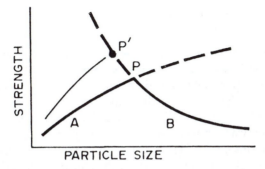

Fig. 2.24 Representation of relationship between strength and particle size for a typical age-hardening alloy: (A) particles sheared by dislocations; (B) particles not sheared (i.e. bypassed) by dislocations (from Nicholson R. B., *Strengthening Methods in Crystals*, Kelly, A. and Nicholson, R. B., Eds, Elsevier, Amsterdam, p. 535, 1971).

(curve B) of precipitates, as shown schematically in Fig. 2.24. In theory, the intersection point at P represents the maximum strength that can be developed in the alloy.

Accompanying the formation of the intermediate precipitate is the development of wider, precipitate-free zones adjacent to grain bound-aries as shown in Fig. 2.15a. These zones are relatively weak with respect to the age-hardened matrix and may deform preferentially leading to high stress concentrations at triple points (Fig. 2.22b) which, in turn, may cause premature cracking (Section 2.5).

The most interesting situation arises if precipitates are present which can resist shearing by dislocations and yet be too closely spaced to allow bypassing by dislocations. In such a case, the motion of dislocation lines would only be possible if sections can pass over or under individual particles by a process such as cross-slip. High levels of both strengthening and work-hardening would then be expected. Normally such precipitates are too widely spaced for this to occur. However, some success has been achieved through the following strategies in stimulating formation of dispersions of precipitates that resist cutting by dislocations:

1. Duplex ageing treatments first below, and then above the GP zones solvus temperature which enable finer dispersions of intermediate precipitates to be formed in some alloys (e.g. Section 3.4.5).
2. Co-precipitation of two phases, one which forms as finely dispersed zones or particles that contribute mainly to raising yield strength and the other as larger particles that resist shearing by dislocations so that plastic deformation is distributed more uniformly (e.g. Fig. 2.48b).
3. Co-precipitation of two or more intermediate phases, each of which forms on different crystallographic planes so that dislocation mobility is again reduced (e.g. Section 3.4.6).
4. Nucleation of uniform dispersions of intermediate precipitates by the addition of specific trace elements (e.g. Fig. 3.18).

An increase in the volume fraction of precipitate particles raises both curves A and B in Fig. 2.24 resulting in higher strength. Similarly, a decrease in the particle size of precipitates that are still capable of resisting shearing by dislocations will also raise the peak strength by moving P to P'. The critical size, D_c, of particles capable of resisting shearing varies with different precipitate phases and depends on crystal structure and morphology. For example, the T_1 phase (Al$_2$CuLi) that forms on the $\{111\}_\alpha$ planes in certain artificially aged, lithium-containing alloys has a smaller value of D_c than the phase θ' (Al$_2$Cu) that forms on the $\{100\}_\alpha$ planes (Table 2.3).

Modelling concepts are now permitting further refinements to be made in the understanding of the microstructural design of high-strength aluminium alloys. While it is agreed that the desired microstructure to obtain high strength combined with a high resistance to fracture is one that consists of a small volume fraction of very fine, hard particles, it is also recognised that a common feature in high-strength aluminium alloys is the presence of shear-resistant, plate-shaped precipitates that form on the $\{100\}_\alpha$ or $\{111\}_\alpha$ matrix planes, or rod-shaped precipitates that form in the $<100>_\alpha$ directions. Less attention has been paid to a quantitative analysis of the effects of particle shape and orientation because of a lack of appropriate versions of the Orowan equation that relate the critical resolved shear stress due to dispersion hardening to precipitate characteristics.

The version of the Orowan equation currently accepted for spherical particles is:

$$\Delta\tau = 2\left\{\frac{Gb}{4\pi\sqrt{1-v}}\right\}\left\{\frac{1}{\lambda}\right\}\left\{\ln\frac{D}{r_0}\right\}$$

Where $\Delta\tau$ = increment in critical resolved shear stress due to dispersion strengthening, v = Poissons ratio, G = shear modulus, b = Burgers vector of

Table 2.3 Probable precipitation processes in aluminium alloys of commercial interest

Alloy	Precipitates	Remarks
Al–Cu	GP zones as thin plates on $\{100\}_\alpha$ θ'' (formerly GP zones [2])	Usually single layers of copper atoms on $\{100\}_\alpha$. Coherent, probably two layers of copper atoms separated by three layers of aluminium atoms. May be nucleated at GP zones (Fig. 2.11).
	θ' tetragonal Al_2Cu $a = 0.404$ nm $c = 0.580$ nm	Semi-coherent plates nucleated at dislocations. Form on $\{100\}_\alpha$.
	θ body-centred tetragonal Al_2Cu $a = 0.607$ nm $c = 0.487$ nm	Incoherent equilibrium phase. May nucleate at surface of θ'.
Al–Mg (>5%)	Spherical GP zones	GP zones solvus below room temperature if <5%Mg and close to room temperature in compositions between 5 and 10%Mg.
	β' hexagonal $a = 1.002$ nm $c = 1.636$ nm	Probably semi-coherent. Nucleated on dislocations. $(0001)_\beta//(001)_\alpha:[01\bar{1}0]_\beta//[110]_\alpha$.
	β face-centered cubic Mg_5Al_8 (formerly Mg_2Al_3) $a = 2.824$ nm	Incoherent, equilibrium phase. Forms as plates or laths in grain boundaries and at a surface of β' particles in matrix. $(111)_\beta//(001)_\alpha:[110]_\beta//[010]_\alpha$.
Al–Si	Silicon diamond cubic $a = 0.542$ nm	Silicon forms directly from SSSS.

(Continued)

57

Table 2.3 (*Continued*)

Alloy	Precipitates	Remarks
Al–Cu–Mg	Disordered clusters of Cu & Mg atoms	Form rapidly in most compositions and promote hardening. May be an early stage of GP (Cu, Mg) zones.
	GP (Cu,Mg) zones as rods along $<100>_\alpha$	May form from clusters. Stable to relatively high temperatures. Also known as GPB zones.
	S' orthorhombic Al_2CuMg $a = 0.404$ nm $b = 0.925$ nm $c = 0.718$ nm	Semi-choherent and nucleated at dislocations (Fig. 2.13). Forms as laths on $\{210\}_\alpha$ along $<001>_\alpha$. Very similar to equilibrium S.
	S orthorhomic Al_2CuMg $a = 0.400$ nm $b = 0.923$ nm $c = 0.714$ nm	Incoherent equilibrium phase, probably transforms from S'. Note that precipitates from the Al–Cu system can also form in compositions with high Cu:Mg ratios.
Al–Mg–Si	Clusters & co-clusters of Mg & Si atoms GP zones	Form rapidly but cause little hardening (Fig. 2.32). GP zones solvus occurs at temperatures that are normally higher than the ageing temperatures. Zones seem spherical but structure not well defined.
	β" monoclinic $a = 1.534$ nm $b = 0.405$ nm $c = 0.683$ nm $\beta = 106°$	Coherent needles, lie along $\langle 100\rangle_\alpha$. $(010)_\beta''//(001)_\alpha$; $[001]_\beta''//[310]_\alpha$. Lattice dimensions of monoclinic cell are changed in alloys with high Si contents. Main strengthening precipitate. Forms from GP zones. Composition close to MgSi
	β' hexagonal $a = 0.705$ nm $c = 0.405$ nm	Semi-coherent rods, lie along $\langle 100\rangle_\alpha$. $(001)_\beta'//(100)_\alpha$; $[100]_\beta'//[011]_\alpha$. May form from β"? Composition close to $Mg_{1.7}Si$.
	B' hexagonal $a = 1.04$ nm $c = 0.405$ nm	Semi-coherent laths, lie along $\langle 100\rangle_\alpha$. $(0001)_B'//(001)_\alpha$; $(10\bar{1}0)_B'//[510]_\alpha$. Forms together with β'; favoured by high Si:Mg ratios. Two other (orthorhombic and hexagonal) precipitates have been detected. Composition close to MgSi.

Alloy system	Phase	Lattice parameters	Description
	β face-centred cubic Mg_2Si	$a = 0.639$ nm	Platelets on $\{100\}_\alpha$. May transform directly from β'. $(100)_\beta//(100)_\alpha$; $[110]_\beta//[100]_\alpha$.
Al–Zn–Mg	GP zones: two types		GP[1] Spherical, 1–1.5 nm, ordered. GP [11] thin Zn discs, 1–2 atom layers thick, form on $\{111\}_\alpha$. Partly ordered.
	η' (or M') hexagonal	$a = 0.496$ nm $c = 1.405$ nm	May form from GP zones in alloys with Zn:Mg > 3:1 $(0001)_{\eta'}//(111)_\alpha$; $[11\bar{2}0]_{\eta'}//[11\bar{2}]_\alpha$. Semi-coherent. Disc shaped. $a//<112>_\alpha$, $c//<111>_\alpha$. Composition close to MgZn (e.g. Fig. 2.15).
	η (or M) hexagonal $MgZn_2$	$a = 0.521$ nm $c = 0.860$ nm	Forms at or from η', may have one of nine orientation relationships with matrix. Most common are: $(10\bar{1}0)_\eta//(001)_\alpha$; $(0001)_\eta//(110)_\alpha$ and $(0001)_\eta//(1\bar{1}\bar{1})_\alpha$; $(10\bar{1}0)_\eta//(110)_\alpha$.
	T' hexagonal, probably Mg_{32} (Al, $Zn)_{49}$	$a = 1.388$ nm $c = 2.752$ nm	Semi-coherent. May form instead of η in alloys with high Mg:Zn ratios. $(0001)_{T'}//(111)_\alpha$; $(10\bar{1}1)_{T'}//(11\bar{2})_\alpha$.
	T cubic Mg_{32} (Al, $Zn)_{49}$	$a = 1.416$ nm	May form from η if ageing temperature > 190 °C, or from T' in alloy with high Mg:Zn ratios. $(100)_T//(112)_\alpha$; $[001]_T//[100]_\alpha$.
Al–Li–Mg	δ' cubic Al_3Li	$a = 0.404$ to 0.401 nm	Metastable coherent precipitate with ordered $Cu_3Au(L1_2)$ type superlattice (Fig. 3.36). Low misfit.
	Al_2LiMg cubic	$a = 1.99$ nm	Forms as coarse rods with $\langle110\rangle$ growth directions in alloys with $\geq 2\%$Mg. $(110)_{ppt}//(110)_{A1}$; $[110]_{ppt}//[111]_{A1}$.
Al–Li–Cu	δ' cubic Al_3Li δ cubic AlLi	$a = 0.637$ nm	As for Al–Li and Al–Li–Mg alloys. Nucleates heterogeneously, mainly in grain boundaries. $(100)_\delta//(100)_\alpha$; $(011)_\delta//(111)_\alpha$; $(011)_\delta//(112)_\alpha$.
	T_1 hexagonal Al_2LiCu	$a = 0.497$ nm $c = 0.934$ nm	Thin hexagonal shaped plates with $\{111\}$ habit plane. $(001)_{T_1}//\{111\}_\alpha$; $\langle10\bar{1}0\rangle_{T_1}//\langle110\rangle_\alpha$ (Figs 2.27 and 3.37).
	θ'', θ'		Phases present in binary Al–Cu alloys may also form at low Li:Cu ratios.

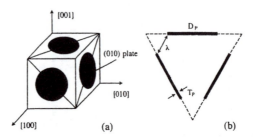

Fig. 2.25 (a) Circular $\{100\}_\alpha$ precipitate plates in a cubic volume of the aluminium matrix and (b) projection of intersected $\{100\}_\alpha$ precipitate plates on a $\{111\}_\alpha$ plane of the matrix (from Nie J.-F. et al, Mater. Sci. Forum, **217–222**, 1257, 1996).

the gliding dislocations, λ = effective inter-particle spacing, Dp = planar diameter of the precipitate particles, and r_0 = core radius of the dislocations. Within this equation, it is λ that varies with shape, orientation and distribution of the particles, and derivation of appropriate versions of the Orowan equation require the calculation of λ for different particle arrays. If it is assumed that $\{100\}_\alpha$ precipitate plates are circular discs of diameter D and thickness T distributed at the centre of each surface of a cubic volume of the matrix (Fig 2.25a), then the intersection of these plates with the $\{111\}_\alpha$ slip plane in the matrix will have a triangular distribution on this slip plane (Fig. 2.25b).

Calculations show that the effective planar inter-particle spacing for the $\{100\}_\alpha$ plates is given by $\lambda = 0.931(0.306\ \pi D\ T/f)^{1/2} - \pi D/8 - 1.061D$ where f = volume fraction of particles. Similar calculations for $\{111\}_\alpha$ precipitate plates show $\lambda = 0.931(0.265\ \pi DT/f)^{1/2} - \pi D/8 - 0.919T$, and for $<100>_\alpha$ rods $\lambda = 1.075D(0.433\ \pi/f)^{1/2} - (1.732D)^{1/2}$. Substitution of these expressions for λ in the Orowan equation shown above enables the critical resolved shear stresses to be determined for model alloys containing these three different precipitates. This analysis shows, quantitatively, that plate-shaped precipitates are more effective barriers to gliding dislocations than either rods or spherical precipitates. Furthermore, the increment of strengthening produced by $\{111\}_\alpha$ plates is invariably larger than that produced by $\{100\}_\alpha$ plates and, for both orientations, this increment becomes progressively larger as the aspect ratio (length to thickness) increases. Above a critical value of aspect ratio, plates on either set of planes form what is essentially a closed network that entraps gliding dislocations. The influences of precipitate orientation and shape are demonstrated in Fig. 2.26 and are in accord with the observed behaviour of high-strength aluminium alloys.

One example of high-strength alloys hardened by precipitates that form on the $\{111\}_\alpha$ planes is those based on the Al–Zn–Mg–Cu system (Table 2.3). Some of these commercial alloys develop yield strengths exceeding 600 MPa which are significantly higher than the yield strengths possible with alloys based on the Al–Cu system in which the precipitates form on the $\{100\}_\alpha$ planes. Another

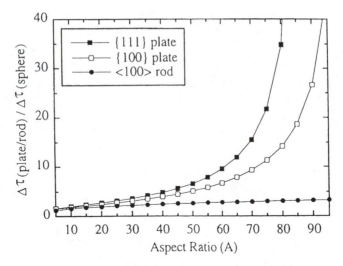

Fig. 2.26 Variation in ratio of critical resolved shear stress $\Delta\tau$ (plate,rod)/$\Delta\tau$ (sphere) with aspect ratio for Orowan strengthening attributable to $\{111\}_\alpha$ and $\{100\}_\alpha$ precipitate plates and $<100>_\alpha$ precipitate rods. Volume fraction of precipitates f = 0.05. (from Nie J.-F. *et al*, *Mater. Sci. Forum*, **217–222**, 1257, 1996).

example of a precipitate which forms on the $\{111\}_\alpha$ planes is the T_1 phase (Al$_2$CuLi) that was mentioned above. This phase has a particularly high aspect ratio and its ability to promote greater hardening than a much higher density of zones of the finer, shearable phase θ'' formed on the $\{100\}_\alpha$ planes is illustrated in an Al–Cu–Li–Mg–Ag–Zr alloy in Fig. 2.27. In this regard, it may be noted that yield stresses exceeding 700 MPa have been recorded for this alloy which are close to the theoretical upper limit for aluminium (approximately 900 MPa).

2.3 AGEING PROCESSES

2.3.1 Precipitation sequences

As mentioned earlier, several aluminium alloys display a marked response to age-hardening. By suitable alloying and heat treatment it is possible to increase the yield stress of high-purity aluminium by as much as 50 times. Details of the precipitates that may be present in alloy systems having commercial significance are shown in Table 2.3. The actual precipitate or precipitates that form in a particular alloy during ageing depends mainly on the ageing temperature. For example, for GP zones to form, ageing must be carried out below the relevant GP zones solvus temperature as mentioned in Section 2.2.2. If intermediate precipitates are formed, they may nucleate from pre-existing GP zones, at the sites of these zones, or independently depending on the alloy concerned. At some ageing temperatures, both GP zones and an intermediate precipitate

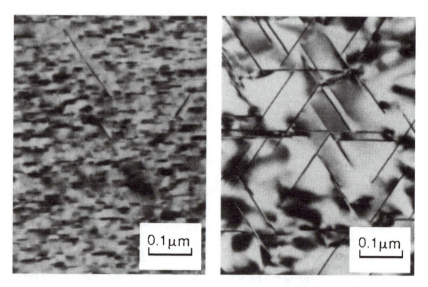

Fig. 2.27 Electron micrographs of the alloy Al–5.3Cu–1.3Li–0.4Mg–0.4Ag–0.16Zr: (a) quenched and aged 8 h at 160 °C showing finely dispersed, coherent θ'' particles and occasional plates of the T_1 phase. Hardness 146 DPN; (b) quenched, cold worked 6% and aged 8 h 160 °C showing a much coarser but uniform dispersion of semi-coherent T_1 plates. Hardness 200 DPN (courtesy S. P. Ringer) $b = <110>_\alpha$.

may be present together. Cold work prior to ageing increases the density of dislocations which may provide sites for the heterogeneous nucleation of specific precipitates.

A partial phase diagram for the Al–Cu system was shown in Fig. 2.1. Al–Mg–Si alloys can be represented as a pseudo-binary Al–Mg$_2$Si system (Fig. 2.28) and sections of the ternary phase diagrams for the important Al–Cu–Mg and Al–Zn–Mg systems are shown in Figs. 2.29 and 2.30. Most commercial alloys based on these systems have additional alloying elements present that modify the respective ternary diagrams and, in Fig. 2.31, an example is shown for the section at 460 °C for Al–Zn–Mg alloys containing 1.5% copper. Since this is close to the usual solution treatment temperature for alloys of this type, it should be noted that some quaternary compositions will not be single phase prior to quenching.

2.3.2 Clustering phenomena

Although the random clustering of solute atoms prior to precipitation in quenched and aged aluminium alloys was detected by small angle x-ray diffraction many years ago, the effects of this phenomenon on subsequent ageing

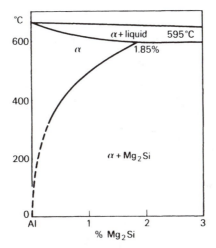

Fig. 2.28 Pseudo-binary phase diagram for Al–Mg₂Si.

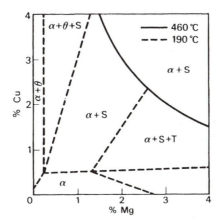

Fig. 2.29 Section of ternary Al–Cu–Mg phase diagram at 460 °C and 190 °C (estimated). θ = Al₂Cu, S = Al₂CuMg, T = Al₆CuMg₄.

processes have been little understood. Now there is evidence that clustering events may promote formation of existing precipitates in an alloy, stimulate nucleation of new precipitates as was discussed in Section 2.2.4, and contribute to the actual age hardening of certain alloys.

In the Al–Mg–Si system in which ageing processes are particularly complex, atom probe studies have shown that the formation of GP zones may be preceded initially by the appearance of individual clusters of magnesium and silicon atoms, followed by the formation of co-clusters of these elements. This

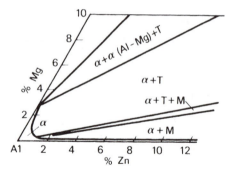

Fig. 2.30 Section of ternary Al–Zn–Mg phase diagrams at 200 °C. M = MgZn$_2$, T = Al$_{32}$ (Mg,Zn)$_{49}$.

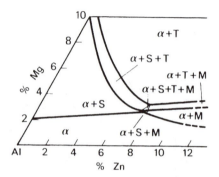

Fig. 2.31 Section of Al–Zn–Mg–Cu phase diagram (1.5% Cu) at 460 °C. S = Al$_2$CuMg, T = Al$_6$CuMg$_4$ + Al$_{32}$ (Mg,Zn)$_{49}$, M = MgZn$_2$ + AlCuMg.

behaviour is demonstrated in Fig. 2.32 which shows atom probe concentration profiles after ageing the alloy Al–1Mg–0.6Si for (a) 0.5 h and (b) 8 h at 70 °C. These profiles are developed by collecting, counting and identifying ions as they are evaporated from the tip of an alloy specimen in a field ion microscope. The clusters will, for example, form during a delay at ambient temperature after quenching and before artificial ageing. As described in Section 3.4.3, their presence may alter the dispersion of precipitates that form on subsequent ageing such that the response to hardening is reduced.

It is well known that age-hardening in most alloys based on the Al–Cu–Mg system occurs in two distinct stages over a wide temperature range (~100 to 240 °C). The first stage, which may account for 60–70% of the total hardening response, is characteristically and uniquely rapid, and may be completed within 60 s. This behaviour is then followed by what, for some compositions, can be a prolonged period (e.g. 100 h) during which the hardness shows little or no change, and after which there is a second rise to a peak value. Previously, this

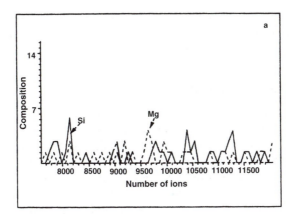

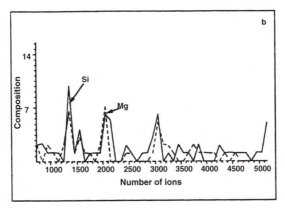

Fig. 2.32 Atom probe profiles showing evidence of clustering of magnesium and silicon atoms in the alloy 6061 (Al–1Mg–0.6Si) quenched and aged (a) 0.5 h and (b) 8 h at 70 °C (from Edwards G. A., et al, Applied Surf. Sci., **76, 77**, 219, 1994).

early hardening phenomenon was attributed to the rapid formation of GP(Cu,Mg) zones (also known as GPB zones). However, recent studies using high resolution electron microscopy and electron diffraction have not detected evidence of these zones until alloys are aged for times that places them further along the hardness plateau. Instead, atom probe field ion microscopy has revealed the presence of a high density (e.g. 10^{19} cm^{-3}) of small, disordered clusters of atoms immediately after rapid early hardening is completed, and the phenomenon has been termed "cluster hardening" to distinguish it from normal precipitation reactions. There is debate whether or not these clusters are actually small GP(Cu,Mg) zones. Also the mechanism by which they can cause hardening remains uncertain. One possibility is that rapid solute/dislocation interactions occur in which the relatively small copper atoms and large magnesium atoms immobilise edge dislocations by segregating preferentially to the respective compression and tension regions. Another suggestion is that hardening may arise because of differences in elastic modulii between aluminium and the Cu–Mg clusters.

2.3.3 Intermediate precipitates

It has generally been accepted that most intermediate precipitates which form in aged aluminium alloys have compositions and crystal structures that differ only slightly from those of the respective equilibrium precipitates. In fact, for Al–Cu–Mg alloys in which the equilibrium precipitate is S (Al_2CuMg), the intermediate precipitate S′ differs so little in its crystallographic dimensions that it is sometimes ignored. However, recent studies of some alloys using one dimensional and three dimensional atom probe field ion microscopy have revealed some unexpected compositional variations between intermediate and equilibrium precipitates.

One example is the Al–Mg–Si system in which the compositions of the intermediate precipitates β″ and β′ were assumed to be the same as the equilibrium precipitate β (Mg_2Si) (Table 2.3). Because of this, the compositions of some commercial alloys have been designed deliberately to have a balanced (2:1) atomic ratio of magnesium and silicon in order to maximize precipitation of β″ and β′ during ageing. Now there is strong experimental evidence that the actual Mg:Si ratios of these intermediate precipitates are close to 1:1. As mentioned in Section 3.4.3, this has opened up the prospect of producing a new range of Al–Mg–Si alloys in which the magnesium content has been reduced to improve their hot working characteristics. Al–Zn–Mg–(Cu) alloys are others in which atom probe studies have shown that the Mg:Zn ratio for the intermediate precipitate η′ differs substantially from that of the equilibrium precipitate η ($MgZn_2$). In this case the Mg:Zn ratio appears to lie in the range 1:1 to 1:1.15 rather than the expected 1:2. This suggests that the composition of η′ is linked more to the pre-existing GP zones than to the equilibrium precipitate η, and supports the suggestion that η′ can nucleate directly from these zones. These new observations about the compositions of some intermediate precipitates means that a substantial number of atom positions in their unit cells must still be occupied by aluminium atoms rather than the respective solute atoms.

2.3.4 Secondary precipitation

For many years there was an implicit acceptance that, once an alloy had been aged at an elevated temperature, its mechanical properties remained stable on exposure for an indefinite time at a significantly lower temperature. However, it was found that highly saturated Al–Zn alloys aged at 180 °C will continue to age and undergo what has been termed "secondary precipitation" if cooled and then held at ambient temperature. Similar behaviour has also been observed in highly saturated lithium-containing aluminium alloys aged first at 170 °C and then exposed at temperatures in the range 60–130 °C. In this case, there is a progressive increase in hardness and mechanical strength accompanied by an unacceptable decrease in ductility and toughness that is attributed to secondary precipitation of the finely dispersed δ' throughout the matrix. More recently, observations on a wide range of aluminium alloys have shown that secondary precipitation is, in fact, a more general phenomenon. This conclusion is supported by results obtained using the technique of positron annihilation spectroscopy which have indicated that vacancies may be retained and remain mobile at ambient temperatures after aged aluminium alloys are cooled from a higher ageing temperature.

Positron annihilation spectroscopy is proving to be a powerful tool for studying the role of lattice defects, such as vacancies, in the decomposition kinetics of aged alloys. This technique involves measurement of the lifetimes of positrons emitted from a radioactive source before there are annihilated by interacting with electrons in the alloy. The annihilation process may be thought of as a chemical reaction that has a rate which is directly proportional to the local electron density. Vacant lattice sites can be detected because they are open volume defects and offer temporary "shelter" to incident positrons, thereby slowing their annihilation rate. As an example, Fig. 2.33 shows the evolution of positron lifetimes during ageing the alloy Al–4Cu–0.3 Mg at 20 °C after first being solution treated, quenched and aged for various times at 180 °C. The increases in positron lifetimes, which are comparatively large for alloys initially aged for short times (30 and 120 s) at 180 °C, but still significant for longer

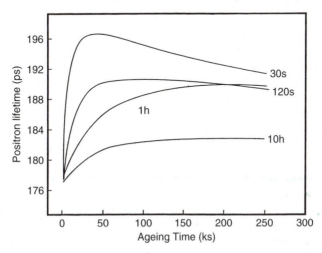

Fig. 2.33 Positron lifetimes during secondary ageing Al–4Cu–0.3 Mg at 20 °C after solution treatment at 520 °C, quenching at 0 °C and first ageing 30 s, 120 s, 1 h or 10 h at 180 °C (from Somoza, A. *et al, Phys. Rev. B*, **61**, 14454, 2000).

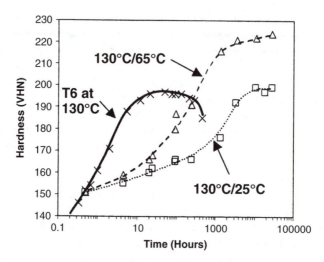

Fig. 2.34 Hardness-time curves for the Al–Zn–Mg–Cu alloy 7075 aged at 130 °C (solid line), and underaged 0.5 h at 130 °C, quenched to 25 °C and held either at this temperature or at 65 °C (From Lumley, R. N. *et al.*, *Mater Sci and Technology*, 2005, 22, 1025).

times of 1 and 10 h, are all taken to indicate that retained vacancies (and solute atoms) are mobile at 20 °C thereby allowing further ageing to occur.

Detailed studies of secondary precipitation in a wide range of aluminium alloys have revealed that the levels of residual or "free" solute that remain in solid solution until the alloys are in the overaged condition is higher than has generally been accepted. However, as expected from Fig. 2.33, the response to secondary precipitation at the lower ageing temperature is greater if an alloy is first artificially aged for a short time (i.e. underaged). The behaviour is illustrated in Fig. 2.34 for the Al–Zn–Mg–Cu alloy 7075 which normally reaches a peak hardness of 195 DPN if aged at 130 °C for 24 h (T6 temper). If this alloy is first underaged for 0.5 h at 130 °C and quenched to 25 °C, the hardness is 150 DPN. During prolonged secondary ageing at this lower temperature the hardness gradually increases and it becomes equal to the T6 value. Alternately, if 7075 is held at the slightly higher temperature of 65 °C after quenching from 130 °C, the hardness increases faster and reaches the higher value of 225 DPN after 10 000 h.

During secondary ageing at the lower temperature, GP zone formation usually occurs. If, after a dwell period, ageing at the initial elevated temperature is resumed, then the microstructure at peak hardness is refined so that a greater overall response to hardening can be achieved than is possible using a single stage artificial ageing treatment. This effect is shown for the alloy Al–4%Cu in Fig. 2.35 in which the multi-stage ageing schedule is given the designation "T6I6", where "I" means that artificial ageing at an elevated temperature has been interrupted. As shown later in Table 3.7, experimental interrupted ageing

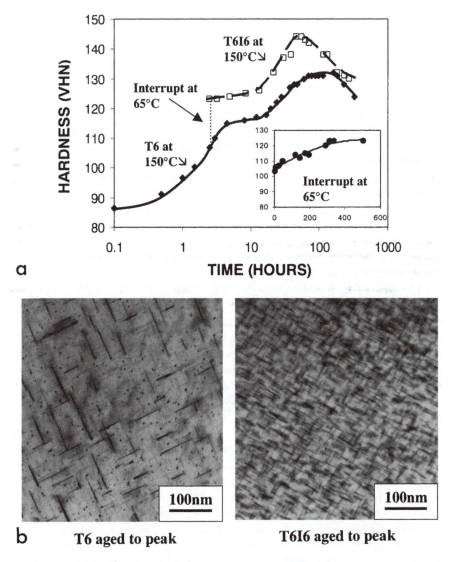

a

Fig. 2.35 (a) Differences in the hardness curves for Al–4%Cu artificially aged at 150 °C, with and without an interrupted period of secondary ageing at 65 °C. The inset plot shows the hardness change during this dwell period at 65 °C. (b) Transmission electron micrographs in the $[001]_\alpha$ direction showing dispersions of the θ' precipitate plates and minor amounts of the θ'' phase in these two aged conditions.

T6 temper: 100 h at 150 °C

T6I6 temper: 3 h at 150 °C, quench, 500 h at 65 °C, and 50 h at 150 °C (from Lumley, R.N. et al, Mater. Sci. and Technology, **19**, 1453, 2003).

cycles have been developed that enable simultaneous increases to be achieved in tensile and fracture toughness properties of a wide range of aluminium alloys.

2.4 CORROSION

2.4.1 Surface oxide film

Aluminium is an active metal which will oxidize readily under the influence of the high free energy of the reaction whenever the necessary conditions for oxidation prevail. Nevertheless, aluminium and its alloys are relatively stable in most environments due to the rapid formation of a natural oxide film of alumina on the surface that inhibits the bulk reaction predicted from thermo-dynamic data. Moreover, if the surface of aluminium is scratched sufficiently to remove the oxide film, a new film quickly re-forms in most environments. As a general rule, the protective film is stable in aqueous solutions of the pH range 4.5–8.5, whereas it is soluble in strong acids or alkalis, leading to rapid attack of the aluminium. Exceptions are concentrated nitric acid, glacial acetic acid and ammonium hydroxide.

The oxide film formed on freshly rolled aluminium exposed to air is very thin and has been measured as 2.5 nm. It may continue to grow at a decreasing rate for several years to reach a thickness of some tens of nanometres. The rate of film growth becomes more rapid at higher temperatures and higher relative humidities, so in water it is many times that occurring in dry air. In aqueous solutions, it has been suggested that the initial corrosion product is aluminium hydroxide, which changes with time to become a hydrated aluminium oxide. The main difference between this film and that formed in air is that it is less adherent and so is far less protective.

Much thicker surface oxide films that give enhanced corrosion resistance to aluminium and its alloys can be produced by various chemical and electro-chemical treatments. The natural film can be thickened some 500 times, to say 1–2 μm, by immersion of components in certain hot acid or alkaline solutions. Although the films produced are mainly Al_2O_3, they also contain chemicals such as chromates which are collected from the bath to render them more corrosion resistant. A number of proprietary solutions are available and the films they produce are known generally as conversion coatings. Even thicker, e.g. 10–20 μm, surface films are produced by the more commonly used treatment known as anodizing. In this case the component is made the anode in an electrolyte, such as an aqueous solution containing 15% sulphuric acid, which produces a porous Al_2O_3 film that is subsequently sealed, i.e. rendered non-porous, by boiling in water. Both conversion and anodic coatings can be dyed to give attractive colours and the latter process is widely applied to architectural products.

It should be noted that chromate conversion coatings are widely used in corrosion protection schemes for aluminium alloys in aircraft structures and other applications. However, it has been recognized recently that chromates may present a health hazard which has led to an interest in other, non-toxic, coating processes. Promising results have been reported for cerium-rich coatings which can be applied by several methods. A durable cerium oxide/hydroxide film replaces natural Al_2O_3 and protection is afforded by partial or complete suppression of the reduction of oxygen at cathodic sites which normally occurs during electrolytic corrosion.

Chemicals known as inhibitors may be added to potentially corrosive liquid environments for the purpose of minimizing or preventing corrosion of aluminium and its alloys. Inhibitors may be classified as being anodic, cathodic, or mixed depending on whether they mainly affect the anodic, cathodic, or both anodic and cathodic corrosion processes. Anodic inhibitors stifle the anodic reaction, usually by depositing on the surface sparingly soluble substances as a direct anodic product. There is often no change in the surface appearance. Chromates are commonly used for this purpose and are generally effective. However, they can cause problems if present in insufficient amounts because they may decrease the surface area under attack without decreasing appreciably the amount of metal dissolution. Such a situation may lead to intensified local attack, e.g. by pitting. Cathodic inhibitors are safer in this respect. They serve to stifle the cathodic reaction either by restricting access to oxygen, or by "poisoning" local spots favourable for cathodic hydrogen evolution. They form a visible film on the aluminium and are usually less efficient than anodic inhibitors as they do not completely prevent attack. Examples are phosphates, silicates and soluble oils. The choice and concentration of an inhibitor depends on several factors such as the compositions of the alloy and the liquid environment to which it is to be exposed, the temperature, and the rate of movement of the liquid. An inhibitor that offers protection in one environment may increase it in another.

2.4.2 Contact with dissimilar metals

The electrode potential of aluminium with respect to other metals becomes particularly important when considering galvanic effects arising from dissimilar metal contact. Comparisons must be made by taking measurements in the same solution and Table 2.4 shows the electrode potentials with respect to the 0.1 M calomel electrode (Hg–$HgCl_2$, 0.1 M KCl) for various metals and alloys immersed in an aqueous solution of 1 M NaCl and 0.1 M H_2O_2. The value for aluminium is -0.85 V whereas aluminium alloys range from -0.69 V to -0.99 V. Magnesium which has an electrode potential of -1.73 V is more active than aluminium whereas mild steel is cathodic having a value of -0.58 V.

Table 2.4 suggests that sacrificial attack of aluminium and its alloys will occur when they are in contact with most other metals in a corrosive environment. However, it should be noted that electrode potentials serve only as a guide to the possibility of galvanic corrosion. The actual magnitude of the

Table 2.4 Electrode potentials of various metals and alloys with respect to the 0.1 M calomel electrode in aqueous solutions of 53 g l^{-1} NaCl and 3 g l^{-1} H$_2$O$_2$ at 25 °C (from *Metals Handbook*, Volume 1, American Society for Metals, Cleveland, Ohio, 1961)

Metal or alloy		Potential (V)
Magnesium		-1.73
Zinc		-1.10
Alclad 6061, Alclad 7075		-0.99
5456, 5083		-0.87
Aluminium (99.95%), 5052, 5086	aluminium	-0.85
3004, 1060, 5050	alloys*	-0.84
1100, 3003, 6063, 6061, Alclad 2024		-0.83
2014–T4		-0.69
Cadmium		-0.82
Mild steel		-0.58
Lead		-0.55
Tin		-0.49
Copper		-0.20
Stainless steel (3xx series)		-0.09
Nickel		-0.07
Chromium		-0.49 to $+0.18$

*Compositions corresponding to the numbers are given in Tables 3.2 and 3.4

galvanic corrosion current is determined not only by the difference in electrode potentials between the particular dissimilar metals but also by the total electrical resistance, or polarization, of the galvanic circuit. Polarization itself is influenced by the nature of the metal/liquid interface and more particularly by the oxides formed on metal surfaces. For example, contact between aluminum and stainless steels usually results in less electrolytic attack than might be expected from the relatively large difference in the electrode potentials, whereas contact with copper causes severe galvanic corrosion of aluminium even though this difference is less.

Galvanic corrosion of aluminium and its alloys may be minimized in several ways. If contact with other metals cannot be avoided, these should be chosen so that they have electrode potentials close to aluminium; alternatively it may be possible to locate a dissimilar metal joint away from the corrosive environment. If not then complete electrical isolation of aluminium and the other metal must be arranged by using non-conducting washers, sleeves or gaskets. When paint is used for protection, it should be applied to the cathodic metal and not the aluminium. This practice is required because pinholes that may form in a paint film on aluminium can lead to pitting attack because of the large cathode to anode area ratio. In a closed loop system, such as used for cooling automobile engines, a mixed anodic and cathodic inhibitor should be added to the cooling water.

2.4.3 Influence of alloying elements and impurities

Alloying elements may be present as solid solutions with aluminium, or as micro-constituents comprising the element itself, e.g. silicon, a compound between one or more elements and aluminium (e.g. Al_2CuMg) or as a compound between one or more elements (e.g. Mg_2Si). Any or all of the above conditions may exist in a commercial alloy. Table 2.5 gives values of the electrode potentials of some aluminium solid solutions and micro-constituents.

In general, a solid solution is the most corrosion resistant form in which an alloy may exist. Magnesium dissolved in aluminium renders it more anodic although dilute Al–Mg alloys retain a relatively high resistance to corrosion, particularly to sea water and alkaline solutions. Chromium, silicon and zinc in solid solution in aluminium have only minor effects on corrosion resistance although zinc does cause a significant increase in the electrode potential. As a result, Al–Zn alloys are used as clad coatings for certain aluminium alloys (see Section 3.1.5) and as galvanic anodes for the cathodic protection of steel structures in sea water. Copper reduces the corrosion resistance of aluminium more than any other alloying element and this arises mainly because of its presence in micro-constituents. However, it should be noted that when added in small amounts (0.05–0.2%), corrosion of aluminium and its alloys tends to become more general and pitting attack is reduced. Thus, although under corrosive conditons the overall weight loss is greater, perforation by pitting is retarded.

Table 2.5 Electrode potentials of aluminum solid solutions and micro-constituents with respect to the 0.1 M calomel electrode in aqueous solutions of 53 g l^{-1} NaCl and 3 g l^{-1} H_2O_2 at 25 °C (from *Metals Handbook, Volume I*, American Society for Metals, Cleveland, Ohio, 1961)

Solid solution or micro-constituent	Potential (V)
Mg_5Al_8	−1.24
Al–Zn–Mg solid solution (4% $MgZn_2$)	−1.07
$MgZn_2$	−1.05
Al_2CuMg	−1.00
Al–5% Mg solid solution	−0.88
$MnAl_6$	−0.85
Aluminium (99.95%)	−0.85
Al–Mg–Si solid solution (1% Mg_2Si)	−0.83
Al–1% Si solid solution	−0.81
Al–2% Cu supersaturated solid solution	−0.75
Al–4% Cu supersaturated solid solution	−0.69
$FeAl_3$	−0.56
$CuAl_2$	−0.53
$NiAl_3$	−0.52
Si	−0.26

Micro-constituents are usually the source of most problems with electro-chemical corrosion as they lead to non-uniform attack at specific areas of the alloy surface. Pitting and intergranular corrosion are examples of localized attack (Fig. 2.36), and an extreme example of this is that components with a marked directionality of grain structure show exfoliation (layer) corrosion (Fig. 2.37). In exfoliation corrosion, delamination of surface grains or layers occurs under forces exerted by the voluminous corrosion products.

Fig. 2.36 Microsection of surface pits in a high-strength aluminium alloy. Note that intergranular stess-corrosion cracks are propagating from the base of these pits ($\times$ 100).

Fig. 2.37 Microsection showing exfoliation (layer) corrosion of an aluminium alloy plate ($\times$ 100).

Iron and silicon occur as impurities and form compounds most of which are cathodic with respect to aluminium. For example, the compound Al_3Fe provides points at which the surface oxide film is weak, thereby promoting electrochemical attack. The rate of general corrosion of high-purity aluminium is much less than that of the commercial-purity grades which is attributed to the smaller size and number of these cathodic constituents throughout the grains. However, it should be noted that this may be a disadvantage in some environments as attack of high-purity aluminium may be concentrated in grain boundaries. Nickel and titanium also form cathodic phases although nickel is present in very few alloys. Titanium, which forms Al_3Ti, is commonly added to refine grain size (Section 3.1.3) but the amount is too small to have a significant effect on corrosion resistance. Manganese and aluminium form Al_6Mn, which has almost the same electrode potential as aluminium, and this compound is capable of dissolving iron which reduces the detrimental effect of this element. Magnesium in excess of that in solid solution in binary aluminium alloys tends to form the strongly anodic phase Mg_5Al_8 which precipitates in grain boundaries and promotes intercrystalline attack. However, magnesium and silicon, when together in the atomic ratio 2:1, form the phase Mg_2Si which has a similar electrode potential to aluminium.

Where a basic alloy is vulnerable to corrosive attack, it is possible to provide surface protection for wrought products such as sheet, plate, and, to a lesser extent, tube and wire by means of metallurgically bonded, thin layers of pure aluminium or an aluminium alloy. Such alloys are commonly those based on the Al–Cu–Mg and Al–Zn–Mg–Cu systems and products are said to be alclad (Fig. 3.10). In a corrosive environment, the cladding will anodic with respect to the core and provide sustained electrochemical protection at abraded or corroded areas, as well as at exposed edges.

2.4.4 Crevice corrosion

If an electrolyte penetrates a crevice formed between two aluminium surfaces in contact, or between an aluminium surface and a non-metallic material such as a gasket or washer, localised corrosion may occur by etching or pitting. The oxygen content of the liquid in the crevice is consumed by the reaction at the aluminium surfaces and corrosion will be inhibited if replenishment of oxygen by diffusion into the crevice is slow. However, if oxygen remains plentiful at the mouth of the crevice, a localised electrolytic cell will be created in which the oxygen-depleted region becomes the anode. Furthermore, once crevice attack has been initiated, this anodic area becomes acidic and the larger external cathodic area becomes alkaline. These changes further enhance local cell action and more corrosion occurs in the crevice, particularly in a submerged situation.

A common example of crevice corrosion occurs when water is present in the restricted space between layers of aluminium sheets or foil in close contact in stacks or coils. This may take place by condensation during storage if the metal temperature falls below the dew point, or by the ingress of rain when being transported. Irregular stain patches may form which impair the surface appearance. In severe cases, the corrosion product may cement two surfaces together and make separation difficult.

A special form of crevice corrosion may occur on an aluminium surface that is covered by an organic coating. It takes the form of tracks of thread-like filaments and has the name filiform corrosion. The tracks proceed from one or more places where the coating is breached for some reason and the corrosion products raise bulges in the surface. The amount of aluminium consumed is small and filiform corrosion only assumes practical importance if the metal is of thin cross-section. Filiform corrosion occurs only in the atmosphere and relative humidity is the key factor. It has been observed on lacquered aluminium surfaces in aircraft exposed to marine or high-humidity environments and may be controlled by anodising, chemical conversion coatings or by using chromate-containing primers prior to painting.

2.4.5 Cavitation corrosion

Protective films on the surfaces of aluminium and its alloys may be removed by mechanical actions of many sorts such as turbulent effects arising from moving fluid. Electrolytic reactions may then occur which can proceed without inhibition. If voids (gas bubbles) form in the turbulent liquid because the pressure falls below the vapour pressure, then cavitation corrosion may take place. Collapse of these voids at the metal surface allows the sudden release the latent heat of vaporisation, which may dislodge a protective film during service and even alter the state of work hardening of the metal at the surface. Cavitation corrosion therefore combines electrochemical action with mechanical damage, the relative proportion of each being controlled by the severity of the turbulence and the aggressiveness of the environment.

Weight loss in standard tests on aluminium alloys has been found to decrease with as strength (and hardness) increased, However, compared with other non-ferrous metals and alloys, aluminium and its alloys do not perform well under cavitation conditions. For example, common wrought aluminium alloys, that are considered to be relatively resistant to corrosion, have been found to suffer weight losses 100 to 200 times greater than aluminium bronze under cavitation conditions in fresh water.

2.4.6 Waterline corrosion

This form of corrosion can affect semi-submerged structures, such as ships, whereby the zone very close to the air/water boundary can suffer differential corrosion that is sometimes severe. With aluminium alloys, waterline corrosion may arise because of a difference in the chloride level between the sea water at the air/water parting line and that contained in the meniscus formed by capillary action in which the chlorides become concentrated by evaporation. This effect is weak in water that is in motion because the meniscus is constantly being renewed. Although aluminium alloys used for the hulls of ships and other semi-submerged structures are not very sensitive to waterline corrosion, this region should be painted to avoid the risk of attack. If the water is stagnant, painting is essential.

2.4.7 Metallurgical and thermal treatments

Treatments that are carried out to change the shape and achieve a desired level of mechanical properties in aluminium alloys may also modify corrosion resistance, largely through their effects on both the quantity and the distribution of micro-constituents. In this regard, the complex changes associated with ageing or tempering treatments are on a fine scale and these are considered in Chapter 3. Both mechanical and thermal treatments can introduce residual stresses into components which may contribute to the phenomenon of stress-corrosion cracking and this is discussed in Section 2.5.4.

If one portion of an alloy surface receives a thermal treatment different from the remainder of the alloy, differences in potential between these regions can result. Welding processes provide an extreme example of such an effect and differences of up to 0.1 V may exist between the weld bead, heat-affected zones and the remainder of the parent alloy.

Most wrought products do not undergo bulk recrystallization during subsequent heat treatment so that the elongated grain structure resulting from mechanical working is retained. Three principal directions are recognized: longitudinal, transverse (or long transverse) and short transverse, and these are represented in Fig. 2.38. This directionality of grain structure is significant in components when corrosion processes involve intercrystalline attack as has been illustrated by exfoliation corrosion. It is particularly important in regard to stress–corrosion cracking, which is discussed in Section 2.5.4.

In certain products such as extrusions and die forgings, working is non-uniform and a mixture of unrecrystallized and recrystallized grain structures may form between which potential differences may exist. Large, recrystallized grains normally occur at the surface (see Fig. 3.9) and these are usually slightly cathodic with respect to the underlying, unrecrystallized grains. Preferential attack may occur if the relatively more anodic internal grains are partly exposed as may occur by machining.

2.5 MECHANICAL BEHAVIOUR

The principal microstructural features that control the mechanical properties of aluminium alloys are as follows:

1. Coarse intermetallic compounds (often called constituent particles) that form interdendritically by eutectic decomposition during ingot solidification. One group comprises virtually insoluble compounds that usually contain the

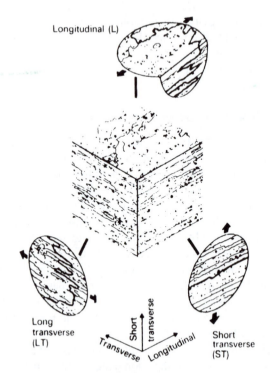

Fig. 2.38 The three principal directions with respect to the grain structure in a wrought aluminium alloy. Note the appearance of cracks that may form when stressing in these three directions (from Speidel, M. O. and Hyatt, M. V., *Advances in Corrosion Science and Technology*, Plenum Press, New York, 1972).

impurity elements iron or silicon and examples are $Al_6(Fe,Mn)$, Al_3Fe, $\alpha Al(Fe,Mn,Si)$ and Al_7Cu_2Fe. The second group, which are known as the soluble constituents, consists of equilibrium intermetallic compounds of the major alloying elements. Typical examples are Al_2Cu, Al_2CuMg, and Mg_2Si. Both types of particles form as lacy networks surrounding the cast grains and one purpose of the process referred to as preheating or ingot homogenization (Section 3.1.4) is to dissolve the soluble constituents. During subsequent fabrication of the cast ingots, the largest of the remaining particles usually fracture, which reduces their sizes to the range 0.5–10 μm and causes them to become aligned as stringers in the direction of working or metal flow (Fig. 2.39)

Constituent particles serve no useful function in high-strength wrought alloys and they are tolerated in most commercial compositions because their removal would necessitate a significant cost increase. They do, however, serve a useful purpose in certain alloys such as those used for canstock (Section 3.6.5).

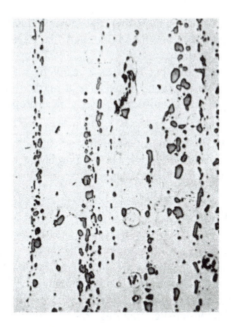

Fig. 2.39 Aligned stringers of coarse intermetallic compounds in a rolled aluminium alloy ($\times$ 250).

2. Smaller submicron particles, or dispersoids (typically 0.05–0.5 μm) that form during homogenization of the ingots by solid state precipitation of compounds containing elements which have modest solubility and which diffuse slowly in solid aluminium. Once formed, these particles resist either dissolution or coarsening. The compounds usually contain one of the transition metals and examples are $Al_{20}Mn_3Cu_2$, $Al_{12}Mg_2Cr$ and Al_3Zr. They serve to retard recrystallization and grain growth during processing and heat treatment of the alloys concerned. Moreover, they may also exert an important influence on certain mechanical properties through their effects both on the response of some alloys to ageing treatments, and on dislocation substructures formed as a result of plastic deformation.
3. Fine precipitates (up to 0.1 μm) which form during age-hardening and normally have by far the largest effect on strengthening of alloys that respond to such treatments.
4. Grain size and shape. The most significant microstructural feature that differentiates wrought products such as sheet from plate, forgings and extrusions is the degree of recrystallization. Aluminium dynamically recovers during hot deformation producing a network of subgrains and this characteristic is attributed to its relatively high stacking-fault energy. However, thick sections, which experience less deformation, usually do not undergo bulk recrystallization during processing so that an elongated grain structure is retained (Fig. 2.38)

5. Dislocation substructure, notably that caused by cold working of those alloys which do not respond to age-hardening, and that developed due to service stresses.
6. Crystallographic textures that form as a result of working and annealing, particularly in rolled products. They have a marked effect on formability (Section 2.1.4) and lead to anisotropic mechanical properties.

Each of these features may be influenced by the various stages involved in the solidification and processing of wrought and cast alloys and these are discussed in detail in Chapters 3 and 4. Here it is relevant to consider how these features influence mechanical behaviour.

2.5.1 Tensile properties

Aluminium alloys may be divided into two groups depending upon whether or not they respond to precipitation hardening. The tensile properties of commercial wrought and cast compositions are considered in Chapters 3 and 4 respectively. For alloys that do respond to ageing treatments, it is the finely dispersed precipitates that have the dominant effect in inhibiting dislocation motion, thereby raising yield and tensile strengths. For the other group, the dislocation substructure produced by cold-working in the case of wrought alloys and the grain size of cast alloys are of prime importance.

Coarse intermetallic compounds have relatively little effect on yield or tensile strength but they can cause a marked loss of ductility in both the cast and wrought products. The particles may crack at small plastic strains forming internal voids which, under the action of further plastic strain, may coalesce leading to premature fracture.

As mentioned above, the fabrication of wrought products may cause highly directional grain structures. Moreover, the coarse intermetallic compounds and smaller dispersoids also become aligned to form stringers in the direction of metal flow (Fig. 2.39). These microstructural features are known as mechanical fibring and, together with crystallographic texturing (Section 2.1.4), they cause anisotropy in tensile and other properties. Accordingly, measurements are often made in the three principal directions shown in Fig. 2.38. Tensile properties, notably ductility, are greatest in the longitudinal direction and least in the short transverse direction in which stressing is normal to the stringers of intermetallics, e.g. Table 2.6.

2.5.2 Toughness

Early work on the higher strength aluminium alloys was directed primarily at maximizing tensile properties in materials for aircraft construction. More recently, the emphasis in alloy development has shifted away from tensile strength as an over-riding consideration and more attention is being given to the behaviour of alloys under the variety of conditions encountered in service.

Table 2.6 Variation in tensile properties with direction in 76 mm thick aluminium alloy plates (from Forsyth, P.J.E. and Stubbington, A., *Metals Technology*, **2**, 158, 1975)

Alloy direction	0.2% proof stress (MPa)	Tensile strength (MPa)	Elongation (%)
Al–Zn–Mg–Cu (7075)			
Longitudinal (L)	523	570	15.5
Long transverse (LT)	482	552	12.0
Short transverse (ST)	445	527	7.5
ST/L ratio	0.85	0.93	0.48
Al–Cu–Mg (2014)			
Longitudinal (L)	441	477	14.0
Long transverse (LT)	423	471	10.5
Short transverse (ST)	404	449	4.0
ST/L ratio	0.91	0.94	0.29

Tensile strength controls resistance to failure by mechanical overload but, in the presence of cracks and other flaws, it is the toughness (and more particularly the fracture toughness) of the alloy that becomes the most important parameter. In common with other metallic materials, the toughness of aluminium alloys decreases as the general level of strength is raised by alloying and heat treatment. Minimum fracture toughness requirements become more stringent and, in the high-strength alloys, it is necessary to place a ceiling on the level of yield strength that can be safely employed by the designer.

Crack extension in commercial aluminium alloys proceeds by the ductile, fibrous mode involving the growth and coalescence of voids nucleated by cracking or by decohesion at the interface between second phase particles and the matrix (Fig. 2.40). Consequently, the important metallurgical factors are:

1. The distribution of the particles that crack.
2. The resistance of the particles and their interfaces with the matrix to cleavage and decohesion.
3. The local strain concentrations which accelerate coalescence of the voids.
4. The grain size when coalescence involves grain boundaries.

The major step in the development of aluminium alloys with greatly improved fracture toughness has come from the control of the levels of the impurity elements iron and silicon. This effect is shown in Fig. 2.41 for alloys based on the Al–Cu–Mg system and it can be seen that plane strain fracture toughness values may be doubled by maintaining the combined levels of these elements below 0.5% as compared with similar alloys in which this value exceeds 1.0%. As a consequence of this, a range of high-toughness versions of older alloy compositions is now in commercial use in which the levels of impurities have been reduced (see Table 3.4).

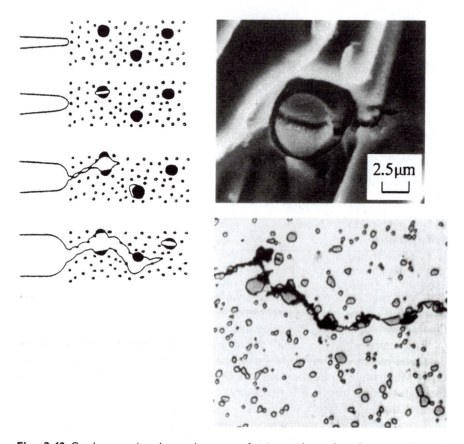

Fig. 2.40 Crack extension by coalescence of microvoids nucleated at particles and dispersoids: (a) schematic showing nucleation of voids due to particle cleavage followed by progressive linkage of advancing crack to these voids (courtesy R. J. H. Wanhill); (b) cleavage of an Al_7Cu_2Fe intermetallic particle with associated void nucleation and crack extension into the adjacent ductile matrix in a high-strength aluminium alloy. Scanning electron micrograph $\times$ 2800 (courtesy R. Gürbüz); (c) crack path in an aluminium casting alloy which has been influenced by the presence of coarse, brittle silicon particles. Optical micrograph $\times$ 700 (courtesy R.W. Coade).

The role of the submicron dispersoids with respect to toughness is more complex as they have both good and bad effects. To the extent that they suppress recrystallization and limit grain growth they are beneficial. The effect of these factors on a range of high-strength sheet alloys based on the Al–Zn–Mg–Cu system is shown in Fig. 2.42. Fine, unrecrystallized grains favour a high energy absorbing, transcrystalline mode of fracture. On the other hand, such particles also nucleate microvoids by decohesion at the interface with the matrix which may lead to the formation of sheets of voids between the larger voids that are

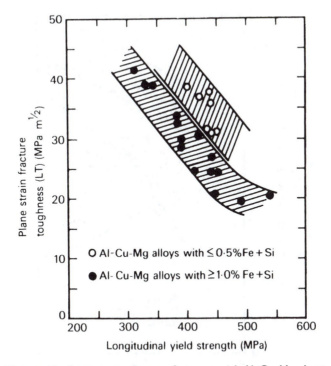

Fig. 2.41 Plane strain fracture toughness of commercial Al–Cu–Mg sheet alloys with differing levels of iron and silicon (from Speidel, M. O., *Proceedings of 6th International Conference on Light Metals*, Leoben, Austria, Aluminium-Verlag, Düsseldorf, 1975).

associated with the coarse intermetallic compounds. In this regard, the effects do vary with different transition metals and there is evidence to suggest that alloys containing zirconium to control grain shape are more resistant to fracture than those to which chromium or manganese has been added to this purpose. This is attributed to the fact that zirconium forms relatively small particles of the compound Al_3Zr, which are around 20 nm in diameter.

The fine precipitates developed by age-hardening are also thought to have at least two effects with regard to the toughness of aluminium alloys. To the extent that they reduce deformation, toughness is enhanced and it has been observed that, for equal dispersions of particles, an alloy with a higher yield stress has greater toughness. At the same time, these fine particles tend to cause localization of slip during plastic deformation, particularly under plane strain conditions, leading to development of pockets of slip or so-called super-bands ahead of an advancing crack. Strain is concentrated within these bands and may cause premature cracking at the sites of inter-metallic compounds ahead of an advancing crack. This effect is usually greatest for alloys aged to peak hardness and the overall result is a net loss of toughness at the highest strength levels.

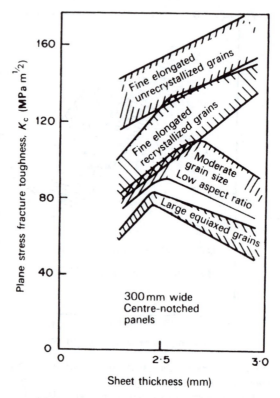

Fig. 2.42 Effect of recrystallization and grain size and shape of various alloys based on the Al–Zn–Mg–Cu system (from Thompson, D. S., *Met. Trans.*, **6A**, 671, 1975).

The fact that toughness does not follow a simple inverse relationship to the strength of age-hardened aluminium alloys is shown by comparing alloys in the under- and over-aged conditions. Toughness is greatest in the underaged condition and decreases as ageing proceeds to peak strength. In some alloys, this condition corresponds to a minimum value of toughness and some improvement may occur on over-ageing which is associated with a reduction in yield strength (e.g. Al–Zn–Mg–Cu alloys: Section 3.4.5). However, where over-ageing leads to the formation of large precipitate particles in grain boundaries, or wide precipitate-free zones then toughness can continue to decrease because each of these features promotes more intergranular fracture. Lithium-containing alloys tend to behave in this way (Section 3.4.6).

2.5.3 Fatigue

It is well known that, contrary to steels, the increases that have been achieved in the tensile strength of most non-ferrous alloys have not been accompanied by proportionate improvements in fatigue properties. This feature is illustrated

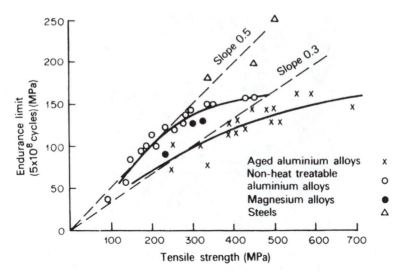

Fig. 2.43 Fatigue ratios (endurance limit:tensile strength) for aluminium alloys and other materials (from Varley, P. C., *The Technology of Aluminium and its Alloys*, Newnes-Butterworths, London, 1970).

in Fig. 2.43 which shows relationships between fatigue endurance limit $(5 \times 10^8$ cycles) and tensile strength for different alloys. It should also be noted that the so-called fatigue ratios are lowest for age-hardened aluminium alloys and, as a general rule, the more an alloy is dependent upon precipitation-hardening for its tensile strength, the lower its ratio becomes.

Detailed studies of the processes of fatigue in metals and alloys have shown that the initiation of cracks normally occurs at the surface. It is here that strain becomes localized due to the presence of pre-existing stress concentrations such as mechanical notches or corrosion pits, coarse (persistent) slip bands in which minute extrusions and intrusions may form, or at relatively soft zones such as the precipitate-free regions adjacent to grain boundaries.

The disappointing fatigue properties of age-hardened aluminium alloys are also attributed to an additional factor, and that is the metastable nature of the metallurgical structure under conditions of cyclic stressing. Localization of strain is particularly harmful because the precipitate may be removed from certain slip bands which causes softening there and leads to a further concentration of stress so that the whole process of cracking is accelerated. This effect is shown in an exaggerated manner in a recrystallized, high-purity alloy in Fig. 2.44. It has been proposed that removal of the precipitate occurs either by over-ageing or re-solution, the latter now being considered to apply in most cases. One suggestion is that the particles in the slip bands are cut by moving dislocations, and re-solution occurs when they become smaller than the critical size for thermodynamic stability.

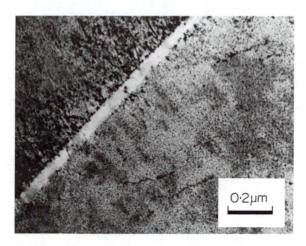

Fig. 2.44 Transmission electron micrograph showing precipitate depletion in a persistent slip band formed by fatigue stressing a high-purity Al–Zn–Mg alloy (courtesy A. Stubbington, copyright HMSO).

The fatigue behaviour of age-hardened aluminium alloys should therefore be improved if fatigue deformation could be dispersed more uniformly. Factors which prevent the formation of coarse slip bands should assist in this regard. Thus it is to be expected that commercial-purity alloys should perform better than equivalent high-purity compositions because the presence of inclusions and intermetallic compounds would tend to disperse slip. This effect is demonstrated for the Al–Zn–Mg–Cu alloy 7075 in Fig. 2.45 (a) which shows fatigue (S/N) curves for commercial-purity and high-purity compositions. Tests were carried out on smooth specimens prepared from the alloys which had been aged under similar conditions to produce fine, shearable precipitates. It will be noted that the fatigue performance of the commercial-purity alloys is superior because crack initiation is delayed due to the fact that slip is more uniformly dispersed by dispersoid particles such as $Al_{12}Mg_2Cr$. These particles are absent in the high-purity alloy. However, the fatigue performance characteristics are reversed if tests are carried out on pre-cracked specimens (Fig. 2.45 b). In this condition, the rate of fatigue crack growth is faster in the commercial-purity alloy because voids nucleate more readily at dispersed particles which are within the plastic zone of the advancing crack (Fig. 2.40).

Thermomechanical processing whereby plastic deformation before, or during, the ageing treatment increases the dislocation density, has also been found to improve the fatigue performance of certain alloys although this effect arises in part from an increase in tensile properties caused by such a treatment (Fig. 2.46). It should be noted, however, that the promising results mentioned above were obtained for smooth specimens. The improved fatigue behaviour has not been

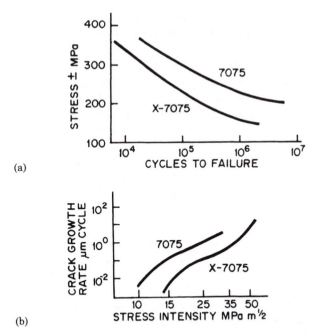

(a)

(b)

Fig. 2.45 (a) Fatigue (S/N) curves for Al–Zn–Mg–Cu alloys 7075 and X7075. X7075 is a high-purity version of the commercial alloy 7075 (from Lütjering, G., *Micromechanisms in Particle-hardened Alloys*, Martin, J. W., Cambridge University Press, 142, 1980); (b) Fatigue crack growth rate curves for alloys 7075 and X7075 (from Albrecht, J. et al., *Proc. 4th Inter. Conf. on Strength of Metals and Alloys*, J. de Physique, Paris, 463, 1976).

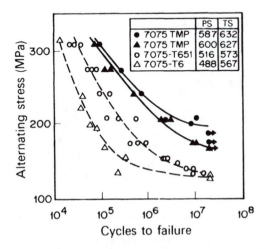

Fig. 2.46 Effect of thermomechanical processing (TMP) on the unnotched fatigue properties of the commercial Al–Zn–Mg–Cu alloy 7075. PS = proof stress (MPa), TS = tensile strength (MPa) (from Ostermann, F. G., *Met. Trans.*, **2A**, 2897, 1971).

sustained for severely notched conditions and it seems that the resultant stress concentrations over-ride the more subtle microstructural effects that have been described.

Alloys which are aged at higher temperatures, and thus form relatively more stable precipitates, might also be expected to show better fatigue properties and this trend is observed. For example, the fatigue performance of the alloys based on the Al–Cu–Mg system is generally better than that of Al–Zn–Mg–Cu alloys, although this effect is again greatly reduced for notched conditions.

The fact that the microstructure can have a greater influence upon the fatigue properties of aluminium alloys than the level of tensile properties has been demonstrated for an Al–Mg alloy containing a small addition of silver. It is well known that binary Al–Mg alloys such as Al–5Mg, in which the magnesium is present in solid solution, display a relatively high level of fatigue strength. The same applies for an Al–5Mg–0.5Ag alloy in the as-quenched condition and Fig. 2.47 shows that the endurance limit after 10^8 cycles is ± 87 MPa which approximately equals the 0.2% proof stress. This result is attributed to the interaction of magnesium atoms with dislocations which minimizes formation of coarse slip bands during fatigue. The silver-containing alloy responds to age-hardening at elevated temperatures due to the formation of a finely dispersed precipitate, and the 0.2% proof stress may be raised to 200 MPa after ageing for one day at 175 °C. However, the endurance limit for 10^8 cycles is actually decreased to ± 48 MPa due to the localization of strain in a limited number of coarse slip bands (Fig. 2.48a).

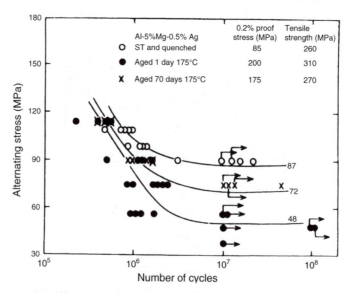

Fig. 2.47 Fatigue (S/N) curves for the alloy Al–5Mg–0.5Ag in different conditions (from Boyapati, K. and Polmear, I. J., *Fatigue of Engineering Materials and Structures*, **2**, 23, 1979).

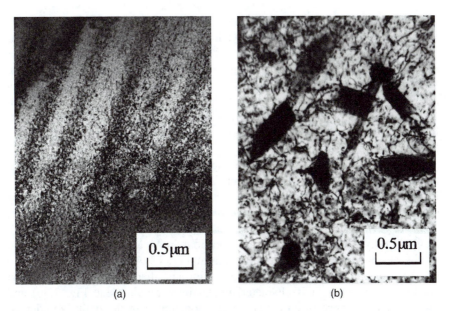

Fig. 2.48 (a) Coarse slip bands containing a high density of dislocations. Alloy Al–5Mg–0.5Ag aged one day at 175 °C and tested at a stress of ±75 MPa for 1.4 × 10⁶ cycles; (b) large particles of a second precipitate, that have formed in the alloy aged 70 days at 175 °C, which have dispersed dislocations (see arrow) generated by fatigue stressing for 10⁷ cycles at a stress of ±75 MPa.

Continued ageing of the alloy at 175 °C causes only slight softening (0.2% proof stress 175 MPa after 70 days) although large particles of a second precipitate are formed (Fig. 2.48b) which have the effect of dispersing dislocations generated by cyclic stressing. As a result, fatigue properties are improved and the endurance limit for 10^8 cycles is raised to ±72 MPa (Fig. 2.47). These particles serve the same role as the submicron particles in the commercial alloy 7075 (Fig. 2.45) but they have formed by a precipitation process. This again suggests the desirability of having a duplex precipitate structure; fine particles to give a high level of tensile properties, and coarse particles to improve fatigue strength.

2.5.4 Stress-corrosion cracking

Stress-corrosion cracking (SCC) may be defined as a phenomenon which results in brittle failure in alloys, normally considered ductile, when they are exposed to the simultaneous action of surface tensile stress and a corrosive environment, neither of which when operating separately could cause major damage. It involves time-dependent interactions between the microstructure of a susceptible alloy, mechanical deformation, and local environmental conditions. A threshold stress is needed for crack initiation and growth, the level of which is normally well below that required to cause yielding. Specific corrosive environments can be quite mild, e.g. water vapour, although in the case of aluminium alloys it is common for halide ions to be present. The relative importance of each of the two factors, stress and corrosion (i.e. electrochemistry) varies with different alloy systems. For aluminium alloys, there is general agreement that electro-chemical factors predominate and it has been on this basis that new compositions and tempers have been developed which provide improved resistance to SCC (e.g. Section 3.4.5).

Only aluminium alloys that contain appreciable amounts of solute elements, notably copper, magnesium, silicon, zinc and lithium may be susceptible to SCC. In practical terms, the commonly used commercial alloys in which SCC may occur are those based on the systems Al–Cu–Mg (2xxx series), Al–Mg (5xxx) containing more than 3% magnesium, and Al–Zn–Mg and Al–Zn–Mg–Cu (7xxx). SCC has only been observed on rare occasions in Al–Mg–Si alloys (6xxx) and is absent in commercial-purity aluminium (1xxx), Al–Mn and Al–Mn–Mg (3xxx) alloys, and in Al–Mg alloys containing less than 3% magnesium. When cracking does occur, it is characteristically intergranular (e.g. Fig. 2.36) and involves the presence of an active anodic constituent in the grain boundaries. Alloys are usually most susceptible in the recrystallized condition and it is for this reason that compositions, working procedures, and heat treatment temperatures for wrought alloys are normally adjusted to prevent recrystallization. It should be noted, however, that the resistance of a particular wrought alloy to SCC will now vary depending upon the direction of stressing with respect to the elongated grain structure. Maximum susceptibility occurs if stressing is normal to the grain direction, i.e. in the short transverse direction of components, because the crack path along grain boundaries is so clearly defined (Fig. 2.38).

Residual stresses are introduced into aluminium alloy products when they are solution treated and quenched. The quenching operation places surface regions of a component into compression and the centre into tension. If the compressive surface stresses are not disturbed during subsequent fabrication procedures, then a component will actually have an enhanced resistance to SCC because a sustained tensile stress is required to initiate and propagate this type of cracking. On the other hand, if the central regions are exposed (e.g. by machining away sections in a component that has not been stress relieved), then the internal residual tensile tresses will be additive to any tensile stresses imposed in service, thereby increasing the probability of SCC. As mentioned in Section 3.1.6, it is for this reason that aluminium alloy products of constant cross-section (rolled plate and extrusions) are usually given a mechanical stretch of 1 to 3% after quenching which greatly reduces the levels of residual stress.

In precipitation-hardened aluminium alloys aged at elevated temperatures, resistance to SCC varies with ageing condition and generally has an inverse relationship with strength similar to that shown in Fig. 2.49. Resistance to cracking is significantly higher in the lower strength, underaged and overaged

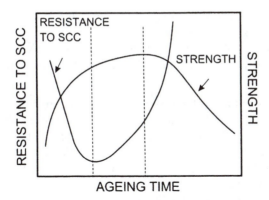

Fig. 2.49 General relationship between resistance to SCC and strength during the ageing of aluminium alloys at elevated temperatures.

conditions. It also tends to increase as the ageing temperature is raised. Much attention has therefore been given to devising new ageing tempers and thermomechanical practices whereby high strength can be combined with an acceptable resistance to SCC, some of which are discussed in Section 3.4.5.

Considerable effort has been directed towards understanding the mechanism of SCC in aluminium alloys and significance has been attached to the following microstructural features.

1. Precipitate-free zones adjacent to grain boundaries (Figs 2.15a and 2.22b).

In a corrosive medium, it is considered that either these zones or the grain boundaries will be anodic with respect to the grain centres. Moreover, strain is likely to be concentrated in the zones because they are relatively soft.
2. Nature of the matrix precipitate.

Maximum susceptibility to cracking occurs in alloys when GP zones are present. In this condition, deformation tends to be concentrated in discrete slip bands similar in appearance to those shown in Fig. 2.48a. It is considered that stress is generated where these bands impinge upon grain boundaries which can contribute to intercrystalline cracking under stress-corrosion conditions (Fig. 2.22a)
3. Dispersion of precipitate particles in grain boundaries.

In some aged aluminium alloys, it has been shown that SCC occurs more rapidly when particles in grain boundaries are closely spaced.
4. Solute concentrations in the region of grain boundaries.

Differences in solute levels that arise during ageing are thought to modify local electrochemical potentials. Moreover, it has been observed that a higher magnesium content develops in these regions. This results in an adjacent oxide layer with an increased MgO content which, in turn, is a less effective barrier against environmental influences.

5. Hydrogen embrittlement that may occur due to the rapid diffusion of atomic hydrogen along grain boundaries.
6. Chemisorption of atom species at the surface of cracks which may lower the cohesive strength of the interatomic bonds in the region ahead of an advancing crack.

Recent experimental work has shown that stress-corrosion cracking at grain boundaries occurs in a brittle and discontinuous manner and there is clear evidence that hydrogen diffuses there, even in the absence of stress (e.g. Fig. 2.50). It thus seems that the presence of hydrogen does play a vital part in SCC due to one or both of mechanisms (5) and (6). However, the overall process of SCC is complex and it seems probable that one or more of the other microstructural factors are also involved. The relative importance of each of these factors may depend upon the particular combination of alloy and environment.

A useful measure of the relative susceptibility of alloys to SCC can be obtained by using pre-cracked specimens that are subjected to sustained tensile loading in an appropriate corrosive environment. The rate of crack growth (da/dt) is monitored as a function of the instantaneous value of stress intensity (K) and plots of the data take the form shown in Fig. 2.51. Three distinct regimes of crack growth are evident. In region I, da/dt is strongly dependent on K whereas, in region II, da/dt is virtually independent of the prevailing value of K. It is within this region that environmental factors predominate and, in some alloys, a steady state or plateau velocity is recorded (see also Figs. 3.22 and 3.33). In region III, da/dt again becomes strongly dependent of until final overload failure occurs. At very low values of K, da/dt becomes vanishingly

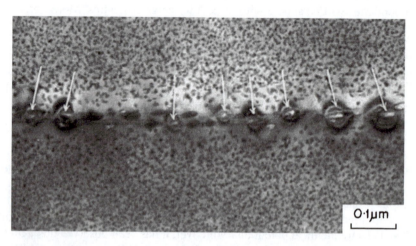

Fig. 2.50 Transmission electron micrograph showing hydrogen bubble development at precipitate particles in a grain boundary of a thin foil of an artificially aged Al–Zn–Mg alloy exposed to laboratory air for three months (from Scamans, G. M., J. Mater. Sci., **13**, 27, 1978).

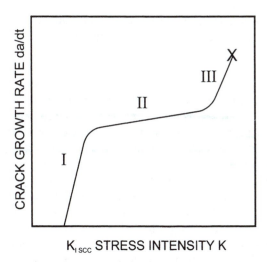

K_{Iscc} STRESS INTENSITY K

Fig. 2.51 Typical relationship between rate of crack growth and stress intensity during SCC of a pre-cracked specimen of a susceptible alloy loaded in tension and exposed to a corrosive environment.

small and the value (often referred to as K_{Iscc}) is considered to be the threshold level below which SCC does not occur in the particular environment.

2.5.5 Corrosion fatigue

Under conditions of simultaneous corrosion and cyclic stressing (corrosion fatigue), the reduction in strength is greater than the additive effects if each is considered either separately or alternately. Although it is often possible to provide adequate protection for metallic parts which are stressed under static conditions, most surface films (including naturally protective oxides) can be more easily broken or disrupted under cyclic loading. Corrosion fatigue is accentuated by the rapid movement of the corroding medium over the surface, since protecting layers that might form will be washed away. Localised attack of the aluminium surface, such as by pitting corrosion, will also provide stress concentrators that may greatly reduce fatigue life. Contrary to failures by stress-corrosion cracking in aluminium alloys, which are invariably intercrystalline, corrosion fatigue cracks are characteristically transcrystalline in their mode of propagation.

In general, the reduction in a fatigue strength of a material in a particular corrosive medium will be related to the corrosion resistance of the material in that medium. Under conditions of corrosion fatigue, all types of aluminium alloys exhibit about the same percentage reduction in strength when compared with their fatigue strength in air. For example, under freshwater conditions the fatigue strength at 10^8 cycles is about 60% of that in air, and in NaCl solutions it is normally between 25 and 35% of that in air. Another general observation is that the corrosion fatigue strength of a particular aluminium alloy appears to be virtually independent of its metallurgical condition.

2.5.6 Creep

Creep fracture, even in pure metals, normally occurs by the initiation of cracks in grain boundaries. The susceptibility of this region to cracking in age-hardened aluminium alloys is enhanced because the grains are harder and less willing to accommodate deformation than the relatively softer precipitate-free zones adjacent to the boundaries (Figs 2.15a and 2.22b). Moreover, the strength of the grain boundaries may be modified by the presence there of precipitate particles.

Precipitation-hardened alloys are normally aged at one or two temperatures which allow peak properties to be realized in a relatively short time. Continued exposure to these temperatures normally leads to rapid over-ageing and softening and it follows that service temperatures must be well below the final ageing temperature if a loss of strength due to over-ageing is to be minimized. For example, the alloy selected for the structure and skin of the supersonic Concorde aircraft, which was normally required to operate in service at 100–110 °C, is aged at around 190 °C.

Creep resistance in aluminium alloys is promoted by the presence of submicron intermetallic compounds such as Al_9FeNi or other fine particles that are stable at the required service temperatures (normally below 200 °C). Fine, stable particles or fibres can be introduced by a variety of novel processing methods and the products that result can show much improved creep resistance when compared with conventional age-hardened alloys. These new materials are discussed in Chapter 7.

Recent work has also shown that aged aluminium alloys may show superior creep resistance if they are tested in the underaged rather than fully hardened condition (T6 temper, Section 3.2.2). As demonstrated in Fig. 2.52 (a) with an experimental Al–Cu–Mg–Ag alloy tested at 150 °C and a stress of 300 MPa, this beneficial effect of underageing is manifest in significantly reduced rates of secondary creep. In this case, the creep rate has been reduced to one third of the value for same alloy tested in the fully hardened T6 condition. With the commercial Al–Cu–Mg–Mn alloy 2024 (Table 3.4), underageing has nearly doubled the time to failure from 260 h for the T6 condition to 480 h (Fig. 2.52 (b)). It is also interesting to note these results were obtained despite the fact that, in the underaged condition, both alloys had levels of yield stress significantly lower than the T6 values.

Microstructural studies have suggested that the enhanced resistance of aluminium alloys to creep in the underaged condition is a consequence of the presence of "free" solute in solid solution that is not yet committed to the precipitation process. This free solute is available to retard the motion of dislocations during creep deformation through the formation of solute atmospheres, or by facilitating dynamic precipitation within the matrix.

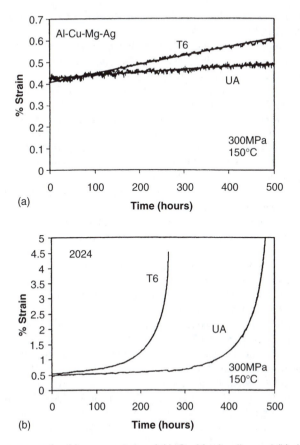

Fig. 2.52 Creep curves for (a) an experimental Al–Cu–Mg–Ag alloy and (b) the commercial alloy 2024, both tested in the underaged and fully hardened (T6) conditions at 150 °C and a stress of 300 MPa (after Lumley, R. N. *et al*, *Acta Mater.*, **50**, 3597, 2002).

FURTHER READING

Mondolfo, L.F., *Aluminium* Alloys: *Structure and Properties*, Butterworths, London, 1976

Murakami, Y., Aluminium-based alloys, *Materials Science and Technology: A Comprehensive Treatment*, Vol. **8**, Matucha, K.H., Ed., VCH, Weinheim, Germany, 217, 1996

Humphreys, F.J. and Hatherly, M. *Recrystallization and Related Annealing Phenomena*, Pergamon, Elsevier Science, Oxford, UK, 1996

Vasudevan, A.K. and Doherty, R.D. Eds., *Treatise on Materials Science and Technology*, Vol. **31**,: *Aluminium Alloys – Contemporary Research and Applications*, Academic Press, New York, 1989

Polmear, I.J., Aluminium alloys: a century of age hardening, *Proc. 9th Inter. Conf. on Aluminium Alloys*, IMEA, 1, 2004

Ardell, A.J., Precipitation hardening, *Metall. Trans. A*, **16A**, 2131, 1985

Lorimer, G.W., in *Precipitation Processes in Solids*, Russell, K.C. and Aaronson, H.I., Eds. Chap. 3, Met. Soc. AIME, New York, 1978

Martin, J.W. *Precipitation Hardening*, 2nd Edn, Butterworth-Heinemann, Oxford, 1998

Lloyd, D.J., Precipitation hardening, *Proc. 7th Inter. Conf. on Strength of Metals*, McQueen, H.J. *et al*, Eds., Pergamon Press, Toronto, **3**, 1745, 1985

Shercliff, H.R. and Ashby, M.F., A process model for age hardening of aluminium alloys, *Acta Mater.*, **38**, 1789 and 1803, 1990

Bratland, D.H., Grong, O., Shercliff, H., Myhr, O.R., and Tjotta, S., Modelling of precipitation reactions in industrial processing, *Acta Mater.*, **45**, 1, 1997

Martin, J.W., *Micromechanisms in Particle-Hardened Alloys*, CUP, Cambridge, UK, 1980

Polmear, I.J. and Ringer, S.P., Evolution and control of microstructure in aged aluminium alloys, *J. Japan Inst. Light Metals*, **50**, 633, 2000

Polmear, I.J. Control of precipitation processes and properties in aged aluminium alloys by microalloying, *Mater. Forum*, **23**, 117, 1999

Ringer, S.P. and Hono, K., Microstructural evolution and age hardening in aluminium alloys: atom probe field ion microscopy and transmission electron microscopy studies, *Mater. Characterization*, **44**, 101, 2000

Hornbogen, E. and Starke, E.A. Jr., Theory assisted design of high strength low alloy aluminium, *Acta Metall. Mater.*, **41**, 1, 1993

Nie, J.-F. and Muddle, B.C., Microstructural design of high-strength aluminium alloys, *J. Phase Equilibria*, **19**, 543, 1998

Dupasquier, A., Kögel, and Somoza, A., Studies of light alloys by positron annihilation techniques, *Acta Mater.*, **52**, 4707, 2004

Hollingsworth, E.H. and Hunsicker. H.Y., Corrosion of aluminium and aluminium alloys, *Metals Handbook*, 10th Edn., ASM, Metals Park, Ohio, USA, 583, 1987

Cramer, S.D. and Corvino Jr, B.S., Eds., *Metals Handbook Volume 13A, Corrosion: Fundamentals, Testing and Protection*, ASM International, Ohio, USA, 2003

Staley, J.T., Metallurgical aspects affecting strength of heat-treated products used in the aerospace industry, *Proc. 3rd Int. Conf. on Aluminium Alloys*, Vol. **3**, Arnberg *et al*. Eds., Trondheim, Norway, NTH/SINTEF, 107, 1992

Kaufman, J.G., *Fracture Resistance of Aluminium Alloys*, ASM International, Materials Park, Ohio, USA, 2001

Kobayashi, T., *Strength and Toughness of Materials*, Springer-Verlag, Tokyo, 2004

Sanders, T.H. and Staley, J.T., Fatigue and fracture research on high-strength aluminium alloys, *Fatigue and Microstructure*, ASM, Cleveland, Ohio, USA, 467, 1979

Parkins, R.N., Current understanding of stress-corrosion cracking, *JOM*, **48**, No. 12, 12, 1992

3

WROUGHT ALUMINIUM ALLOYS

As a general average, 75 to 80% of aluminium is used for wrought products, e.g. rolled plate (>6 mm thickness), sheet (0.15–6 mm), foil (<0.15 mm), extrusions, tube, rod, bar and wire. These are produced from cast ingots, the structures of which are greatly changed by the various working operations and thermal treatments. Each class of alloys behaves differently, with composition and structure dictating the working characteristics and subsequent properties that are developed. Relationships between the various wrought alloy systems are summarised concisely in Fig. 3.1. Before considering each system individually, it is desirable to examine how wrought alloys are produced and processed.

3.1 PRODUCTION OF WROUGHT ALLOYS

3.1.1 Melting and casting

Ingots are prepared for subsequent mechanical working by first melting virgin aluminium, scrap and the alloying additions, usually in the form of a concentrated hardener or master alloy, in a suitable furnace. A fuel-fired reverberatory type is most commonly used. The main essentials in promoting ingot quality are a thorough mixing of the constituents together with effective fluxing, degassing, and filtering of the melt before casting in order to remove dross, oxides, gases and other non-metallic impurities. Hydrogen is the only gas with measurable solubility in aluminium, the respective equilibrium solubilities in the liquid and solid states at the melting temperature and a pressure of one atmosphere being 0.68 and 0.036 cm^3 per 100 g of metal—a difference of 19 times. Atomic size factors require that hydrogen enters solution in the atomic form and the gas is derived from the surface reaction of aluminium with water vapour:

$$2Al + 3H_2O \rightarrow Al_2O_3 + 3H_2$$

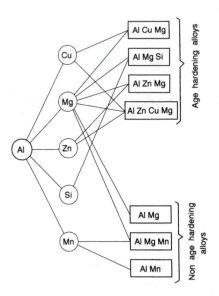

Fig. 3.1 Representation of alloying relationships of the principal wrought aluminium alloys (courtesy D. G. Altenpohl).

The standard free energy change for this reaction is very high (equilibrium constant = 8×10^{40} at 725 °C) so that, for practical purposes, all traces of water vapour contacting the metal are converted to hydrogen. The main sources of water vapour are the furnace atmosphere in contact with the molten metal, hydrated surface contaminants and residual moisture in launders and moulds.

During solidification, excess hydrogen is rejected from solution and it recombines as molecular gas which may be entrapped in the solid structure leading to porosity, notably in interdendritic regions. In order to obtain an ingot that is free from gas porosity, it has been found necessary to reduce the hydrogen content of the molten metal to less than 0.15 cm^3 per 100 g. Although hydrogen can escape from molten aluminium by evaporation from the surface, the process is slow and it is common practice to purge by bubbling an insoluble gas through the melt in the reverberatory furnace prior to pouring. Partial pressure requirements cause hydrogen to diffuse to these bubbles and be carried out of the metal. According to circumstances either nitrogen, argon, chlorine, mixtures of these gases or a solid chlorinated hydrocarbon is used. Chlorine is normally present because it serves the important additional function of increasing the surface tension between inclusions and the melt so that they tend to rise to the top and can be skimmed off. However, the use of chlorine does present environmental problems and an example of an alternative method is the continuous, fumeless, in-line degassing (FILD) process developed by the former British Aluminium Company. In this process, the molten aluminium from the melting furnace enters a vessel through a liquid flux cover (KCl + NaCl eutectic with small amounts of CaF$_2$). It is then degassed with nitrogen under conditions which do not increase the inclusion content and passes through a bed of balls of Al_2O_3 which become coated with flux and serve to reduce further the content of non-metallic inclusions in the aluminium. This process eliminates the capital and operating costs of fume treatment, as well as the need to hold the aluminium in a furnace whilst degassing is carried out. Loss of metal as dross is also reduced. Other degassing regimes have been developed including an Alcoa process that involves injecting a finely dispersed mixture of argon-chlorine bubbles into the molten metal by means of a spinning nozzle. Chlorine levels can be kept at a low level (1–10%) which reduces emissions while maintaining ingot quality. Removal of oxide skins and other undesirable particulate matter is also achieved by filtering the molten metal through a glass cloth on a porous ceramic foam block prior to casting.

The production of a uniform ingot structure is desirable and this is promoted by direct-chill (DC), semi-continuous processes. Most commonly, ingots are cast by the vertical process in which the molten alloy is poured into one, or more fixed, water-cooled moulds having retractable bases (Fig. 3.2a). The process of solidification is accomplished in two stages: formation of solid metal at the chilled mould wall and solidification of the remainder of the billet cross-section by the removal of heat by submould, spray cooling. Ingots may be rectangular in section for rolling (Fig. 3.3) and weigh as much as 15 tonnes. Round ingots for extrusion are now being DC cast with diameters exceeding 2 m. Several may be produced at once. Similar sections of generally smaller dimensions may be cast in a horizontal (Ugine) arrangement also shown in Fig. 3.2b, although control of microstructure, notably grain size, is more difficult in this process. Maximum section thicknesses are around 650 mm and cooling rates usually lie in the range 1–5 °C s^{-1}.

A problem with these DC processes is that the surfaces of the ingots tend to be rippled in contour (Fig. 3.3) due to stick-slip contact as they move past the sides of the mould when solidification first occurs. Surface tears and

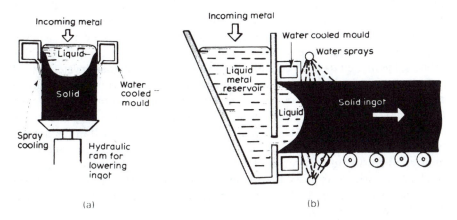

Fig. 3.2 Direct-chill casting processes: (a) vertical; (b) horizontal.

Fig. 3.3 DC cast ingots for subsequent rolling into plate, sheet or foil (courtesy Australian Aluminium Council).

microstructural inhomogeneities such as inverse segregation also tend to occur in the surface regions which may cause edge cracking during rolling. For these reasons, it is necessary to machine or scalp the surfaces of DC ingots prior to rolling or extrusion which adds cost to the overall operation.

For vertically cast DC ingots, improved surface quality may be obtained by reducing the rate of heat flow from the solidifying billet to the mould through the use of a narrow stream of high-pressure air directed along the metal/mould interface. Scalping may be avoided altogether by using a novel method of casting in an electromagnetic field which was first invented in the Soviet Union. Further developments have occurred in the United States and elsewhere. In this process, the molten metal is contained and shaped by an internally water-cooled inductor which repels the liquid metal and prevents contact with the sides of the mould. Metal solidification occurs only on the aluminium bottom block or table (Fig. 3.2a). Subsequent solidification is rapid and is achieved by the direct application of water from the coolant jacket on to the ingot shell in the normal way. A very smooth ingot surface is achievable together with a finer and more uniform microstructure. Direct rolling to produce difficult products such as canstock (Section 3.6.5) is possible without scalping which offers economic advantages that outweigh the higher capital investment and energy costs associated with electromagnetic casting.

3.1.2 Continuous casting with moving moulds

The concept of the moving mould has revolutionized the casting of some lower-strength aluminium alloys as it is now possible to produce continuous shapes in sizes close to the requirements of the final wrought products, thereby reducing the investment in the heavy equipment required for subsequent mechanical working. Basic methods for casting bar and sheet are shown in Fig. 3.4 and each offers considerable economic advantage over earlier practices involving extrusion or rolling of large ingots to small section sizes. Developments with aluminium alloys are now being adapted for the continuous casting of other metals.

The application of continuous casting is expanding, particularly with the production of thin slab and sheet. Two thin slab casting processes are in commercial operation. One involves a block caster, first developed by Alusuisse in Switzerland, in which two sets of externally cooled cast iron, steel or copper blocks rotate continuously in opposite directions on caterpillar tracks, to form the mould cavity into which the molten metal is introduced through an insulated nozzle. The other system has a similar configuration except that flexible steel belts are used (Hazelett Process; Fig. 3.5). Molten metal is introduced into the caster via a holding tundish and through a wide ceramic nozzle placed accurately between the moving belts so that turbulence in minimized. The rotating belts are held in tension to form the top and bottom surfaces, and proprietary coatings may be applied continuously to provide specific characteristics to the mould/metal interface. At the same time, gas mixtures are injected to minimize oxidation and the belts are water cooled to enhance heat transfer from the solidifying strip or slab. Side chains made from small rectangular steel blocks that move with the belts are spaced to control the width. The strip or slab may range

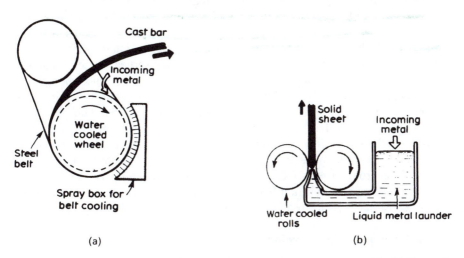

(a)   (b)

Fig. 3.4 Processes for continuously casting bar and sheet in moving moulds: (a) Properzi process; (b) Hunter process.

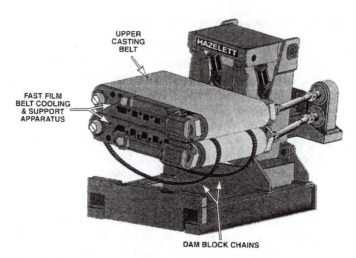

Fig. 3.5 Schematic diagram showing a Hazelett twin belt continuous caster (Hazelett, D. N. and Szczypiorski, W.S., The Hazelett strip casting process for aluminium packaging applications, *Aluminium*, **79**, 11, 2003).

in thickness from around 12 to 75 mm, and up to a maximum designed width of 2.3 m. The mould section is around 2 m long and cooling rates for a 19 mm slab vary for 11 to 40 °C s^{-1} at the surface, to 4 °C s^{-1} in the centre. Production rates are generally in the range 21 to 24.5 tonnes per metre of width per hour. As a contrast, twin roll casting of sheet occurs at much faster cooling rates that may be in the range 10^2–10^3 °C s^{-1}. In this case, sheet thicknesses are commonly 3 to 10 mm so the minimal final rolling is required.

Rate of solidification has an important impact on the quality of an ingot, slab or sheet. Faster rates are generally desirable as they lead to finer dendrite arm spacings, which, in turn, reduces microsegregation to the interdendritic regions. Dendrite cell sizes typically vary from 30–70 μm for vertical DC cast ingots to 5–10 μm for twin roll cast sheet. Faster casting rates also reduce the sizes of intermetallic compounds and promote finer and more uniform grain sizes. Another consequence of the faster solidification rates associated with continuous casting processes is that higher concentrations (supersaturations) of solute elements may be retained in solid solution. This applies particularly to the slowly diffusing, transition metal elements such as iron and mechanical properties which differ from those of DC cast products, even though the compositions are similar. In some situations a different alloy, or one with a lower solute content, can be substituted when a product is prepared by continuous casting.

3.1.3 Grain refinement by inoculation

Because fine grain size is so desirable, it is usual to add small amounts of master alloys of Al–Ti, Al–Ti–C or Al–Ti–B to the melt before casting to promote further refinement. Al–Ti–B is the most widely used alloy with Ti:B ratios vary-ing from 3:1 to 50:1, although Al–5Ti–1B is now the favoured composition. These grain refiners used to be introduced to the melt in the form of tablets made from titanium and boron salts. Now they are prepared as small ingots that are added to the melt in the furnace, or as rods which are fed automatically into the launder and dissolved by the molten metal as it passes from the furnace to the casting station. This latter method is preferred because less time is available for the effects of the grain refiner to fade before the ingots have solidified.

Despite the commercial importance of this practice, the actual mechanisms of grain refinement by these inoculants has remained uncertain and two theories have been keenly debated for many years. One is based on the heterogeneous nucleation of aluminium alloy grains at insoluble particles of TiC in the melt since this compound has a crystal structure similar to aluminium. The second theory attributes grain refinement to the following peritectic reaction involving particles of the compound $TiAl_3$:

liquid aluminium + $TiAl_3$ → α-aluminium alloy solid solution (Fig. 2.2)

The newly formed solid solution will coat (envelop) the particles of $TiAl_3$, giving nuclei on which grains may then grow. Under certain conditions, these particles adopt a petal-like form (Fig. 3.6) and the crystal facets at each petal tip

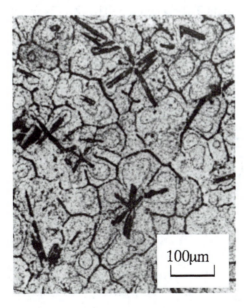

Fig. 3.6 Petal-like form of $TiAl_3$ particles in a matrix of α-aluminium solid solution. Note that several grains have been nucleated at each particle (from St John, D. H. and Hogan, L. M., *J. Australian Inst. Metals*, **22**, 160, 1977).

can be the point of origin of a separate α-aluminium dendrite without the need of the intermediate stage of forming a peritectic envelope. Multiple nucleation of as many as twelve new grains may occur on each particle, although the average is nearer to eight.

Although some doubts remain about both mechanisms, it is generally accepted that conditions exist under which each may apply. Epitaxial nucleation of single α-aluminium grains on TiC particles may be favoured in melts in which the titanium content is below that needed to initiate the peritectic reaction, whereas multiple nucleation of grains may occur by this reaction in the presence of higher levels of titanium.

Boron additions lead to the formation of particles of the compound TiB_2 which has a favourable {0001} lattice parameter such that it provides an interface on which α–aluminium grains can nucleate during solidification in the absence of $TiAl_3$, i.e. at excess titanium levels below 0.15% (Fig. 2.2). However, $TiAl_3$ is a more potent nucleating agent and this is evidence that the primary role of TiB_2 is to provide a surface on which titanium forms a thin coating of $TiAl_3$. Subsequently this compound may grow to form several needles similar to those shown in Fig. 3.6. These effects are demonstrated in Fig. 3.7 in an experiment in which TiB_2 particles were deliberately introduced into an Al–Si alloy containing an excess of titanium.

Some solutes can interfere with (poison) the ability of TiB_2 to facilitate grain refinement. Zirconium is one example and it is considered that atomic substitution occurs with TiB_2 becoming $(Ti_{1-x},Zr_x)B_2$ which changes its lattice parameters. Silicon, which is present in large quantities in several casting

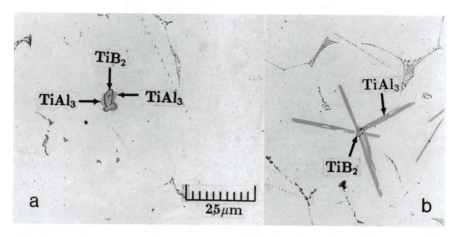

Fig. 3.7 (a, on left) Segregation of Ti_3Al at surface of a TiB_2 particle in a cast Al–Si alloy and (b, on right) growth of $TiAl_3$ needles (courtesy P. S. Mohanty).

alloys, may also lead to coarser grain sizes if added to alloys are refined with Al–Ti–B. Two suggestions have been proposed to explain this behaviour. One is that silicon increases dendrite growth rates as grains are forming. The other proposal is that the TiB_2 particles may become covered by $TiSi_2$ thereby rendering them incapable of nucleating α-aluminium grains.

Most studies of grain refinement in aluminium alloys have been focused on the role of inoculants. However, since it is found that casting alloys are more difficult to grain refine than wrought alloys, it appears that alloy chemistry may also be important. This suggestion is supported by the fact that segregation of titanium from the melt to the surface of TiB_2 particles is essential for effective nucleation of fine grains. As solidification occurs, it is known that solute elements partition to the adjacent liquid which lowers the local freezing temperature and may result in constitutional undercooling of this zone. Such an effect will restrict growth of the existing solid crystals and may facilitate grain refinement by allowing time for more nucleating events to occur in this undercooled zone.

3.1.4 Homogenization of DC ingots

Before DC ingots are fabricated into semi-finished forms it is necessary to homogenize at a temperature commonly in the range of 450–600 °C. This treatment has the following objectives:

1. Reduction of the effects of microsegregation.
2. Removal of non-equilibrium, low melting point eutectics that may cause cracking during subsequent working.
3. Controlled precipitation of excess concentrations of elements that are dissolved during solidification.

Homogenization mainly involves diffusion of alloying elements from grain boundaries and other solute-rich regions to grain centres. The time required depends on diffusion distances, i.e. grain size (or dendrite arm spacing) and the rates of diffusion of the alloying elements. In this regard the mean distance, x, a particular atom may travel in time t is given by $x = \sqrt{Dt}$, where D is the diffusion coefficient. D is strongly temperature dependent and values for different elements may vary by several orders of magnitude. Examples for some elements dissolved in aluminium at 500 °C are: Mn 5×10^{-13} cm^2 s^{-1}, Cu 5×10^{-10} cm^2 s^{-1}, Mg and Si 1×10^{-9} cm^2 s^{-1}. Obviously, the larger the dendrite cell size in an ingot, the greater the diffusion distances involved. For example, if x is doubled in the above expression, time t must be quadrupled. On the other hand, raising temperatures increases diffusion rates. As a rough guide, increasing the homogenisation temperature by 50 °C reduces the furnace time to approximately one third of that needed at the lower temperature. In practice, homogenization times usually vary from 6–24 h depending upon casting conditions and the alloy system.

Homogenization is particularly important for the higher-strength alloys as it serves to precipitate and redistribute submicron inter-metallic compounds such as

Al_6Mn, $Al_{12}Mg_2Cr$, and Al_3Zr which were discussed in Section 2.4. As mentioned in Section 3.1.2, transition metals may supersaturate in aluminium during the cooling of DC cast ingots and, more particularly, continuously cast slab or strip. It is necessary to promote their precipitation as uniformly dispersed compounds in order to control grain structure during subsequent fabrication and heat treatment. Moreover, it is now realized that they may exert a marked influence on various mechanical properties through their effects on the dislocation substructures formed by deformation and on the subsequent response to ageing treatments.

Regulation of these various functions requires a careful choice of conditions for homogenizing ingots of different alloys. When precipitation of these compounds is involved, both time and temperature are significant and the rate of heating to the homogenization temperature is of crucial importance. Relatively slow rates, e.g. 75 °C h^{-1}, are necessary to promote nucleation and growth of a fine and uniform dispersion of the compounds. It has been found that the compounds are actually nucleated heterogeneously at the surfaces of precipitate particles which form and grow to relatively large sizes during the slow heating cycle. Once formed, the submicron compounds remain stable at the homogenization temperature, whereas the precipitates are redissolved. This effect is illustrated for an Al–Zn–Mg alloy in Fig. 3.8. In this case, the precipitate η-($MgZn_2$) has provided an interface for the subsequent heterogeneous nucleation of Al_6Mn in the ingot

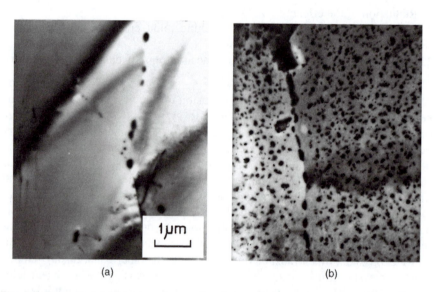

(a)   (b)

Fig. 3.8 Transmission electron micrographs showing formation of submicron particles of Al_6Mn in an Al–Zn–Mg alloy containing 0.3Mn. The ingots were homogenized at 500 °C for 24 h following: (a) a fast heating rate (500 °C h^{-1}); (b) a slow heating rate (50 °C h^{-1}) (from Thomas, A. T., *Proceedings of 6th International Conference on Light Metals*, Leoben, Austria, Aluminium-Verlag, Düsseldorf, 1975).

that was heated at the slower rate. Control of rates of cooling from the homogenization temperature may also be necessary and sometimes a second isothermal treatment is used to promote precipitation of the submicron particles.

Homogenization treatments are also important because they may change the dispersion and nature of large, primary intermetallic compounds that form in the interdendrite regions during casting. One example which is important in producing alloys for canstock (Section 3.6.5) is the reaction of dissolved silicon with the compound $Al_6(Fe,Mn)$ converting it to the harder phase $\alpha\text{-}Al_{12}(Fe,Mn)_3Si$.

3.1.5 Fabrication of DC ingots

The next stage in the production of wrought alloys is the conversion of the ingot into semi-fabricated forms. Most alloys are first hot-worked to break down the cast structure with the aim of achieving uniformity of grain size as well as constituent size and distribution. Casting pores are also closed and are usually welded up. Processing by cold-working may follow, particularly for sheet, although it may be necessary to interrupt processing to give intermediate annealing treatments. Annealing usually involves heating the alloy to a temperature of 345–415 °C and holding for times ranging from a few minutes up to 3 h, depending upon alloy composition and section size. Cold-working is also important as the means of strengthening by work hardening those alloys which do not respond to age-hardening heat treatments.

To produce sheet or plate, modern hot mills usually commence with a reversing roughing mill in which the rectangular DC cast ingot, 200 to 600 mm thick, is rolled down with heavy reductions to a slab gauge of 15 to 35 mm. Hot-rolling may then continue in a series of mills arranged in tandem resulting in a final thickness of 2.5–8 mm, after which the product is coiled. Grains become elongated in the rolling direction and, depending on the composition of the alloy, the temperature and the reduction made each pass, recovery and partial recrystallization will occur during the hot rolling operation. With plate, the degree of working is initially uneven throughout the thickness, decreasing from surface to centre. Uniformity of work can be improved by increasing the reduction for each pass or by preforging or pressing the cast ingot before rolling. The problem of differential working throughout a section also applies with forgings and several techniques are used to control grain flow. In addition, it is necessary to exercise particular control of reheating cycles for alloys containing elements that assist in inhibiting recrystallization, so that coarse grains do not form in critically strained regions.

Extrusion is second to rolling for making semi-fabricated products from aluminium and its alloys. Essentially, the process involves holding a round, pre-heated ingot in a container in a hydraulic press and forcing the ingot through a steel die opening to form elongated shapes or tubes with constant cross-sections. Examples of extruded profiles are shown in Fig. 3.23. Most

Fig. 3.9 Section through an extruded bar etched to show coarse recrystallized grains around the periphery (courtesy of A. T. Thomas) ($\times$ 1.5).

sections are extruded straight, although increasing use is being made of curved profiles that are produced by placing one or more guiding devices at the front of the exit side of the die. As with rolled products, a major objective with extrusions is to avoid forming coarse, recrystallized grains which tend to form around the periphery of sections that are heavily worked as the alloy flows through and past the edges of the die (Fig. 3.9). This effect predominates at the back end of an extruded length because of the nature of flow during extrusion, and is detrimental for several reasons, notably that the longitudinal tensile strength can be reduced by 10–15%. However, the heavy deformation associated with the process of extrusion is beneficial in producing a highly refined microstructure elsewhere in a section.

Most aluminium alloys are also amenable to forging by hammering or pressing, usually at high temperature (e.g. 450 °C), which is the next most common way of forming wrought products. There are two main types of forging machines: a hammer or drop hammer that strikes the alloy with rapid blows, and a press that squeezes metal slowly into the desired shape. Tooling comprises two types: open dies for forging between open tools, and more costly closed dies which surround and shape the product.

With sheet alloys used for some building materials and for high-strength aircraft panels, it is usual to roll-clad the surfaces with high-purity aluminium or an Al–1Zn alloy as a protection against atmospheric corrosion. This is arranged by attaching the cladding plates to each side of the freshly scalped ingot at the first rolling pass, taking care that the mating surfaces are clean. Good bonding is achieved and each clad surface is normally 5% of the total thickness of the composite sheet (Fig. 3.10). The cladding layer is selected to be at least 100 mV more anodic than the core. Normally 1xxx alloys are chosen to protect 2xxx alloy cores, and 7072 (Al–1Zn) for 3xxx, 5xxx, 6xxx, or 7xxx cores.

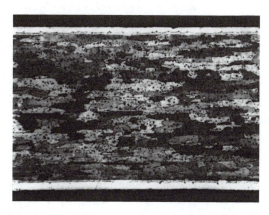

Fig. 3.10 Section of a high-strength alloy sheet roll-clad with pure aluminium (courtesy D. W. Glanvill) ($\times$ 50).

3.1.6 Thermal treatment

The function of thermal treatment is to develop a desired balance of mechanical properties required for consistent service performance. It will be clear from the above considerations that such consistency presupposes attainment of a satisfactory uniformity of microstructure in the preceding stages of production of the wrought material. Also, annealing treatments, if required, are normally carried out within the temperature range 350–420 °C. Attention must now be given to the processes of solution treatment, quenching and ageing, which are the three stages in strengthening alloys by precipitation hardening.

Solution treatment Since the main purpose of this treatment is to obtain complete solution of most of the alloying elements, it should ideally be carried out at a temperature within the single phase, equilibrium solid solution range for the alloy concerned, (α_{Al} in Fig. 2.1). However, it is essential that alloys are not heated above the solidus temperature which will cause overheating, i.e. liquation (melting) of compounds and grain boundary regions with a subsequent adverse effect on ductility and other mechanical properties. Special problems are sometimes encountered with alloys based on the Al–Cu–Mg system in which adequate solution of the alloying elements is possible only if solution treatment is carried out within a few degrees of the solidus, and special care must be taken with control of the furnace temperature. An example of liquation along grain boundaries in an Al–Cu–Mg alloy is shown in Fig. 3.11.

Further precautions are necessary in solution treatment of hot-worked products, again to prevent the growth of coarse, recrystallized grains. Unnecessarily high temperatures and excessively long solution treatment times are to be avoided, particularly with extrusions or with forgings produced from extruded stock. Differential working that is common in such products is the reason why they are so sensitive to grain growth in localized regions.

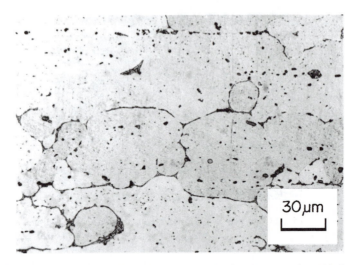

Fig. 3.11 Liquation along grain boundaries caused by overheating of an Al–Cu–Mg alloy during solution treatment.

The consequences of the introduction of hydrogen into molten aluminium through the surface reaction with water vapour were mentioned in Section 3.1.1. Such a reaction may also occur with solid aluminium during solution treatment leading to the adsorption of hydrogen atoms. These atoms can recombine at internal cavities to form pockets of molecular gas. Localized gas pressures can develop which, bearing in mind the relatively high plasticity of the metal at the solution treatment temperatures, may lead in turn to the irreversible formation of surface blisters (Fig. 3.12). Sources of the internal cavities at which blisters may

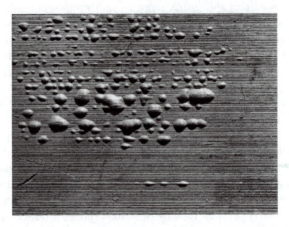

Fig. 3.12 Blisters on the surface of an aluminium alloy component solution treated in a humid atmosphere (courtesy A. T. Thomas) ($\times$ 3).

form are unhealed porosity from the original ingot, intermetallic compounds that have cracked during fabrication and, possibly, clusters of vacant lattice sites that may have formed when precipitates or compounds are dissolved. In these cases, the presence of blisters, while spoiling surface appearance, may have little effect upon mechanical properties of the components. However, blistering is often associated with overheating because the hydrogen can readily collect at liquated regions and this is a more serious problem requiring rejection of the affected material.

As it is difficult to eliminate internal cavities in wrought products, it is imperative that the water vapour content of furnace atmospheres be minimized. Where this is not possible, the introduction of a fluoride salt into the furnace during the heat treatment of critical components can be beneficial by reducing the surface reaction with water vapour.

Finally, reference should be made to the solution treatment of roll-clad sheet. Here it is necessary to avoid dissolved solute atoms, e.g. copper, diffusing from the alloy core through the high-purity aluminium or Al–Zn cladding, thereby reducing its effect in providing protection against corrosion. Such a risk is particularly serious in thinner gauges of sheet. Strict control of temperature and time is necessary and such times should be kept to a minimum consistent with achieving full solution of the alloying elements in the core. The rate of heating to the solution treatment temperature is also important and it is customary to use a mixed nitrate salt bath for clad material because this rate is much faster than that found in air furnaces.

Quenching After solution treatment, aluminium alloy components must be cooled or quenched, usually to room temperature, which is a straightforward operation in principle since the aim is simply to achieve a maximum supersaturation of alloying elements in preparation for subsequent ageing. Cold water quenching is very effective for this and is frequently necessary in order to obtain adequate cooling rates in thicker sections. However, rapid quenching distorts thinner products such as sheet and introduces internal (residual) stresses into thicker products which are normally compressive at the surface and tensile in the core.

Residual stresses may cause dimensional instability, particularly when components have an irregular shape or when subsequent machining operations expose the underlying tensile stresses. What is also serious is that the level of residual stresses may approach the yield stress in some high-strength alloys which, when superimposed, upon normal assembly and service stresses, can cause premature failure. For products of regular section such as sheet, plate, and extrusions, the level of residual quenching stresses can be much reduced by stretching after quenching although this technique has limited use with sheet because it may cause an unacceptable reduction in thickness or gauge. In such cases roller levelling or flattening in a press may be a more satisfactory operation. The ageing treatment also allows some relaxation of stresses and reductions of between 20 and 40% have been measured.

Quenching stresses will be reduced if slower rates of cooling are used and this alternative is particulary important in the case of forgings. Some alloys may be quenched with hot or boiling water, water sprayed, or even air-cooled after solution treatment, and still show an acceptable response to subsequent age hardening. The extent to which slower quenching rates can be tolerated is controlled by what is known as the quench sensitivity of the alloy concerned. During slow cooling there is a tendency of some of the solute elements to precipitate out as coarse particles which reduces the level of supersaturation and hence lowers the subsequent response of the alloys to age hardening. As is demonstrated in Fig. 3.13, this effect is more pronounced in highly concentrated alloys. It is also aggravated by the presence of submicron intermetallic compounds which provide interfaces for the heterogeneous nucleation of large precipitates (Fig. 3.14) during cooling. This behaviour is in fact the reverse of that occurring during the heating cycle when homogenizing ingots (Fig. 3.8). Of the dispersoids commonly present, the chromium-containing phase $(Al_{12}Mg_2Cr)$ causes the greatest increase in quench sensitivity, whereas the zirconium-containing phase (Al_3Zr) has a much less deleterious effect in this regard (Fig. 3.15). The manganese-containing compound Al_6Mn has an effect intermediate between these two.

Microstructural changes may also occur in the region of grain boundaries which are a consequence of slow quenching. In particular, segregation to the grain boundaries of solute elements, such as copper, may cause reduced toughness and higher susceptibility to intergranular corrosion in service.

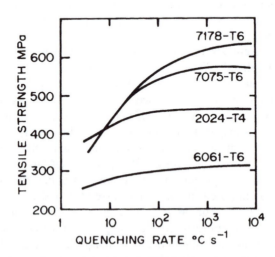

Fig. 3.13 Tensile strengths of various commercial alloys as a function of quenching rate in the critical temperature range 400–290 °C. Alloy compositions are shown in Table 3.4 (from Van Horn, K. R. (Ed.), *Aluminium*, Vol. 1, Chapter 5, American Society for Metals, Cleveland, Ohio, 1967).

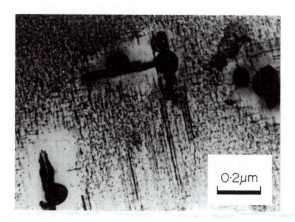

Fig. 3.14 Heterogenous nucleation of large needles of Mg_2Si phase on manganese-bearing intermetallic compounds in a slowly quenched and aged Al–Mg–Si alloy. This has caused denudation of the fine precipitate in surrounding regions of the matrix. (Harding, A. R., *Aluminium Transformation Technology and Applications*, Pampillo, C. A. *et al.* (Eds), American Society for Metals, Cleveland, Ohio, 211, 1980).

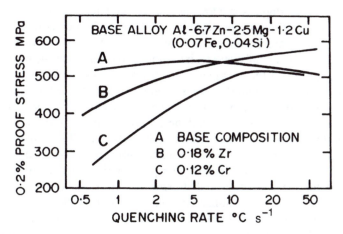

Fig. 3.15 The effect of minor additions of chromium and zirconium on the quench sensitivity of an alloy similar to 7075 (Table 3.4). Quench rates are shown in the critical temperature range 400–290 °C (from Spangler, G. E. *et al.*, *Aluminium Transformation Technology and Applications in 1981*, Pampillo, C. A. *et al.* (Eds), American Society for Metals, Cleveland, Ohio, 1982).

The critical temperature range over which alloys display maximum quench sensitivity is 290–400 °C and specialized techniques have been devised which allow rapid cooling through this range but still permit a reduction in residual stresses. One method which has been used for certain high-strength

Al–Zn–Mg–Cu alloys is to quench into a fused salt bath at 180 °C and hold for a time before cooling to room temperature. A second technique involves the use of proprietary organic liquids having an inverse solubility/temperature relationship. Whereas immersion of a hot body in boiling water generates a tenacious blanket of steam around the body, thereby reducing the cooling rate in the critical temperature range, the organic liquids are formulated to have the reverse effect. Initially the cooling rate is reduced by localized precipitation of a solute in the quenching medium after which it increases through the critical temperature range as the precipitate redissolves. The overall cooling rate is comparatively uniform and a desirable combination of stress relief and high level of mechanical properties has been reported.

Ageing Age-hardening is the final stage in the development of properties in the heat-treatable aluminium alloys. Metallurgical changes that occur during ageing have been discussed in Chapter 2. Some alloys undergo ageing at room temperature (natural ageing) but most require heating for a time interval at one or more elevated temperatures (artificial ageing) which are usually in the range 100–190 °C. Ageing temperatures and times are generally less critical than those in the solution treatment operation and depend upon the particular alloys concerned. Where single-stage ageing is involved, a temperature is selected for which the ageing time to develop high-strength properties is of a convenient duration, e.g. 8 h corresponding to a working day or 16 h for an overnight treatment. Usually the only other stipulation is to ensure that the ageing time is sufficient to allow for the charge to reach the required temperature.

Multiple ageing treatments are sometimes given to certain alloys which have desirable effects on properties such as the stress-corrosion resistance. Such treatments may involve several days at room temperature followed by one or two periods at elevated temperatures. If alloys are slowly quenched after solution treatment, room temperature incubation may be critical because the lower supersaturation of vacancies alters precipitation kinetics. Sufficient time is required for the formation and growth of GP zones, particularly if these zones are to transform to another precipitate on subsequent elevated temperature ageing (Section 2.2.2). Similar considerations apply with respect to the rate of heating to the ageing temperature.

A recent development has been the recognition that alloys may undergo secondary precipitation and further hardening at relatively low temperatures after first being artificially aged at a higher temperature (see Section 2.3.4). Response to these effects depends on alloy composition, the duration of artificial ageing, and the cooling rate to the lower temperature. Alloy properties may therefore continue to change with time, particularly in an alloy is first underaged condition at an elevated temperature. Furthermore, if an alloy in this underaged condition, is then quenched and artificial ageing is resumed after a dwell period at a lower temperature, then the overall level of age hardening may exceed what may be achieved following a standard, single stage (T6) temper. The effects of these interrupted ageing treatments on the tensile and fracture toughness properties of selected alloys are described in Section 3.4.2.

Thermomechanical processing Some alloys show an enhanced response to hardening if they are cold-worked after quenching and prior to ageing and such treatments have been used for many years (T8 temper, see Section 3.2.2). Practices that combine plastic deformation with heat treatment are known generally as thermomechanical processing or thermomechanical treatments (TMT). There are two types which have been applied to some high-strength aluminium alloys.

1. Intermediate TMT (ITMT), in which the processing, usually during rolling, is controlled so as to result in recrystallized but very fine grain structure prior to solution treatment. One example is shown in Fig. 3.63.
2. Final TMT (FTMT), in which deformation is applied after solution treatment and may involve cold or warm working before, during, or after ageing. The purpose is to increase the dislocation density and to stabilize this configuration by precipitates that are nucleated along the lines. A schematic representation of one type of FTMT is shown in Fig. 3.32. The combination of precipitation and substructure hardening may enhance the strength and toughness of some alloys, and its effect on unnotched fatigue properties of Al–Zn–Mg–Cu alloys was illustrated in Fig. 2.46.

3.2 DESIGNATION OF ALLOYS AND TEMPERS

3.2.1 Nomenclature of alloys

The selection of aluminium alloys for use in engineering has often been difficult because specification and alloy designations have differed from country to country. Moreover, in some countries, the system used has been simply to number alloys in the historical sequence of their development rather than in a more logical arrangement. For these reasons, the introduction of an International Alloy Designation System (IADS) for wrought products in 1970 and its gradual acceptance by most countries is to be welcomed. The system is based on the classification used for many years by the Aluminium Association of the United States and it is used when describing alloys in this book.

The IADS gives each wrought alloy a four digit number of which the first digit is assigned on the basis of the major alloying element(s) (Fig. 3.16). Hence there are the 1xxx series alloys which are unalloyed aluminium (with 99% aluminium minimum), the 2xxx series with copper as the major alloying element, 3xxx series with manganese, 4xxx series with silicon, 5xxx series with magnesium, 6xxx with magnesium and silicon and 7xxx series with zinc (and magnesium) as the major alloying elements. The 8xxx series is used for compositions not covered by the above designations and several different alloys are

ALUMINIUM ALLOY AND TEMPER DESIGNATION SYSTEMS

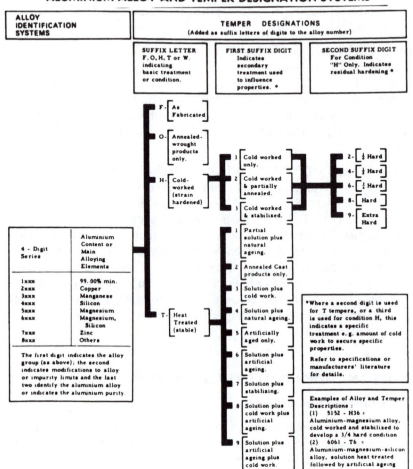

Fig. 3.16 Aluminium alloy and temper designation systems (courtesy Institute of Metals and Materials, Australia).

mentioned in some other sections of Chapter 3. It should also be noted that wrought Al–Si alloys of the 4xxx series are used mainly for welding and brazing electrodes and brazing sheet rather than for structural purposes.

The third and fourth digits are significant in the 1xxx series but not in other alloys. In the 1xxx series, the minimum purity of the aluminium is denoted by these digits, e.g. 1145 has a minimum purity of 99.45%; 1200 has a minimum purity of 99.00%. In all other series, the third and fourth digits have little meaning and serve only to identify the different aluminium alloys in the series. Hence 3003, 3004 and 3005 are distinctly different Al–Mn alloys just as 5082 and 5083 denote two types of Al–Mg alloys (Table 3.2). The second digit indicates purity or alloy modifications. If the second digit is zero, it indicates the original alloy; integers 1 to 9, which are assigned consecutively, indicate alloy modifications. A close relationship usually exists between the alloys, e.g. 5352 is closely related to 5052 and 5252 (Table 3.2); just as 7075 and 7475 differ only slightly in composition (Table 3.4). A prefix X is used to denote that an alloy is at an experimental stage of its development.

In Britain, it has been traditional to use three principal types of specifications to which aluminium and its alloys have been supplied:

1. BS (British Standard) specifications for general engineering use.
2. BS Specifications for aeronautical use (designated as the L series).
3. DTD (Directorate of Technical Development) specifications issued by the Ministry of Technology for specialized aeronautical applications.

In addition there are several supplementary engineering specifications which cover other specialized alloys or those with limited use, while electrical applications are covered by a further six specifications.

The general engineering series is specified BS 1470–75, the six standards covering the different forms: plate, sheet, and strip (BS 1470); drawn tube (BS 1471); forging stock and forgings (BS 1472); rivet, bolt and screw stock (BS 1473); bars, extruded round tube and sections (BS 1474); wire (BS 1475). Every composition is denoted by a number which always indicates the same chemical composition irrespective of form or condition. Pure aluminium (99.99% minimum content) is numbered 1 and the other grades by suffix 1A, 1B, and 1C. The alloys follow from 2 onwards with numerous omissions corresponding to obsolete alloys or numbers not now used. Non-heat treatable alloys are prefixed with the letter N and the heat-treatable alloys with H. The various grades of aluminium (series 1) which are not heat-treatable do not, however, carry the N prefix. Other letters and figures indicate the form and condition of the material.

3.2.2 Temper or heat-treatment nomenclatures

In order to specify the mechanical properties of an alloy and the way these properties were achieved, a system of temper nomenclature has also been adopted as part of the IADS. This takes the form of letters and digits that are added as suffixes to the alloy number. The system deals separately with the non-heat-treatable, strain-hardening alloys on the one hand and heat-treatable alloys on the other. The essential features of the system are outlined in Fig. 3.16 although recourse to detailed specifications or manufacturer's literature is recommended, particularly when several digits are included in the temper designation.

Alloys supplied in the as-fabricated or annealed conditions are designated with the suffixes F and O respectively. Strain-hardening is a natural consequence of most working and forming operations on aluminium alloys. For the various grades of aluminium (1xxx series) and the non-heat-treatable Al–Mn (3xxx) and Al–Mg (5xxx) alloys, this form of hardening augments the strengthening that arises from solid solution and dispersion hardening. Strain-hardened alloys are designated with the letter H. The first suffix digit indicates the secondary treatment: 1 is cold-worked only; 2 is cold-worked and partially annealed; 3 is cold-worked and stabilized. The second digit represents hardening and the severely cold-worked or fully hard condition is designated H18. This corresponds to the tensile strength achieved by a 75% reduction in the original cross-sectional area following a full anneal (0 temper). The H12 temper represents a tensile strength quarter way between that of the 0 temper and the H18 temper (i.e. quarter hard) while the H14 and H16 tempers are half and three-quarters of H18 respectively. Standards do not specify the actual cold reductions required for the H12, H14 and H16 conditions and each individual mill determines its own cold rolling practices to fulfil these temper requirements. A combination of strain-hardening and partial annealing is used to produce the H2 series of tempers. In these the products are cold-worked more than is required to achieve the desired mechanical properties and then reduced in strength by partial annealing. The H3 tempers apply only to Al–Mg alloys which have a tendency to soften with time at room temperature after strain-hardening. This may be avoided by heating for a short time at an elevated temperature (120–175 °C) to ensure completion of the softening process. Such treatment provides stable mechanical properties and improves working characteristics. It should also be noted that an H4 temper has been introduced which applies to products that are strain-hardened and then subjected to some partial annealing during a subsequent paint baking or lacquering operation.

A series of H temper designations having three digits has been assigned to wrought products as follows.

H111 applies to products which are strain-hardened less than the amount required for a controlled H11 temper.

H112 applies to products which are strain-hardened less than the amount incidental to the shaping process. No special control is exerted over the amount of strain-hardening or thermal treatment, but there are mechanical property limits and mechanical property testing is specified.

H121 applies to products which are strain-hardened less than the amount required for a controlled H12 temper.

H311 applies to products which are strain-hardened less than the amount required for a controlled H31 temper.

H321 applies to products which are strain-hardened less than the amount required for a controlled H32 temper.

In addition, temper designations H323 and H343 have been assigned to wrought products containing more than 4% magnesium and apply to products that are specially fabricated to have acceptable resistance to stress-corrosion cracking.

A different system of nomenclature applies for heat-treatable alloys. Tempers other than 0 are denoted by the letter T followed by one or more digits. The more common designations are: T4 which indicates that the alloy has been solution treated, quenched, and naturally aged; T5 in which the alloy has been rapidly cooled following processing at an elevated temperature, e.g. by extrusion, and then artificially aged; and T6 which denotes solution treatment, quenching and artificial ageing. For the T8 condition in which products are cold-worked between quenching and artificial ageing to improve strength, the amount of cold-work is indicated by a second digit, e.g. T85 means 5% cold-work. It should also be noted that the letter W may be used to indicate an unstable temper when an alloy ages spontaneously at room temperature after solution treatment. This designation is specific only when the period of natural ageing is indicated, e.g. $W\frac{1}{2}h$.

Several designations involving additional digits have been assigned to stress-relieved tempers of wrought products.

Tx51—stress-relieved by stretching. Applies to products that are stress-relieved after quenching by stretching the following amounts: plate 0.5–3% permanent set; rod, bar and shapes 1–3% permanent set. These products receive no further straightening after stretching. Additional digits are used in the designations for extruded rod, bar, shapes and tube as follows: Tx510 applies to products that receive no further straightening after stretching; Tx511 applies where minor straightening after stretching is necessary to comply with standard tolerances for straightness.

Tx52—stress-relieved by compressing. Applies to products that are stress-relieved after quenching by compressing to produce a nominal permanent set of 2.5%.

Tx53—stress-relieved by thermal treatment.

In the above designations the letter x represents digits, 3, 4, 6 or 8, whichever is applicable.

In cases where wrought products may require heat treatment by the user, the following temper designations have been assigned.

T42—solution treated, quenched and naturally aged.
T62—solution treated, quenched and artificially aged.

Compositions and tensile properties of selected aluminium alloys are included in Tables 3.2 to 3.5. As an illustration of the variation in tensile properties that may be obtained when the same alloy is prepared under different temper conditions, Table 3.1 shows typical results for the commonly used Al–Mg–Si alloy 6063.

In the British system of nomenclature, the form of the product is denoted by the letters: B for bar and screw stock; C for clad plate, sheet or strip; E for bars,

Table 3.1 Tensile properties of alloy 6063 in different temper conditions (from Australian Aluminium Development Council Handbook, 1994)

Temper condition	0.2% proof stress (MPa)	Tensile strength (MPa)	Elongation (% in 50 mm)
0	50	90	30
T1	90	150	20
T31		180	12
T32		205	8
T33	240	260	20
T34	325	340	15
T4	90	170	22
T5	180	220	12
T6	215	240	12
T81	230	250	11
T82	255	260	9
T83	270	280	8
H112	90	150	20
H14	95	160	18
H18	150	200	8

extruded round tube and sections; F for forgings and forging stock; G for wire; J for longitudinally welded tube; R for rivet stock; S for plate, sheet and strip; T for drawn tube. The condition of the product with regard to strain hardening or heat treatment is denoted by suffixes: M as manufactured; O annealed; H1 to H8 the degrees of strain hardening in increasing order of strength, with two additional categories, H68 applicable only to wire and H9 for extra-hard electrical wire. The heat treatment tempers are as follows.

TB—solution treated, quenched and naturally aged (formerly designated W).
TD—solution treated, cold-worked and naturally aged (applicable only to wires and formerly designated WD).
TE—artificially aged for after cooling from a high-temperature forming process (formerly P).
TF—solution treated, quenched and artificially aged (formerly WP).
TH—solution treated, quenched, cold-worked and artificially aged (applicable only to wire and formerly designated WDP).

The combinations of letters and numbers enables an aluminium alloy to be identified and its form and condition described. For example HE9–TF is heat-treatable composition 9 (Al–Mg–Si) in the form of bar, extruded round tube or section, in the fully heat-treated condition, i.e. solution treated, quenched and artificially aged. The equivalent designation in the IADS would be 6063–T6, which does not define the form of the product.

Table 3.2 Compositions of selected non-heat-treatable wrought aluminium alloys

IADS designation	Si		Fe	Cu	Mn	Mg	Zn	Cr	Ti	Other
1100	1.0 Si	+	Fe	0.05–0.20	0.05		0.10			Al min 99.0
1200	1.0 Si	+	Fe	0.05	0.05		0.10		0.05	Al min 99.0
1145	0.55 Si	+	Fe	0.05	0.05	0.05	0.05		0.03	Al min 99.45
1199	0.006		0.006	0.006	0.002	0.006	0.006		0.002	Al min 99.99
3003	0.60		0.70	0.05–0.20	1.0–1.5		0.10			
3004	0.30		0.70	0.25	1.0–1.5	0.8–1.3	0.25			
3005	0.60		0.70	0.30	1.0–1.5	0.20–0.6	0.25	0.10		
3105	0.60		0.70	0.30	0.3–0.8	0.20–0.8	0.40	0.20	0.10	
5005	0.30		0.70	0.20	0.20	0.50–1.1	0.25	0.10		
5050	0.40		0.70	0.20	0.10	1.1–1.8	0.25	0.10		
5052	0.25		0.40	0.10	0.10	2.2–2.8	0.10	0.15–0.35		
5454	0.25		0.40	0.10	0.50–1.0	2.4–3.0	0.25	0.05–0.20	0.20	
5456	0.25		0.40	0.10	0.50–1.0	4.7–5.5	0.25	0.05–0.20	0.20	
5056	0.30		0.40	0.10	0.05–0.2	4.5–5.6	0.10	0.05–0.20		
5082	0.20		0.35	0.15	0.15	4.0–5.0	0.25	0.15	0.10	
5182	0.20		0.35	0.15	0.20–0.50	4.0–5.0	0.25	0.10	0.10	
5083	0.40		0.40	0.10	0.40–1.0	4.0–4.9	0.25	0.05–0.25	0.15	
5383	0.25		0.25	0.20	0.70–1.0	4.0–5.2	0.40	0.25	0.15	0.20 Zr
5059	0.45		0.50	0.25	0.60–1.2	5.0–6.0	0.20	0.25	0.20	0.05–0.20 Zr
5086	0.40		0.50	0.10	0.20–0.7	3.5–4.5	0.25	0.05–0.25	0.15	
8001	0.17		0.45–0.70	0.15				0.05		0.09–1.3 Ni
8006	0.40		1.2–2.0	0.30	0.30–1.0	0.10	0.10		0.10	
8010	0.40		0.35–0.7	0.10–0.30	0.10–0.8	0.10–0.50	0.40	0.20		
8011	0.50–0.9		0.6–1.0	0.10	0.10	0.05	0.10	0.05	0.08	
8280	1.0–2.0		0.70	0.7–1.3	0.10		0.05			5.5–7.0 Sn 0.20–0.7 Ni
8081	0.7		0.70	0.7–1.3	0.10		0.05		0.10	18–22 Sn

Compositions are in % maximum by weight unless shown as a range or a minimum

121

Table 3.3 Tensile properties of selected non-heat-treatable wrought aluminium alloys

IADS designation	Temper	0.2% proof stress (MPa)	Tensile strength (MPa)	Elongation (% in 50 mm)	Typical applications
1100	0	35	90	35	Sheet, plate, tube, wire, spun hollow-ware, food equipment
	H18	150	165	5	
1145	0	35	75	40	Foil, sheet, semi-rigid containers
	H18	115	145	5	
1199	0	10	45	50	Electrical and electronic foil
	H18	110	115	5	
3003	0	25	110	30	Sheet, plate, foil, rigid containers, cooking utensils, tubes
	H18	185	200	4	
3004	0	70	180	20	Sheet, plate, rigid containers
	H38	250	280	5	
3005	0	55	130	25	Higher-strength foil, roofing sheet
	H18	225	240	4	
3105	0	55	120	24	General sheet material, plate, extrusions
	H18	195	215	3	
5005	0	40	125	30	General sheet material, high-strength foil, electrical conductor wire
	H18	195	200	4	
	H38	185	200	5	
5050	0	55	145	24	Sheet and tube, rip tops for cans
	H38	200	220	6	
5052	0	90	195	25	Sheet, plate, tubes, marine fittings
	H38	255	270	7	

Alloy	Temper				Applications
5454	0	120	250	22	Special-purpose sheet, plate, extrusions, pressure vessels, marine applications such as hulls and superstructures, dump-truck bodies, cryogenic structures
	H34	240	305	10	
5456	0	160	310	24	
	H24	280	370	12	
5056	0	150	290	35	
	H38	345	415	15	
5083	0	145	290	22	
	H343	280	360	8	
5383	0	145	290	17	
	H321	220	305	10	
5086	0	115	260	22	
	H34	255	325	10	
8001	0	40	110	30	Sheet, tubing for water-cooled nuclear reactors
	H18	185	200	4	

Table 3.4 Compositions of selected heat-treatable wrought aluminium alloys

IADS designation	Si	Fe	Cu	Mn	Mg	Zn	Cr	Ti	Other
2001	0.20	0.20	5.2–6.0	0.15–0.50	0.20–0.45	0.10	0.10	0.20	0.05Zr, 0.003 max. Pb
2011	0.40	0.70	5.0–6.0			0.30			0.2–0.6 Bi;0.2–0.6 Pb
2014	0.50–1.2	0.70	3.9–5.0	0.40–1.2	0.20–0.8	0.25	0.10	0.15	0.2Zr + Ti
2017	0.20–0.8	0.70	3.5–4.5	0.4–1.0	0.4–0.8	0.25	0.10	0.15	0.2Zr + Ti
2618	0.10–0.25	0.9–1.3	1.9–2.7		1.3–1.8	0.10		0.04–0.10	0.9–1.2 Ni
2219	0.20	0.30	5.8–6.8	0.20–0.40	0.02	0.10		0.02–0.10	0.05–0.15V,0.10–0.25Zr
2519	0.25	0.30	5.3–6.4	0.10–0.50	0.05–0.40	0.1		0.10	0.05–0.15V,0.10–0.25Zr
2021	0.20	0.30	5.8–6.8	0.20–0.40	0.02	0.10		0.02–0.10	0.10–0.25Zr,0.05–0.20Cd
2024	0.50	0.50	3.8–4.9	0.30–0.9	1.2–1.8	0.25	0.10	0.15	0.20Zr + Ti
2124	0.20	0.30	3.8–4.9	0.30–0.9	1.2–1.8	0.25	0.10	0.15	0.20Zr + Ti
2324	0.10	0.12	3.8–4.4	0.30–0.6	1.2–1.6	0.20		0.10	
2025	0.50–1.2	1.0	3.9–5.0	0.40–1.2	0.05	0.25	0.10	0.15	
2036	0.50	0.50	2.2–3.0	0.10–0.40	0.30–0.6	0.25	0.10	0.15	
2048	0.15	0.20	2.8–3.8	0.20–0.6	1.2–1.8	0.25		0.10	
6005	0.60–0.9	0.35	0.10	0.10	0.40–0.6	0.10		0.10	
6022	0.80–1.5	0.05–0.2	0.01–0.11	0.02–0.1	0.45–0.7	0.25	0.1	0.15	
6060	0.30–0.6	0.10–0.3	0.10	0.10	0.35–0.6	0.15	0.05	0.10	
6063	0.20–0.6	0.35	0.10	0.10	0.45–0.9	0.10	0.10	0.10	
6463	0.20–0.6	0.15	0.20	0.05	0.45–0.9	0.05			
6061	0.40–0.8	0.70	0.15–0.40	0.15	0.8–1.2	0.25	0.04–0.35	0.15	
6151	0.6–1.2	1.0	0.35	0.20	0.45–0.8	0.25	0.15–0.35	0.15	
6351	0.7–1.3	0.50	0.10	0.40–0.8	0.40–0.8	0.20		0.20	
6262	0.40–0.8	0.70	0.15–0.40	0.15	0.8–1.2	0.25	0.04–0.14	0.15	0.40–0.7Bi;0.40–0.7Pb
6009	0.6–1.6	0.50	0.15–0.6	0.2–0.8	0.40–0.8	0.25	0.10	0.10	
6010	0.8–1.2	0.50	0.15–0.6	0.2–0.8	0.6–1.0	0.25	0.10	0.10	

Alloy									
6111	0.60–1.1	0.40	0.50–0.9	0.10–0.45	0.50–1.0	0.15		0.10	0.10
6013	0.60–1.0	0.50	0.60–1.1	0.20–0.8	0.80–1.2	0.25		0.10	0.10
6016	1.0–1.5	0.50	0.20	0.20	0.25–0.6	0.20		0.10	0.15
6017	0.55–0.7	0.15–0.3	0.05–0.2	0.10	0.45–0.6	0.05		0.10	0.05
6082	0.70–1.3	0.50	0.10	0.40–1.0	0.60–1.0	0.20		0.25	0.10
7001	0.35	0.040	1.6–2.6	0.20	2.6–3.4	6.8–8.0		0.18–0.35	0.20	0.10–0.20Zr
7004	0.25	0.35	0.05	0.20–0.7	1.0–2.0	3.8–4.6		0.05	0.05	0.08–0.20Zr
7005	0.35	0.40	0.10	0.20–0.7	1.0–1.8	4.0–5.0		0.06–0.20	0.01–0.06	0.08–0.20Zr
7009	0.20	0.20	0.6–1.3	0.10	2.1–2.9	5.5–6.5		0.10–0.25	0.20	0.25–0.40Ag
7010	0.10	0.15	1.5–2.0	0.30	2.2–2.7	5.7–6.7		0.05		0.11–0.17Zr
7016	0.10	0.12	0.45–1.0	0.03	0.8–1.4	4.0–5.0			0.03	
7017	0.35	0.45	0.20	0.05–0.5	2.0–3.0	4.0–5.2		0.35	0.15	0.10–0.25Zr, 0.15 min. Mn + Cr
7020	0.35	0.40	0.20	0.05–0.5	1.0–1.4	4.0–5.0		0.10–0.35		0.08–0.20Zr, 0.08–0.25 Zr + Ti
7039	0.30	0.40	0.10	0.10–0.40	2.3–3.3	3.5–4.5		0.15–0.25	0.10	
7049	0.25	0.35	1.2–1.9	0.20	2.0–2.9	7.2–8.2		0.10–0.22	0.10	
7449	0.12	0.15	1.4–2.1	0.20	1.8–2.7	7.5–8.7				
7050	0.12	0.15	2.0–2.6	0.10	1.9–2.6	5.7–6.7		0.04	0.06	0.08–0.15Zr
7055	0.12	0.15	2.0–2.6	0.10	1.8–2.3	7.6–8.4		0.05	.06	0.08–0.25Zr
7075	0.40	0.50	1.2–2.0	0.30	2.1–2.9	5.1–6.1		0.18–0.28	0.20	0.25Zr + Ti
7475	0.10	0.12	1.2–1.9	0.06	1.9–2.6	5.2–6.2		0.18–0.25	0.06	
7178	0.40	0.50	1.6–2.4	0.30	2.4–3.1	6.3–7.3		0.18–0.35	0.20	
7085	0.06	0.08	2.0–2.6	0.04	1.2–1.8	7.0–8.0		0.04	0.06	0.08–0.15Zr
7090†	0.12	0.15	0.6–1.3	–	2.0–3.0	7.3–8.7		–	–	1.0–1.9Co, 0.20–0.50O
7091†	0.12	0.15	1.1–1.8	–	2.0–3.0	5.8–7.1		–	–	0.20–0.6Co, 0.20–0.50O

Compositions are in % maximum by weight unless shown as a range or a minimum

†Alloys prepared by powder metallurgy techniques

Table 3.5 Tensile properties and applications of selected heat-treatable wrought aluminium alloys

IADS designation	Temper	0.2% proof stress (MPa)	Tensile strength (MPa)	Elongation (% in 50 mm)	Typical applications
2011	T6	295	390	17	Screw machine parts
2014	T6	410	480	13	Aircraft structures
2017	T4	275	425	22	Screw machine fittings
2618	T61	330	435	10	
2219	T62	290	415	10	Aircraft parts and structures for use at elevated temperatures. 2219 is weldable
	T87	395	475	10	
2024	T4	325	470	20	Aircraft structures and sheet. Truck wheels
	T6	395	475	10	
	T8	450	480	6	
2124	T8	440	490	8	Aircraft structures
2025	T6	255	400	19	Forgings, aircraft propellors
2036	T4	195	340	24	Automotive body panels
2048	T85	440	480	10	Aircraft structures
6005	T5	270	305	12	General purpose extrusions
6060	T5	195	215	17	
6016	T4	105	210	26	Automobile body sheet
6063	T6	215	240	12	Architectural extrusions, pipes
6061	T6	275	310	12	Welded structures
6151	T6	295	330	17	Medium-strength forgings
6009	T4	130	250	24	
	T6	325	345	12	Automobile body sheet
6010	T4	170	290	24	
6111	T4	160	290	25	
6013	T4	185	315	25	Aircraft sheet
6082	T5	260	300	15	Extrusions
	T6	285	315	12	Plate

Alloy	Temper				
7004	T6	340	400	12	
7005	T53	345	395	15	Medium-strength welded structures
7016	T6	315	360	12	
7020	T4	225	340	18	
	T6	310	370	15	
7039	T61	345	415	13	
7001	T6	625	675	9	
7009	T6	470	535	12	
7010	T6	485	545	12	
7049	T73	470	530	11	Aircraft stuctures
7050	T736	510	550	11	
7075	T6	500	570	11	
	T73	430	500	13	
	T76	470	540	12	
7475	7651	560	590	12	
7178	T6	540	610	10	
7055	T7751	610	630	12	
7085	T7651	475	510	7	
7090†	T7E71	580	620	9	Aircraft parts
7091†	T7E69	545	590	11	

†Alloys prepared by powder metallurgy techniques

3.3 NON-HEAT-TREATABLE ALLOYS

Wrought compositions that do not respond to strengthening by heat treatment mainly comprise the various grades of aluminium as well as alloys with manganese or magnesium as the major additions, either singly or in combination (Table 3.2). Approximately 95% of all aluminium flat rolled products (sheet, plate and foil) are made from these three alloy groups. Packaging, transportation and building and construction represent the largest usage of non-heat treatable (NHT) sheet and the three major criteria for selecting and developing these products are:

(i) structural – based on strength and durability
(ii) formability – based on complexity and productivity in making the final part
(iii) surface quality – based on finishing characteristics, reflectivity or general appearance.

Each product has a particular balance of requirements.

NHT alloys can provide a very wide range of tensile properties and Fig. 3.17 shows that yield strengths between 20 and 500 MPa are possible from the 1xxx, 3xxx and 5xxx classes of alloys fabricated in tempers ranging from the annealed "O" to the severely cold worked H19 conditions. Strength is developed by stain-hardening, usually by cold-working during fabrication, in association with

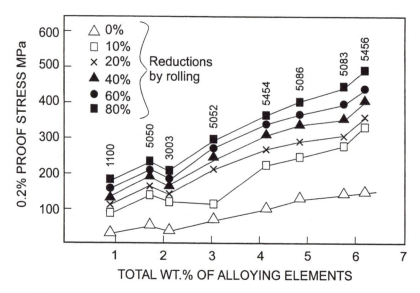

Fig. 3.17 Attainable yield strengths of NHT alloys with different levels of cold work. (courtesy R. E. Sanders Jr., Alcoa, USA).

dispersion-hardening (Al–Mn) and/or solid solution hardening (Al–Mg) or both (Al–Mn–Mg). The miscellaneous alloys in the 8xxx series mostly do not respond to heat treatment and are used for specific applications such as bearings and bottle caps. An Al–1Zn alloy designated 7072 serves as cladding to protect a number of other alloys, e.g. 2219, 7075, from corrosion (Section 3.1.5) and for some fin stock. Several Al–Si alloys in the 4xxx series are available but they are used mainly for welding electrodes, e.g. 4043, or as brazing rods, e.g. 4343. Typical mechanical properties are given in Table 3.3 although it should be noted that the alloys are mainly used for applications in which other properties are the prime consideration. The commercial forms in which they are available and some typical applications are also included in Table 3.3. All are readily weldable and have a high resistance to corrosion in most media.

Some of the essential features of the individual classes of alloys are discussed below. More details concerning special products and applications, e.g. electrical conductors, are considered in Section 3.6.

3.3.1 Super-purity and commercial-purity aluminium (1xxx series)

This group includes super-purity (SP) aluminium (99.99%) and the various grades of commercial-purity (CP) aluminium containing up to 1% of impurities or minor additions (e.g. 1145, 1200). The materials have been utilized as wrought products since the industry was first developed and the CP grades are available in most forms. Tensile properties are low and annealed SP aluminium has a proof stress of only 7–11 MPa. Applications include electrical conductors for which there are several compositions (see Section 3.6.9), chemical process equipment, foil (Section 3.6.4), architectural products requiring decorative finishes, and lithographic plates.

3.3.2 Al–Mn and Al–Mn–Mg alloys (3xxx series)

In general, the 3xxx series of alloys is used when moderate strength combined with high ductility and excellent corrosion resistance is required. Commercial Al–Mn alloys contain up to 1.25% manganese although the maximum solid solubility of this element in aluminium is as high as 1.82%. This limitation is imposed because the presence of iron as an impurity reduces the solubility and there is a danger that large, primary particles of Al_6Mn will form with a disastrous effect on local ductility. The only widely used binary Al–Mn alloy is 3003 which is supplied as sheet. The presence of finer manganese-containing intermetallic compounds confers some dispersion-hardening, and the tensile strength of annealed 3003 is typically 110 MPa compared with 90 MPa for CP aluminium (1100), with corresponding increases in the work-hardened tempers. This alloy is commonly used for foil, cooking utensils and roofing sheet.

The addition of magnesium provides solid solution strengthening and the dilute alloy 3105 (Al–0.55Mn–0.5Mg), which is readily fabricated, is widely used in a variety of strain-hardened tempers. Higher levels of manganese and magnesium are present in 3004 which raises the tensile strength to 180 MPa in the annealed condition. This alloy is used for manufacturing beverage cans, which is currently the largest single use of the metals aluminium and magnesium (Section 3.6.5).

3.3.3 Al–Mg alloys (5xxx series)

Aluminium and magnesium form solid solutions over a wide range of compositions and wrought alloys containing from 0.8% to slightly more than 5% magnesium are widely used. Strength values in the annealed condition range from 40 MPa yield and 125 MPa tensile for Al–0.8Mg (5005) to 160 MPa yield and 310 MPa tensile for the strongest alloy 5456 (Table 3.3). Elongations are relatively high and usually exceed 25%. The alloys work harden rapidly at rates that increase as the magnesium content is raised (Fig. 2.6). As shown in Fig. 3.17, fully work hardened 5456 (H19 temper) may have a yield strength close to 500 MPa although ductility and formability are limited.

The alloys may exhibit some instability in properties which is manifest in two ways. If the magnesium content exceeds 3–4%, there is a tendency for the β-phase, Mg_5Al_8, to precipitate in slip bands and grain boundaries which may lead to intergranular attack and stress-corrosion cracking in corrosive conditions. Precipitation of β occurs only slowly at ambient temperatures but is accelerated if the alloys are in a heavily worked condition, or if the temperature is raised. Small additions of chromium and manganese, which are present in most alloys and raise the recrystallization temperatures, also increase tensile properties for a given magnesium content. This offers the prospect of using alloys with reduced magnesium contents if precipitation of β is to be avoided. For example, alloy 5454 contains 2.7% Mg, 0.7% Mn and 0.12% Cr and has tensile properties similar to those expected from a binary alloy having as much as 4% magnesium.

The second problem is that work-hardened alloys may undergo what is known as age softening at ambient temperatures. Over a period of time, the tensile properties fall due to localized recovery within the deformed grains and, as mentioned in Section 3.2.2, a series of H3 tempers has been devised to overcome this effect. These tempers involve cold-working to a level slightly greater than desired and then stabilizing by heating to a temperature of 120–150 °C. This lowers the tensile properties to the required level and stabilizes them with respect to time. The treatment also reduces the tendency for precipitation of β in the higher magnesium alloys.

The 5xxx series alloys were first developed in the 1930s when there was a need for sheet materials with higher strengths, weldability, good formability and higher levels of corrosion resistance. The first appears to be the alloy 5052 (Al–2.5Mg–0.25Cr) and since then there has been continuing development of stronger alloys with higher magnesium contents, most of which also contain manganese. A well known example is 5083 (Al–4.5Mg–0.7Mn–0.15Cr) and other compositions based on this alloy are discussed in Section 3.6.3. One more recent change has been the deliberate addition of zinc in amounts ranging up to

1.5% with the aim of increasing strength and corrosion resistance (e.g. alloy 5059). However, there is evidence that high zinc levels may cause localised corrosion in the heat affected zones of welded components.

Because of their relatively high strengths in the annealed condition, Al–Mg alloys with higher levels of magnesium are candidates for the rapidly growing area of automotive sheet (Section 3.6.2). However, the fact they are susceptible to surface markings arising from Lüders band formation means that they may be restricted in use for inner panels. If this problem can be avoided, these alloys are capable of being polished to a bright surface finish, particularly if made from high-purity aluminium, and are used for automotive trim and architectural components.

Al–Mg alloys are widely used for welded applications. In transportation structural plate is used for dump truck bodies, large tanks for carrying petrol, milk and grain, and pressure vessels, particularly where cryogenic storage is involved. Their high corrosion resistance makes them suitable for the hulls of small boats and for the super-structures of ocean-going vessels (Section 3.6.3). These alloys are also widely used for ballistic armour plate in light weight military vehicles.

3.3.4 Miscellaneous alloys (8xxx series)

This series contains several dilute alloys, e.g. 8001 (Al–1.1Ni–0.6Fe) which is used in nuclear energy installations where resistance to corrosive attack by water at high temperatures and pressures is the desired characteristic. Its mechanical properties resemble 3003. Alloy 8011 (Al–0.75Fe–0.7Si) is used for bottle caps because of its good deep drawing qualities and several other dilute compositions are included in the range of electrical conductor materials (Section 3.6.9). This alloy and other dilute compositions containing transition metal elements, such as 8006 (Al–1.6Fe–0.65Mn), are used for producing foil (Section 3.6.5) and finstock for heat exchangers (Section 3.5.2). In each of these alloys, the most important contribution to strengthening comes from dispersion hardening which commonly accounts for 30-40 MPa of the values for proof stress and tensile strength. Solid solution strengthening is the next most important factor providing estimated increments of 2–10 MPa in proof stress of 5–20 MPa in tensile strength.

Alloys such as 8280 and 8081 serve an important role as bearing alloys based on the Al–Sn system that are now widely used in motor cars and trucks, particularly where diesel engines are involved. These alloys are considered in some detail in Section 3.6.7. Some new, lithium-containing alloys, designated 8090 and 8091 are described in Section 3.4.6.

3.4 HEAT-TREATABLE ALLOYS

Wrought alloys that respond to strengthening by heat treatment are covered by the three series 2xxx (Al–Cu, Al–Cu–Mg), 6xxx (Al–Mg–Si) and 7xxx (Al–Zn–Mg, Al–Zn–Mg–Cu). All depend on age-hardening to develop enhanced strength properties and they can be classified into two groups: those that have medium strength and are readily weldable (Al–Mg–Si and Al–Zn–Mg), and the high-strength alloys that have been developed primarily for aircraft construction (Al–Cu, Al–Cu–Mg, and Al–Zn–Mg–Cu), most of which have very limited weldability. Compositions of representative commercial alloys are shown in Table 3.4, and Table 3.5 gives typical properties, the forms in which they are available, together with some common applications.

3.4.1 Al–Cu alloys (2xxx series)

Although the complex changes that occur during the ageing of Al–Cu alloys have been studied in greater detail than any other system, there are actually few commercial alloys based on the binary system. Alloy 2011 (Al–5.5Cu) is used where good machining characteristics are required, for which it contains small amounts of the insoluble elements lead and bismuth that form discrete particles in the microstructure and assist with chip formation. Alloy 2025 is used for some forgings although it has largely been superseded by 2219 (Al–6.3Cu) which has a more useful combination of properties and is also available as sheet, plate and extrusions. Alloy 2219 has relatively high tensile properties at room temperature together with good creep strength at elevated temperatures and high toughness at cryogenic temperatures. In addition, it can be welded and has been used for fuel tanks for storing liquified gases that serve as propellants for missiles and space vehicles. Response to age-hardening is enhanced by strain hardening prior to artificial ageing (T8 temper) and the yield strength may be increased by as much as 35% as compared with the T6 temper (Table 3.5)

Modified versions of 2219 have been developed with the United States in order to meet requirements for increased tensile properties. One such alloy is designated 2021 which, as rolled plate, has a yield strength of 435 MPa, tensile strength of 505 MPa and elongation of 9% with no reported sacrifice of weldability or toughness at low temperatures. Increased strength has been achieved by minor or trace additions of 0.15% cadmium and 0.05% tin which have the well-known effect of refining the size of the semi-coherent θ' precipitate that forms on ageing in the medium temperature range ($\sim$130–200 °C) (see Section 2.2.4). However, the toxicity of cadmium causes concern and production of this alloy has been banned in some countries. A second alloy is 2519, in which a minor amount of magnesium (e.g. 0.2%) modifies the precipitation process, resulting in greater age-hardening.

The role of minor elements in modifying precipitation in an aged alloy based on the aluminium–coppper system has also been exploited in a recent experimental alloy to promote improved strength at room and elevated temperatures. This alloy is based on 2219 but has controlled additions of silver (0.3–0.4%) and magnesium (0.4–0.5%) which change the precipitation process. A typical composition is Al–6.3Cu–0.45Mg–0.3Ag–0.3Mn–0.15Zr. Instead of the phases θ'' and θ' which precipitate on the $\{100\}_\alpha$ planes when

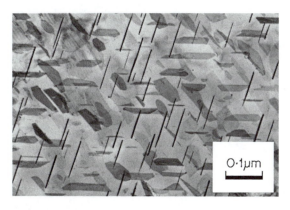

Fig. 3.18 Plates of the Ω phase precipitated on the {111} planes in an Al–Cu–Mg–Ag alloy (courtesy R. J. Chester).

2219 is artificially aged, thin plates of a finely dispersed new phase form on the $\{111\}_\alpha$ planes of the silver-containing alloy (Fig. 3.18). This phase, which has been designated Ω, has an orthorhombic structure (Table 2.2) and has proved to be relatively stable at temperatures up to approximately 200 °C. As shown in Fig. 2.20, nucleation of Ω is facilitated in Al–Cu–Mg alloys having high Cu:Mg ratios due to rapid clustering of magnesium and silver atoms that occurs immediately artificial ageing is commenced. Once Ω precipitates have developed, these two elements are rejected to the Ω/matrix interfaces where they segregate at the broad surfaces of the plates. It is the presence of layers of the silver and magnesium atoms at the Ω plate surfaces that is considered to restrict ledge formation thereby minimizing coarsening at temperatures up to 200 °C.

Alloys based on the Al–Cu–Mg–Ag system develop yield strengths that may exceed 500 Mpa which compares with 290 and 395 Mpa for alloy 2219 in the T6 and T87 conditions, respectively. In Fig. 3.19, accelerated stress-rupture tests suggest that the creep performances of four extruded Al–Cu–Mg–Ag alloys aged to the T6 condition, are generally superior to those of three commercial 2xxx series alloys. As mentioned in Section 2.5.6, there is also evidence that creep resistance can be further improved if aluminium alloys are tested in the underaged condition. As an example, the alloy Al–5.6Cu–0.45Mg–0.40Ag–0.30Mn–0.18Zr, underaged at 185 °C for a time to reach 90% of its T6 0.2% proof stress, has shown zero secondary creep at 130 °C during prolonged exposure for 20 000 h at a relatively high stress of 200 MPa. Several aluminium alloys containing small additions of silver have recently been registered including the nominal compositions 2039 (Al–5.0Cu–0.6Mg–0.38Ag–0.38Mn–0.18Zr) and 2040 (Al–5.1Cu–0.9Mg–0.55Ag–0.68Mn–011Zr–0.0001 Be).

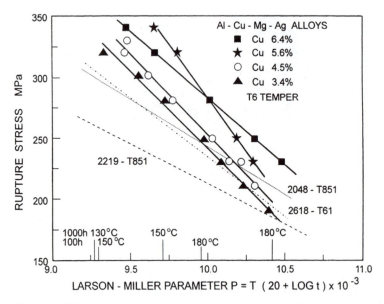

Fig. 3.19 Larson-Miller plots of stress-rupture results for four alloys with compositions Al–3.4 to 6.4Cu–0.45Mg–0.40Ag–0.30Mn–0.18Zr, and for three 2xxx series alloys. Equivalent positions for several temperatures after exposure times of 100 and 1000 h are shown on the horizontal axis (from Polmear I. J. *et al, Mater. Sci. Technol.*, **15**, 861, 1999).

3.4.2 Al–Cu–Mg alloys (2xxx series)

These alloys date from the accidental discovery of the phenomenon of age-hardening by Alfred Wilm, working in Berlin in 1906, who was seeking to develop a stronger aluminium alloy to replace brass for the manufacture of cartridge cases. His work led to the production of an alloy known as Duralumin (Al–3.5Cu–0.5Mg–0.5Mn) which was quickly used for structural members for Zeppelin airships, and later for aircraft (Section 3.6.1). A modified version of this alloy (2017) is still used, mainly for rivets, and several other important alloys have been developed which are now widely used for aircraft construction. An example is 2014 (Al–4.4Cu–0.5Mg–0.9Si–0.8Mn) in which higher strengths have been achieved because the relatively high silicon content increases the response to hardening on artificial ageing. Typical tensile properties are: 0.2% proof stress 320 MPa and tensile strength 485 MPa. Another alloy, 2024, in which the magnesium content is raised to 1.5% and the silicon content is reduced to impurity levels, undergoes significant hardening by natural ageing at room temperature and is frequently used in T3 or T4 tempers. It also has a high response to artificial ageing, particularly if cold-worked prior to ageing at around 175 °C, e.g. 0.2% proof stress 490 MPa and tensile strength 520 MPa for the T86 temper.

These and other 2xxx alloys in the form of sheet are normally rollclad with aluminium or Al–1Zn in order to provide protection against corrosion, and the tensile properties of the composite product may be some 5% below those for the unclad alloy. Much greater reductions in strength may occur under fatigue conditions. For example, cladding 2014 sheet may reduce the fatigue strength by as much as 50% if tests are conducted in air under reversed plane bending. Differences between clad and unclad alloys are much less under axial loading conditions, or if the materials are tested as part of a structural assembly. Under corrosion-fatigue conditions, the strength of unclad sheet may fall well below that of the clad alloy. The adverse effect of cladding on fatigue strength in air is due mainly to the ease by which cracks can be initiated in the soft surface layers and a number of harder cladding materials are being investigated.

Microstructural features that influence toughness and ductility have been considered in Section 2.5.2. Figure 3.20 shows that, for equal values of yield strength, alloys in the 2xxx series have lower fracture toughness than those of the 7xxx series (Al–Zn–Mg–Cu). This is attributed to the larger sizes of intermetallic compounds in the 2xxx alloys. Improvements in both fracture toughness and ductility can be obtained by reducing the levels of iron and silicon impurities as well as that of copper, all of which favour formation of large, brittle compounds in the cast materials. This has led to the development of alloys such as 2124 (iron 0.3% maximum, silicon 0.2% maximum, as compared with levels up to 0.5% for each of these elements in 2024) and 2048 (copper reduced 3.3% and iron and silicon 0.2% and 0.15% maximum, respectively). Some comparative properties are shown in Fig. 2.41 and Table 3.6. High-toughness

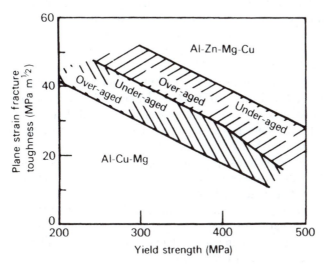

Fig. 3.20 Relationships of plane strain fracture toughness to yield strength for the 2xxx and 7xxx series of alloys (from Develay, R., *Metals and Materials*, **6**, 404, 1972).

Table 3.6 Effect of purity on the fracture toughness of some high-strength wrought aluminium alloys (from Speidel, M. O., *Met. Trans.*, **6A**, 631, 1975)

Alloy and temper	%Fe maximum	%Si maximum	0.2% Proof stress (MPa)	Tensile strength (MPa)	Fracture toughness (MPa m$^{1/2}$)	
					Longitudinal	Short transverse
2024–T8	0.50	0.50	450	480	22–27	18–22
2124–T8	0.30	0.20	440	490	31	25
2048–T8	0.20	0.15	420	460	37	28
7075–T6	0.50	0.40	500	570	26–29	17–22
7075–T73	0.50	0.40	430	500	31–33	20–23
7175–T736	0.20	0.15	470	540	33–38	21–29
7050–T736	0.15	0.12	510	550	33–39	21–29

versions of older alloys such as 2024 are now being used as sheet, plate and forgings in several modern aircraft.

Experimental interrupted ageing treatments involving secondary precipitation (Section 2.3.4) have also been found to increase fracture toughness in the Al–Cu–Mg alloy 2014 and a number of other commercial aluminium alloys. Moreover, these improvements may be accompanied by increases tensile properties despite the fact that toughness and strength are usually inversely related. Results shown in Table 3.7 compare these properties for selected wrought alloys, and the casting alloy 357 (Al–7Si–0.5Mg), that were given standard T6 or interrupted ageing treatments. Increases in fracture toughness by using the latter treatments were as follows : 2014 34.5%; 6061 58.6% and 17.4%; 7050 9.3% and 38.3%; 8090 28.1%; 357 2.0% and 38.8%.

Relationships between rate of growth of fatigue cracks and stress intensity for the alloys 2024–T3 and 7075–T6 are shown in Fig. 3.21. Other 2xxx series alloys show similar rates of crack propagation to 2024–T3 over most of the range of test conditions. In general, these alloys have rates of crack growth that are close to one-third those observed in the 7xxx series alloys.

It is now common to use pre-packed specimens to assess comparative resistance of alloys to stress-corrosion cracking since this type of test avoids uncertainties associated with crack initiation. Relationships between crack growth rate and stress intensity in tests on several 2xxx series alloys are shown in Fig. 3.22. It is clear that there is a large variation in threshold stress intensities for different alloys, while the crack velocity plateaus are similar. Thus the ranking of these alloys is best obtained from measurements of the threshold stress intensities (K_{Iscc} values). In the naturally aged tempers T3 and T4, the 2xxx series are prone to stress-corrosion cracking. The alloy 2014 in the T6 temper is

Table 3.7 Effects of interrupted ageing and secondary hardening on the mechanical properties of selected aluminium alloys (after Lumley, R. N. *et al*, *Mater. Sci. Tech*,. **19**, 1, 2003 and *Proc. 9th Inter. Conf. on Aluminium Alloys*, IMEA, 85, 2004)

Alloy[1] and Treatment[2]	0.2% proof stress (MPa)	tensile strength (MPa)	elongation %	fracture toughness[3] MPa m$^{1/2}$
2014-T6	414	488	5	26.9
2014-T6I6	436	526	10	36.2
6061-T6	267	318	13	36.8
6061-T6I6	299	340	13	58.4
6061-T6I4	302	341	16	43.2
7050-T6	546	621	14	37.6
7050-T6I6	574	639	14	41.1
7050-T6I4	527	626	16	52.0
8090-T6	349	449	4	24.2
8090-T6I6	391	512	5	31.0
357-T6	287	325	7	25.5
357-T6I6	341	375	5	26.0
357-T6I4	280	347	8	35.9

[1]Wrought alloys except for casting alloy 357. Compositions in Tables 3.4 and 4.2 respectively.
[2]T6 involves single stage artificial ageing to peak strength at appropriate temperatures.
  T6I6 is a designation indicating that the T6 temper has been interrupted by quenching and holding the alloys for a prescribed time at 65 °C, after which artificial ageing to peak strength is resumed.
  T6I4 is a designation indicating that T6 temper has been interrupted by quenching and holding the alloys for a long time at 65 °C. In this case, artificial ageing is not resumed.
[3]Fracture toughness tests on wrought alloys in S-L orientation. All tests under plane strain conditions except for alloy 6061.

also susceptible whereas the more recent alloy 2048 in the T8 temper is much more resistant to this form of cracking.

Interest in supersonic aircraft such as Concorde stimulated a need for a sheet alloy with improved creep strength on prolonged exposure, e.g. 50 000 h, at temperatures of around 120 °C. Such a material was developed from a forging alloy known widely as RR58 (2618) which itself had been adapted from an early casting alloy (Al–4Cu–1.5Mg–2Ni) known originally as Y alloy. The alloy 2618 has the nominal composition Al–2.2Cu–1.5Mg–1Ni–1Fe, in which the copper and magnesium contribute to strengthening through age-hardening, whereas nickel and iron form the intermetallic compound Al_9FeNi which causes dispersion hardening and assists in stabilizing the microstructure. One refinement was the addition of 0.2% silicon, which both increases the hardening associated with the first stage in the ageing process (GPB zones) and promotes formation of a more uniform dispersion of the S′ (or S) phase (Table 2.3). Both these changes improve the creep properties of the alloy on long-term exposure at temperatures of 120–150 °C.

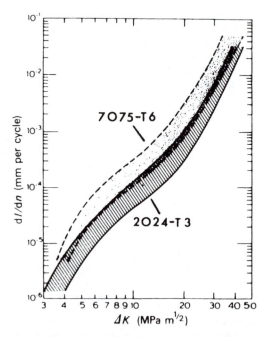

Fig. 3.21 Comparative fatigue crack growth rates for 2024–T3 and 7075–T6 in air of varying humidity (from Hahn, G. T. and Simon, R., *Eng. Fract. Mech.*, **5**, 523, 1973).

Lower strength Al–Cu–Mg alloys are being investigated as possible sheet materials for automotive applications and these are discussed in Section 3.6.2. One example is 2036 which has nominal copper and magnesium contents of 2.5% and 0.45% respectively.

3.4.3 Al–Mg–Si alloys (6xxx series)

Al–Mg–Si alloys are widely used as medium-strength structural alloys which have the additional advantages of good weldability, corrosion resistance, and immunity to stress-corrosion cracking. Just as the 5xxx series of alloys comprise the bulk of sheet products, the 6xxx series are used for the majority of extrusions, with smaller quantities being available as sheet and plate (Fig 3.23). In commercial alloys magnesium and silicon are added either in what are called "balanced" amounts to form quasi-binary Al–Mg_2Si alloys (Mg:Si 1.73:1), or with an excess of silicon above that needed to form Mg_2Si. The commercial alloys may be divided into three groups and a guide to the strength levels that may be attained in the T6 condition is shown as contours drawn on a compositional plot in Fig. 3.24.

The first group comprises alloys with balanced amounts of magnesium and silicon adding up to between 0.8% and 1.2%. These alloys can be readily extruded and offer a further advantage in that they may be quenched at the

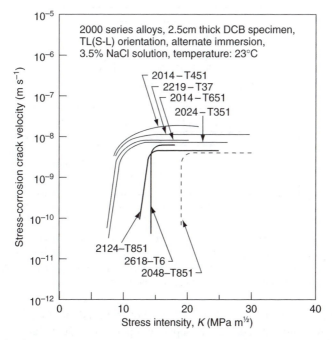

Fig. 3.22 Crack propagation rates in stress-corrosion tests using pre-cracked specimens of 2xxx alloys exposed to an aqueous solution of 3.5% NaCl. Double cantilever beam (DCB) specimens selected from short transverse direction of 25 mm thick plate and wet with the solution twice a day (from Speidel, M. O., *Metall. Trans A*, **6A**, 631, 1975).

Fig. 3.23 Examples of extruded sections made from Al–Mg–Si alloys (courtesy Sumitomo Metal Industries Ltd).

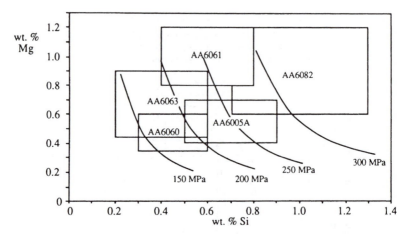

Fig. 3.24 The compositional limits of some common 6xxx alloys, together with contours representing common peak aged (T6) values of yield strength (from Court, S. A., Dudgeon, H. D. and Ricks, R. A., *Proceedings of 4th International Conference on Aluminium Alloys, Atlanta, U.S.A.*, Sanders, T. H. and Starke, E. A. (Eds), 1. 395, 1994).

extrusion press when the product emerges hot from the die, thereby eliminating the need to solution treat as a separate operation. Quenching is normally achieved by means of water sprays, or by leading the product through a trough of water. Thin sections (<3 mm) can be air-cooled. Moderate strength is developed by age-hardening at 160–190 °C and one alloy, 6063, is perhaps the most widely used of all Al–Mg–Si alloys. In the T6 temper, typical tensile properties are 0.2% proof stress 215 MPa and tensile strength 245 MPa. These alloys find particular application for architectural and decorative finishes and, in this regard, they respond well to clear or colour anodizing as well as to the application of other surface finishes. A high-purity version, 6463, in which the iron content is kept to a low level (<0.15%), responds well to chemical brightening and anodizing for use as automotive trim.

The other two groups contain magnesium and silicon in excess of 1.4%. They develop higher strength on ageing and, because they are more quench sensitive, it is usually necessary to solution treat and water quench as separate operations after extrusion. One group, which is particularly popular in North America, has balanced compositions and a common example is 6061 (Al–1Mg–0.6Si) to which is added 0.25% copper to improve mechanical properties together with 0.2% chromium to offset any adverse effect copper may have on corrosion resistance. These alloys are widely used as general-purpose structural materials. The alloys in the other group contain silicon in excess of that needed to form Mg_2Si and the presence of this excess silicon promotes an additional response to age-hardening by both refining the size of the Mg_2Si particles and precipitating as silicon. However, it may also reduce ductility and cause intergranular embrittlement which is attributed in part to the tendency for silicon to segregate to the grain boundaries in these alloys. The presence of chromium (6151) and manganese (6351) helps to counter this effect by promoting fine grain size and inhibiting recrystallization during solution treatment. The alloys are used as extrusions and forgings.

As mentioned in Section 2.3.3, there is strong experimental evidence that the actual atomic ratios Mg:Si in the intermediate precipitates that contribute maximum age hardening in Al–Mg–Si alloys are close to 1:1 rather than the expected 2:1 present in the equilibrium precipitate β (Mg_2Si). It has been suggested, therefore, that the balanced alloys actually have magnesium contents in excess of that needed to promote the required age hardening response, This has led to the design of a new series of alloys in which the magnesium contents have been reduced in order to improve the hot working characteristics and increase productivity without compromising mechanical properties. As an example, the tensile properties of the readily extrudable alloy 6060, with a modified composition (Al–0.35Mg–0.5Si), are found to be comparable with those of the popular alloy 6063 (Al–0.5Mg–0.4Si) whereas extrusion speeds may be up to 20% greater.

Higher strengths may be achieved in Al–Mg–Si alloys by increasing the copper content and the alloy 6013 (Al–1Mg–0.8Si–0.8Cu–0.35Mn) has a 0.2% proof stress of 330 MPa for the T6 temper. This compares with a 0.2% proof stress of 275 MPa for the widely used alloy 6061 which has a maximum copper content of 0.25%. Another alloy is 6111 (Al–0.75Mg–0.85Si–0.7Cu–0.3Mn). These alloys are hardened by the presence of the Q phase ($Al_5Cu_2Mg_8Si_6$) that precipitates in addition to the phases which form in ternary Al–Mg–Si alloys (Table 2.3). The alloys are being promoted for automotive and aircraft applications. Because of the higher copper contents, the alloys may show some susceptibility to intergranular corrosion that is associated with the formation of precipitate-free zones at grain boundaries (see Section 2.2.3) which are depleted of copper and silicon and, as a consequence, are anodic with respect to the grains. This problem can be reduced by using a proprietary T78 temper which involves a controlled amount of overageing.

Al–Mg–Si alloys are normally aged at about 170 °C and the complete precipitation process is now recognized as being perhaps the most complex of all age-hardened aluminium alloys (Table 2.3). Early clustering of silicon atoms has been detected prior to formation of GP zones which can influence subsequent stages of precipitation. For example, during commercial processing, there may be a delay at room temperature between quenching and artificial ageing which may modify the mechanical properties that are developed. In alloys containing more than 1% Mg_2Si, a delay of 24 h causes a reduction of up to 10% in tensile properties as compared with the properties obtained by ageing immediately. However, such a delay can enhance the tensile properties developed in compositions containing less than 0.9% Mg_2Si. These effects have been attributed to clustering of solute atoms and vacancies that occurs at room temperature and to the fact that the GP zones solvus (Section 2.2.2) is above 170 °C for the more highly alloyed compositions. With these alloys, the precipitate which develops directly from the clusters formed at room temperature is coarser than that developed in alloys aged immediately after quenching, with a consequent adverse effect on tensile properties. The reverse occurs in alloys containing less than 0.9% MgSi. The addition of small amounts of copper (e.g. 0.25%) lessens the adverse effects of delays at room temperature by promoting an increased response to artificial ageing.

As with the 2xxx series, there is an Al–Mg–Si alloy (6262) containing additions of lead and bismuth to improve machining characteristics. Although the machinability of this alloy is below the Al–Cu alloy 2011, it is not susceptible to stress-corrosion cracking and is preferred for more highly stressed fittings.

3.4.4 Al–Zn–Mg alloys (7xxx series)

The Al–Zn–Mg system offers the greatest potential of all aluminium alloys for age-hardening although the very high-strength alloys always contain quaternary additions of copper to improve their resistance to stress-corrosion cracking (Section 3.4.5). There is, however, an important range of medium-stength alloys containing little or no copper that have the advantage of being readily weldable. These alloys differ from the other weldable aluminium alloys in that they age-harden significantly at room temperature. Moreover, the strength properties that are developed are relatively insensitive to rate of cooling from high temperatures, and they possess a wide temperature range for solution treatment, i.e. 350 °C and above, with the welding process itself serving this purpose. Thus there is a considerable recovery of strength after welding and tensile strengths of around 320 MPa are obtained without further heat treatment. Yield strengths may be as much as double those obtained for welded components made from the more commonly used Al–Mg and Al–Mg–Si alloys.

Weldable Al–Zn–Mg alloys were first developed for lightweight military bridges but they now have commercial applications, particularly in Europe, e.g. Fig. 3.25. Elsewhere, their use has been less widespread for fear of possible stress-corrosion cracking in the region of welds. Many compositions are now available which may contain 3–7% zinc and 0.8–3% magnesium (Zn + Mg in the range of 4.5–8.5%) together with smaller amounts (0.1–0.3%) of one or more of the elements chromium, manganese and zirconium. These elements are added mainly to control grain structure during fabrication and heat treatment although it has been claimed that zirconium also improves weldability. Minor additions of copper are made to some alloys but the amount is kept below 0.3% to minimize both hot-cracking during the solidification of welds and corrosion in service. Compositions of representative weldable Al–Zn–Mg alloys are shown in Table 3.8 for different categories of tensile strength.

Improvements in resistance to stress-corrosion cracking have come through control of both composition and heat-treatment procedures. With respect to composition, it is well known that both tensile strength and susceptibility to cracking increase as the Zn + Mg content is raised and it is necessary to seek a compromise when selecting an alloy for a particular application. It is generally accepted that the Zn + Mg content should be less than 6% in order for a weld-

Fig. 3.25 Railway carriage made from a welded Al–Zn–Mg alloy.

Table 3.8 Zinc and magnesium contents and ratios in some Al–Zn–Mg and Al–Zn–Mg–Cu alloys

	Alloy	Zn (%)	Mg (%)	Zn + Mg (%)	Zn/Mg ratio
Medium-strength	7104	4.0	0.7	4.7	5.7
weldable	7008	5.0	1.0	6.0	5.0
Al–Zn–Mg alloys	7011	4.7	1.3	6.0	3.7
	7020	4.3	1.2	5.5	3.6
	7005	4.5	1.4	5.9	3.2
	7004	4.2	1.5	5.7	2.8
	7051	3.5	2.1	5.6	1.7
Higher-strength	7003	5.8	0.8	6.6	7.2
weldable	7046	7.1	1.3	8.4	5.5
Al–Zn–Mg alloys	7039	4.0	2.8	6.8	1.4
	7017	4.6	2.5	7.1	1.8
High-strength	7049	7.7	2.5	10.2	3.1
Al–Zn–Mg–Cu	7050	6.2	2.3	8.5	2.7
alloys	7010	6.2	2.5	8.7	2.5
	7475	5.7	2.3	8.0	2.5
	7001	7.4	3.0	10.4	2.5
	7075	5.6	2.5	8.1	2.2
	7055	8.0	2.05	10.05	3.9
	7085	7.5	1.5	9.0	5.0

able alloy to have a satisfactory resistance to cracking, although the additional controls may be required. It has also been proposed that the Zn:Mg ratio is important and, with respect to these two elements, there is experimental evidence which suggests that the maximum resistance to stress-corrosion cracking

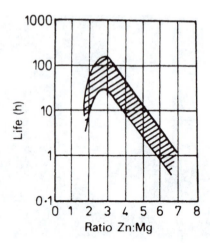

Fig. 3.26 Effect of Zn:Mg ratio on the susceptibility of Al–Zn–Mg alloys to stress-corrosion cracking (from Gruhl, W., *International Congress on Aluminium Alloys in the Aircraft Industry*, Turin, 1976).

occurs if this ratio is between 2.7 and 2.9 (Fig. 3.26). As shown in Table 3.8, few of the existing commercial alloys do in fact comply with this proposed ratio. Small amounts of copper and, more particularly, silver have been shown to increase resistance to SCC but the addition of silver is considered too great a cost penalty for this range of alloys.

Two changes in heat treatment procedures have led to a marked reduction in susceptibility to stress-corrosion cracking in the weldable alloys. One has been the use of slower quench rates, e.g. aircooling, from the solution treatment temperature which both minimizes residual stresses and decreases differences in electrode potentials throughout the microstructure. This practice has also had implications with regard to composition as there has been a tendency to use 0.08–0.25% zirconium to replace chromium and manganese for the purpose of inhibiting recrystallization because this element has the least effect on quench sensitivity (Fig. 3.15). This characteristic is thought to arise because zirconium forms small, insoluble particles of Al_3Zr whereas chromium and manganese combine with some of the principal alloying elements to form $Al_{12}Mg_2Cr$ and $Al_{20}Cu_2Mn_3$ respectively, thereby removing them from solid solution. The other change has been to artificially age the alloys, sometimes to the extent of using a duplex treatment of the T73 type.

The Al–Zn–Mg alloys are normally welded with an Al–4.5Mg + Mn filler wire although some compositions are available which contain both zinc and magnesium. Although problems with intercrystalline SCC of welded structures in service are now comparatively rare, when this does occur the cracks normally form close to the weld bead/parent alloy interface in what has become known as the 'white zone' (Fig. 3.27). This is a region within the parent metal that has

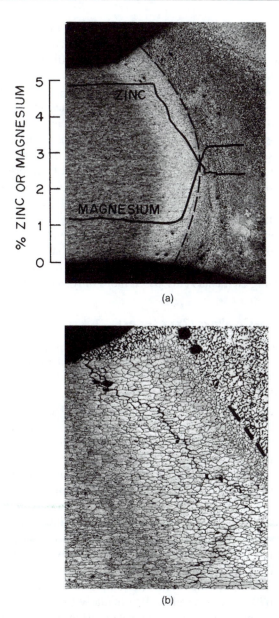

(a)

(b)

Fig. 3.27 (a) Section showing interface between parent metal and weld bead. The 'white zone' of the parent metal is revealed by deep etching with 20% HNO_3. The compositions of the parent metal and filler wire were Al–4.9Zn–1.2Mg and Al–4.5Mg. Variations in the zinc and magnesium contents were determined by microprobe analysis × 15; (b) intercrystalline cracking within the white zone × 40 (from Cordier, H. and Polmear, I. J., *Proceedings of Eurocor* **77**, Soc. Chemical Industry, London, 1979).

undergone partial liquation and in which the zinc and magnesium contents vary considerably (Fig. 3.27). It is also one into which elements added to the filler wire can diffuse, at least in part, but the influence of filler composition on SCC has not been studied in detail.

For the majority of welded aluminium alloys, the weld beads are electro-negative with respect to the adjacent heat-affected zones and parent metal. Thus the weld beads act sacrificially (i.e. they are preferentially corroded) thereby protecting the surrounding, more vulnerable regions from corrosive attack. This situation is in fact reversed with the 7xxx series of alloys as it is the weld bead that is electropositive. The surrounding regions therefore require special protection and traditional methods have involved the use of paints and sprayed coatings. More recently, success has been achieved by the novel technique of completing the welding operation with a capping pass using a filler wire containing the elements tin, indium or gallium. These elements lower the elec-trode potential of the aluminium in the weld bead so that it becomes electroneg-ative (i.e. protective) with respect to the surrounding regions. The role of these so-called activating elements is discussed more fully in Section 3.6.10.

3.4.5 Al–Zn–Mg–Cu alloys (7xxx series)

These alloys have received special attention because it has long been realized that they have a particularly high response to age-hardening. For example, Rosenhain and his colleagues at the National Physical Laboratory in Britain in 1917 obtained a tensile strength of 580 MPa for a composition Al–20Zn–2.5Cu–0.5Mg–0.5Mn when the value for Duralumin was 420 MPa. However, this alloy and others produced over the next two decades proved to be unsuitable for structural use because of a high susceptibility to stress-corrosion cracking. Because of the critical importance of Al–Zn–Mg–Cu alloys for air-craft construction, this problem has been the subject of continuing research and development and will now be considered in some detail.

Military needs in the late 1930s and 1940s for aircraft alloys having higher strength/weight ratios eventually led to the introduction of several Al–Zn–Mg–Cu alloys of which 7075 is perhaps the best known. Later this alloy and equivalent materials such as DTD 683 in Britain were also accepted for the construction of most civil aircraft. Stronger alloys, e.g. 7178–T6, tensile strength 600 MPa, were introduced for compressively stressed members and another alloy 7079–T6 was developed particularly for large forgings for which its lower quench sensitivity was an advantage. However, continuing problems with stress-corrosion cracking, notably in an early alloy 7079–T6, and deficien-cies in other properties stimulated a need for further improvements. Some air-craft constructors, in fact, reverted to using the lower-strength alloys based on the Al–Cu–Mg system even though a significant weight penalty was incurred.

Until then it had been the usual practice to cold-water quench components after solution treatment which could introduce high levels of residual stress. An

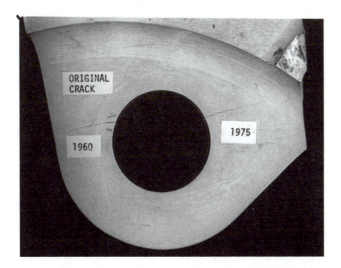

Fig. 3.28 Stress-corrosion cracks in a cold-water quenched Al–Zn–Mg–Cu alloy forging.

example is shown in Fig. 3.28 in which machining of the end lug of a cold-water quenched aircraft forging exposed the underlying residual tensile stresses that contributed to stress-corrosion cracking. It is interesting to note that, although cracking within the bore occurred when the forging was in service, the cracks in the sides of the lugs propagated subsequently after the forging had been removed from service and exposed to corrosive atmospheres on separate occasions many years apart. Because of the problem of quenching stresses some attempt was made, at least in Britain, to use chromium-free alloys for forgings and other components that could not be stress-relieved. Such alloys could accommodate a milder quench, e.g. in boiling water, without suffering a reduced response to ageing.

Another early practice was to give a single ageing treatment at temperatures in the range 120–135 °C at which there was a high response to hardening due to precipitation of GP zones (Fig. 3.29). It was known that ageing at a higher temperature of 160–170 °C, at which the phase η' (or η) formed, did result in a significant increase in resistance to stress-corrosion cracking but tensile properties were much reduced (see curve a in Fig. 3.29). Subsequently, a duplex ageing treatment designated the T73 temper was introduced in which a finer dispersion of the η' (or η) precipitate could be obtained through nucleation from pre-existing GP zones. As shown in Fig. 3.29 (curve b) tensile properties of 7075–T73 are about 15% below those for the T6 temper but now the resistance to stress-corrosion cracking is greatly improved. For example, tests on specimens loaded in the short transverse direction have shown the alloy 7075 aged to the T73 temper to remain uncracked at stress levels of 300 MPa whereas, in the T6 condition, the same alloy failed at stresses of only 50 MPa in the same environment. Confidence in the T73 temper has been demonstrated by the use of 7075–T73 for critical aircraft components such as the large die-forging shown in Fig. 3.30.

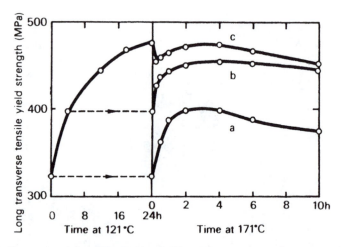

Fig. 3.29 Comparison of yield strengths (0.2% proof stress) of 7075 plate resulting from isothermal (171 °C) (curve a) and two-stage (121/171 °C) (curve b) precipitation heat treatments. The enhanced yield strength for the alloy 7050 given a two-stage treatment is shown in curve c (from Hunsicker, H. Y., *Rosenhain Centenary Conference on the Contribution of Physical Metallurgy to Engineering Practice*, The Royal Society, London, 1976).

Fig. 3.30 Die-forged 7075–T73 integral centre engine support and vertical stabilizer spar for McDonnell-Douglas DC-10 aircraft. Four similar forgings are used in each aircraft (from Hunsicker, H. Y., *Rosenhain Centenary Conference on the Contribution of Physical Metallurgy to Engineering Practice*, The Royal Society, London, 1976).

Another duplex ageing treatment, designated T76, has been applied successfully to 7xxx alloys to increase resistance to exfoliation (layer) corrosion (Fig 2.37).

The use of alloys given the T73 temper has required that some aircraft components be redesigned and weight penalties have resulted when replacing alloys aged to the T6 temper. For this reason, much research has been directed to the development of alloys that could combine a high resistance to stress-corrosion cracking with maximum levels of tensile properties. Some success was achieved with the addition of 0.25–0.4% silver as this element modifies the precipitation process in alloys based on the Al–Zn–Mg–Cu system enabling a high response to age-hardening to be achieved in a single ageing treatment at 160–170 °C. One German commercial high-strength alloy designated AZ74 (7009) contains this element. More recently, an alloy 7050 (Al–6.2Zn–2.25Mg–2.3Cu–0.12Zr) was developed in the United States in which the level of copper normally present in alloys such as 7075 was raised in order to increase the stengthening associated with the second stage of the T73 treatment (curve c in Fig. 3.29). Upper limits must be placed on the copper content in order to mini-mize the presence of large intermetallic compounds, such as Al_2CuMg, that adversely affect toughness. Modelling indicates that the sum of Cu + Mg should be <3.6 for alloys containing around 6%Zn. Modern derivatives of 7050 are 7150 which has marginally higher amounts of Zn and Mg, and the stronger alloys 7055 and 7449, both of which have significantly higher levels of zinc (8% compared with 6.2%).

Another heat treatment has been developed which enables alloys such as 7075 to exhibit the high level of tensile properties expected of the T6 condition combined with SCC resistance equal to the T73 condition. This is known as ret-rogression and re-ageing (RRA) which involves the following stages for 7075.

1. Apply T6 treatment i.e. solution treatment at 465 °C, cold water quench, age 24 h at 120 °C.
2. Heat for a short time (e.g. 5 min) at a temperature in the range 200–280 °C and cold-water quench.
3. Re-age 24 h at 120 °C.

Tensile property changes are shown schematically in Fig. 3.31. It is generally accepted that the RRA treatment results in a microstructure having a matrix simi-lar to that obtained for a T6 temper, combined with grain boundary regions char-acteristic of a T73 temper in that precipitates are larger and more widely sepa-rated. Reduced susceptibility to stress-corrosion cracking has been attributed to this latter change (Section 2.5.4). However, because of the short time interval ini-tially proposed for stage (2), e.g. 1–5 min, an RRA treatment was difficult to apply to thick sections. More recently, the time and temperature conditions for this stage of the process have been optimized (e.g. 1 h at 200 °C) and the RRA treatment has been given the commercial temper designation T77. It has been applied to the highly alloyed composition 7055 (Al–8Zn–2.05Mg–2.3Cu–0.16Zr) that has been used for structural members of the wings of the Boeing 777 passen-ger aircraft (Section 3.6.1).

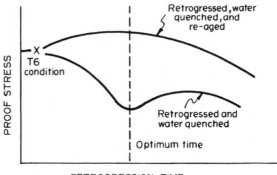

Fig. 3.31 Schematic representation of the effects of retrogression and reageing on the proof stress of an alloy such as 7075 (from Kaneko, R. S., *Met. Prog.*, **41**(4), 1980).

As mentioned in Section 3.1.6, thermomechanical processing offers another method of optimizing the properties of age-hardened alloys. Figure 3.32 shows, schematically, a way of combining ageing with deformation with the objective of achieving dislocation strengthening to compensate for the loss in strength that normally occurs during the over-ageing part of the T73 treatment mentioned above. Deformation is achieved by warm working after a T6 ageing treatment and before over-ageing. It has been reported that thermomechanical processing of some 7xxx alloys in this way can result in an improvement of some 20% in strength with no loss in toughness, or vice versa. However,

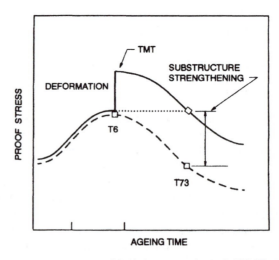

Fig. 3.32 Schematic representation of final thermomechanical (FTMT) strengthening of a 7xxx alloy (from Paton, N. E. and Sommer, A. W., *Proceedings of 3rd International Conference on Strength of Metal and Alloys*, Metals Society, London, **1**, 101, 1973).

commercial implementation has presented difficulties because of problems in controlling temperature and level of deformation that has to be carried out on material which is already in the age-hardened condition.

Other compositional changes have been made. One example is a reduction of the levels of the impurities iron and silicon in alloys such as 7075 (Fe + Si 0.90% maximum) to a combined maximum of 0.22% in the higher purity alloy 7475 (Table 3.4). 7475 also has the manganese content reduced from 0.30% maximum to 0.06% maximum whereas the content of the other alloying elements is essentially the same as 7075. The size and number of intermetallic compounds that assist crack propagation are much reduced in 7475 and, for similar T735 heat treatments and tensile properties, the fracture toughness (K_{1c}) values may exceed 50 MPa $m^{1/2}$ whereas a typical figure for 7075 is 30 MPa $m^{1/2}$. There is a cost penalty of around 10% for producing 7475 but it is widely used in aircraft.

Another modification has been to use 0.08–0.25% zirconium as a recrystallization inhibitor in place of chromium or manganese, or in combination with smaller amounts of these elements, in order to reduce quench sensitivity so that slower quench rates can be used without loss of strength on subsequent ageing. Examples are the alloys 7050, 7150, 7055, and the British alloy 7010. This is also particularly important for alloys used for thick plate since cost reductions are possible if large built-up assemblies can be replaced by machined mono-lithic structures.

Overall, the progress that has been achieved in combating stress-corrosion cracking in the 7xxx series materials through alloy development and changes in heat treatment can be appreciated from results shown in Fig. 3.33. Contrary to the 2xxx series alloys (Fig. 3.22) improvements have come from large changes to the plateau values for crack growth rate rather than from increases in levels of the threshold stress intensity needed to initiate cracks.

Reference has already been made to the fact that 7xxx series alloys tend to show higher values of fracture toughness than the 2xxx series (e.g. Fig. 3.20). Further improvements have been achieved which are shown in Fig. 3.34. Here the inverse relationship between fracture toughness and tensile yield strength is shown for the older alloys 7075–T651 and 7178–T651. Modifications to alloying additions, purity and heat treatment have allowed relatively high values of fracture toughness to be sustained in alloys 7150–T651 and 7055–T7751 despite increasing levels of yield strength. One compositional trend has been to raise the Zn:Mg ratios which have increased from 2.25 for the early alloy 7075 to 2.7 for 7050, 3.9 for 7055 and as high as 5.0 for latest alloy 7085. Another trend has been to reduce the allowable limits on manganese and chromium, as well as the impurities iron and silicon, each to below 0.1% or less. In 7085 (Fe <0.08%, Si <0.06%), the consequent reduction in intermetallic compounds in the microstructure this has resulted in a particularly low quench sensitivity which enables very thick sections (e.g. 150 mm) to be heat treated and record tensile properties higher than earlier aircraft alloys such as 7050. This has been

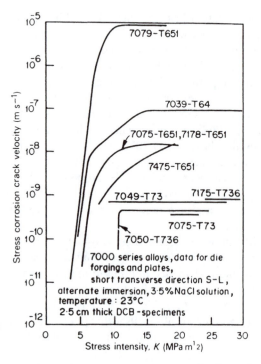

Fig. 3.33 Crack propagation rates in stress-corrosion tests on 7xxx alloys. Test conditions as described for Fig. 3.22. DCB specimens prepared from short transverse directions of die forgings and plates (from Speidel, M. O., *Metall. Trans A*, **6A**, 631, 1975).

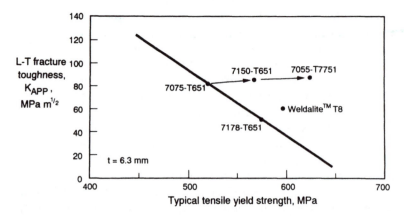

Fig. 3.34 Strength–toughness relationships for 7xxx series alloys in form of rolled plate (from Hyatt, M. V. and Axter, S. E. *Proceedings International Conference on Recent Advances in Science & Engineering of Light Metals*, Japan Institute for Light Metals, Sendai, 274, 1993).

a major factor in the selection of 7085 for the large extruded wing spars and die forged wing rib components for the European A380 Airbus (Section 3.6.1).

As shown in Table 3.7, the alloy 7050 responds to secondary precipitation which has increased both the tensile and fracture toughness properties to levels higher than those obtained using a single stage, T6 temper. Similar responses have been observed with other 7xxx series alloys such as 7075 (Fig. 2.34).

3.4.6 Lithium-containing alloys

Lithium has a high solid solubility in aluminium with a maximum of approximately 4% (16 at %) at 610 °C. This is significant because of the low density of this element (0.54 g cm^{-3}) which means that for each 1% addition, the density of an aluminium alloy is reduced by 3%. Lithium is also unique amongst the more soluble elements in that it causes a marked increase in the Young's modulus of aluminium (6% for each 1% added (Fig. 3.35)). Moreover, binary and more complex alloys containing lithium respond to age-hardening due to precipitation of an ordered, metastable phase δ' (Al$_3$Li) that is coherent and has a particularly small misfit with the matrix (Fig. 3.36). Because of all these features, much attention has been given to developing lithium-containing alloys as a new generation of low-density, high-stiffness materials for use in aircraft structures, and improvements of as much as 25% in specific stiffness are possible. As such, they appear to offer the last opportunity to develop a quite new range of aluminium alloys that can be produced by conventional ingot metallurgy. They may also assist the aluminium industry in responding to the increasing threat imposed by non-metallic composites for use as structural materials for aircraft.

Aged binary Al–Li alloys suffer from low ductility and toughness due, primarily, to severe strain localization that arises because coherent δ' precipitates are readily sheared by moving dislocations (Fig. 3.36b). This can lead to cracking

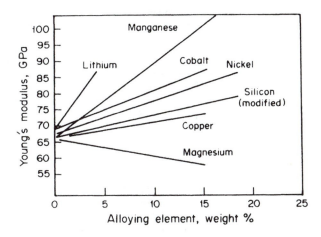

Fig. 3.35 Effects of alloying elements on the Young's modulus of binary aluminium alloys.

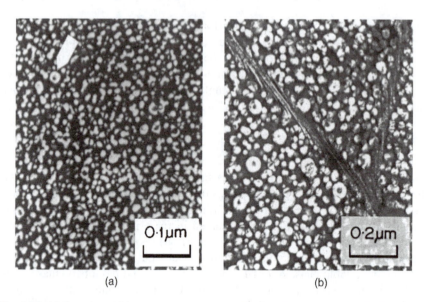

(a)   (b)

Fig. 3.36 (a) Particles of δ' precipitate in an Al–Li–Mg–Zr alloy aged at 190 °C. The arrow indicates a δ' precipitate that has nucleated on an Al$_3$Zr particle (courtesy P. J. Gregson); (b) shearing of δ' precipitates leading to strain localization in a deformed Al–Li–Zr alloy (courtesy D. J. Lloyd).

along grain boundaries as shown schematically in Fig. 2.22a. Accordingly, much of the work on alloy development has concentrated on the investigation of solute elements that will form additional precipitates or dispersoids capable of dispersing dislocations more homogeneously. These developments are summarized below and alloy compositions are shown in Table 3.9.

Table 3.9 Compositions of commercial lithium-containing alloys

| Alloy designation | Major alloying elements | | | | |
	Li	Cu	Mg	Zr	Other
2020	1.3	4.5	–	–	0.5Mn, 0.25Cd
2090	2.2	2.7	–	0.12	–
2091	2.0	2.1	1.5	0.10	–
8090	2.4	1.3	0.9	0.10	–
8091	2.6	1.9	0.9	0.12	–
8092	2.4	0.65	1.2	0.12	–
2094	1.1	4.8	0.5	0.11	0.4Ag
2097	1.5	2.8	0.35 max.	0.12	0.35Mn
2195	1.0	4.0	0.5	0.12	0.4Ag
1420†	2.0	–	5.2	0.11	

†C.I.S. designation

Al–Cu–Li The first commercially used alloy to contain lithium was 2020 (Al–4%Cu–1.1%Li–0.5%Mn–0.2%Cd) which was produced in the United States in the late 1950s. Copper both reduces the solid solubility of lithium so that precipitation of δ' is enhanced and leads to coprecipitation of phases such as G. P. zones and θ' that form in binary Al–Cu alloys. Alloy 2020 combined high room temperature mechanical properties (e.g. 0.2% PS 520 MPa) with good creep properties at temperatures as high as 175 °C, and had an elastic modulus 10% higher than other aluminium alloys. The alloy was used for wing skins for one type of military aircraft but production was subsequently abandoned because of its low toughness properties. More recent Al–Li–Cu alloys have lower copper contents and higher Li:Cu ratios, the best-known example being the United States alloy 2090. In addition to δ', this alloy is hardened by the hexagonal T_1 phase (Al_2CuLi) which forms as thin plates on the {111} planes (Fig. 3.37) and is similar in appearance to the Ω phase mentioned earlier (Fig. 3.18). Plates of the θ' (Al_2Cu) may also be present. Nucleation of T_1 and θ' is difficult and occurs preferentially on dislocation lines so that 2090 is normally cold-worked before artificial ageing (T8 temper) to promote a more uniform distribution of each. Zirconium additions, nominally 0.12%, are also made which form fine dispersoids of Al_3Zr that control recrystallization and grain size. One other effect of this element is that nucleation of some of the δ' precipitates occurs on these dispersoids (see arrow, Fig. 3.36a). Taken overall, zirconium increases tensile properties and toughness.

Al–Li–Mg Russian workers have given particular attention to this system and the properties of the alloy Al–5%Mg–2%Li–0.5%Mn have been evaluated

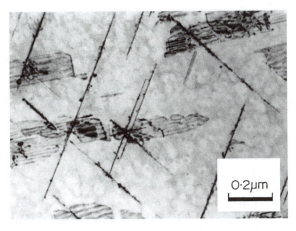

Fig. 3.37 Thin plates of the T_1 (Al_2CuLi) in an Al–Li–Cu–Zr alloy over-aged (500 h) at 170 °C (courtesy P. J. Gregson).

in detail. Magnesium also has the effect of reducing the solubility of lithium but its main role appears to be that of solid solution strengthening. It was hoped that this element may increase the misfit between δ' particles and the matrix but no significant change has been observed. For alloys containing more than 2% magnesium that are aged at relatively high temperatures, a noncoherent, cubic phase Al_2LiMg is formed. However, this phase precipitates preferentially in grain boundaries and has an adverse effect on ductility. Thus the addition of magnesium does not tend to improve the toughness of binary Al–Li alloys. Alloy 1420 has relatively low strength (0.2%PS,280 MPa) but it has high corrosion resistance and has been used successfully for welded sections of at least one advanced military aircraft produced in the former Soviet Union. Higher strengths (0.2% PS, 360 MPa) are developed in a modified alloy 1421. This alloy contains 0.18% of the expensive element scandium which forms fine, stable dispersions of the phase Al_3Sc that is isomorphous with Al_3Zr and promotes additional strengthening and creep resistance.

Al–Li–Cu–Mg In Britain and France, attention has been focused on the development of alloys based on the quaternary Al–Li–Cu–Mg system. Examples are 2091, 8090 and 8091 for which ageing at temperatures close to 200 °C leads to co-precipitation of the semi-coherent phases T_1 and S' in addition to coherent δ'. The S (S') phases (Al_2CuMg) form as laths that are not readily sheared by dislocations, thereby promoting more homogeneous slip. Since these precipitates also nucleate with difficulty, Al–Li–Cu–Mg alloys are again normally used in the T8 temper. In 2003, a so-called third generation Al–Cu–Li–Mg alloy was registered which has lower levels of lithium and magnesium and contains an addition of zinc. This is 2099 (Al–2.7Cu–1.8Li–0.7Zn–0.3Mg–0.3Mn–0.08Zr–0.0001Be) for which it is claimed that difficulties, such as low toughness and anisotropy of mechanical properties traditionally associated with earlier lithium-containing alloys, have been overcome. Weight savings due to the addition of lithium are less and careful analysis will be needed to see if the higher costs involved in its use are justified when possible applications are being considered.

Al–Li–Cu–Mg–Ag The cost of launching payloads into low Earth orbit has been estimated to be approximately $US8000 per kg so that there is a large incentive to reduce the weight of space vehicles through the use of lighter or stronger alloys. Because of this, recent attention has been directed at the prospect of using lithium-containing alloys for the welded fuel tanks in place of the existing alloys 2219 and 2014. Some Al–Cu–Li alloys proved to be weldable and it was then found that minor additions of silver and magnesium could promote large increases in tensile properties in the manner described for Al–Cu alloys in Section 3.4.1. For Al–Cu–Li–Mg–Ag alloys with lithium contents around an optimum level of 1 to 1.3%, the source of this high strength resides in the ability of these minor additions to stimulate nucleation of a fine dispersion of the T_1 phase which co-exists with S' and θ' in the peak hardened condition. Each of these precipitates forms in a different matrix plane and each is resistant to shearing by dislocations. These observations led to the development by Martin Marietta in the United States of an alloy known as Weldalite 049 (Al–6.3Cu–1.3Li–0.4Mg–0.4Ag–0.18Zr) which may exhibit tensile properties exceeding 700 MPa in the T6 and T8 conditions. On a strength:weight basis, an equivalent steel would need to have a tensile strength exceeding 2100 MPa so that Weldalite 049 can be said to be the first ultra-high-strength aluminium alloy to be produced from conventionally cast ingots.

A lower-copper variation of this alloy, 2195–T6, is 5% lighter and 30% stronger than the alloy 2219–T8 that was used for constructing the first version of the large, disposable, external fuel tank used for launching the United States Space Shuttle (Fig. 3.38). This tank is 47 m long, has a diameter of 8.5 m and originally weighed 34 500 kg. A redesign reduced its weight by 4 500 kg and now the adoption of alloy 2195 for the so-called Super Light Weight External Tank has resulted in a further weight reduction of 3 400 kg. This change provides the opportunity to achieve huge cost savings by allowing payloads to be increased, thereby reducing the number of missions needed for the current program of constructing the International Space Station. Each launch costs several hundred million US dollars.

The original objective in developing the Al–Li–Cu and Al–Li–Cu–Mg alloys was as direct substitutes for conventional aircraft alloys with an expected density reduction of around 10% and increase in stiffness of 10 to 15%. These alloys may be classified as high strength (2090, 8091), medium

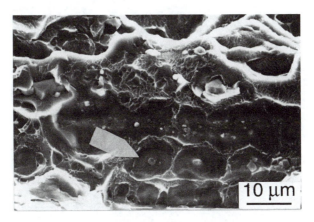

Fig. 3.38 Large external liquid fuel tank for launching the U.S. Space Shuttle to which the orbitor vehicle and solid fuel boosters are attached (courtesy N.A.S.A.).

strength (8090) and damage tolerant (2091, 8090) so far as mechanical properties are concerned. Specific gravities range from 2.53 to 2.60 and elastic
moduli from 78 to 82 GPa. Resistance to fatigue crack growth and stress-
corrosion cracking appears to be equal, or superior to, that displayed by the
existing alloys of the 2xxx and 7xxx series. However, problems with respect
to low toughness in the short transverse (S-L) direction of plate and
extrusions have persisted which appear to be associated with microstructure,
particularly in the region of grain boundaries. Moreover, prolonged exposure
of stressed alloys to elevated temperatures ($>50\,^{\circ}$C) can lead to relatively
high rates of creep crack growth.

One factor that is known to reduce ductility and fracture toughness in
lithium-containing alloys is the presence of alkali metal impurities (notably
sodium and potassium) in grain boundaries. In other aluminium alloys, these
elements are removed into innocuous, solid compounds by elements such as
silicon, whereas they react preferentially with lithium when this element is
present. Observations of the fracture surfaces of partially or fully recrystallized
alloys have revealed the presence of intergranular brittle 'islands', the number
and size of which increases with increasing levels of these impurities above
3–5 ppm (Fig. 3.39). These brittle islands are composed of second-phase particles covered with films of the alkali metals which may be liquid at ambient
temperatures.

Commercial lithium-containing alloys produced by conventional melting
and casting techniques typically contain 3–10 ppm of alkali metal impurities.
The levels of these elements can be reduced to below 1 ppm by vacuum melting
and refining which has been shown to result in a significant improvement in
fracture toughness at room temperature (Fig. 3.40).

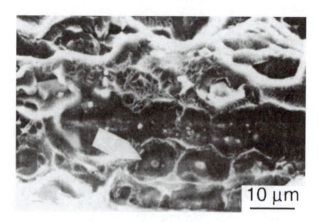

10 µm

Fig. 3.39 Fracture surface of 2090–T8 extrusion with 7.2 ppm alkali metal impurities
showing presence of brittle 'islands' along partially recrystallized grain boundaries (courtesy
S. P. Lynch).

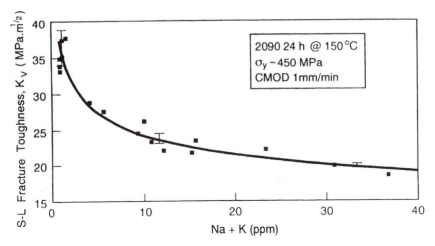

Fig. 3.40 Effect of alkali metal impurities on short transverse (S-L) fracture toughness of a 2090 alloy extrusion. (CMOD denotes 'crack mouth opening displacement') (from Sweet, E. D. *et al., Proceedings 4th International Conference on Aluminium Alloys,* Georgia Institute of Technology, Atlanta, U.S.A., p. 321. 1994).

Susceptibility to intergranular fracture in lithium-containing alloys has also been attributed to other factors including hydrogen embrittlement, the presence of large area fractions of the grain boundary phase δ (AlLi) and soft, precipitate-free zones adjacent to grain boundaries. Recently it was shown that fracture toughness of alloy 8090 aged to peak hardness at 170 °C could be much improved by heating for a short time at a higher temperature (e.g. 5 min at 210 °C), only to decrease again after subsequent prolonged exposure at a lower temperature (e.g. 80 °C). This result has been explained in terms of lithium segregation to grain boundaries as being the key factor influencing fracture behaviour.

Al–Cu–Li and Al–Cu–Li–Mg alloys are available as plate, extrusions and sheet and many trials have been conducted in which they have been substituted for aluminium alloy components in existing aircraft. Extensive databases are now available and approval has been obtained for their use for some structural parts in the latest designs of civil passenger aircraft. However, their acceptance so far has been limited because of concerns about fracture toughness and the fact that their cost has remained at around three times that of existing high-strength alloys, largely due to special requirements needed during melting and casting. One exception has been the adoption of the alloy 8090 for most of the fuselage and main frame of the Westland/Agusta EH101 helicopter (Fig. 3.41) where weight savings of around 200 kg were urgently required.

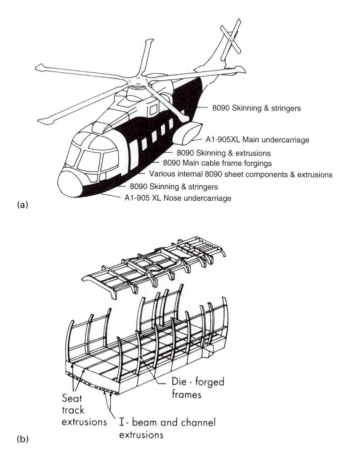

Fig. 3.41 Application of alloy 8090 in (a) the fuselage, and (b) the lift frame of the EH101 helicopter (from Grimes, R., *Metall. and Mater.*, **8**, 436, 1992).

3.5 JOINING

Aluminium can be joined by most methods used for other metals including welding, brazing, soldering, bolting, riveting and adhesive bonding. Welding and brazing will be considered in some detail and the other methods only referred to briefly.

Mechanical joining by fasteners is covered by a range of engineering codes and poses no real technical problems. Aluminium alloy rivets are normally selected so that their mechanical properties closely match those of the material to be joined. Most commercial rivets are produced from one of the following alloys: 1100, 2017, 2024, 2117, 2219, 5056, 6053, 6061, and 7075. They are usually driven cold and some, e.g. 2024, if solution treated and quenched shortly before use, will gradually harden by ageing at room temperature. Alloy

2024–T4 is commonly used for aluminium bolts or screws although steel fasteners are generally cheaper. Such fasteners must be coated to prevent galvanic corrosion of the aluminium, and plating with nickel, cadmium, or zinc is generally used depending upon the corrosive environment. Under severe conditions, stainless steel fasteners are preferable.

Adhesive bonding is particularly suitable for aluminium because of the minimal need for surface preparation. For general bonding with low-strength adhesives, and in field applications, surfaces may be prepared by mechanical abrasion using wire brushes, abrasive cloths or grit or shot blasting. Where higher bond strengths are required, solvent degreasing is essential and this may need to be followed by some form of chemical treatment. This may involve immersion in an acid bath such as an aqueous solution of sodium dichromate and sulphuric acid, or anodizing which thickens the oxide film and provides an excellent surface for bonding. Adhesives include epoxy and phenolic resins, and a range of elastomers.

Soldering involves temperatures below about 450 °C and tends to be troublesome because aggressive chemical fluxes are needed to remove the oxide film. Problems with corrosion may occur if these fluxes are not completely removed, as well as from galvanic effects that can arise because the low-melting-point solders have compositions quite different from the aluminium alloys.

3.5.1 Welding

Oxyacetylene welding was widely applied to the joining of aluminium alloys following the development in 1910 of a flux that removed surface oxide film. However, as such fluxes contain halides, any residues can cause serious corrosion problems and this form of fusion welding has been superseded except for a limited amount of joining of thin sheet.

The process requirements for welding aluminium are:

1. an intense and localized heat source to counter the high thermal conductivity, specific heat and latent heat of this metal
2. the ability to remove the surface oxide film which has a high melting point (about 2000 °C) and may become entrapped to form inclusions in the weld bead
3. a high welding speed to minimize distortion arising because of the relatively large coefficient of thermal expansion of aluminium
4. a low hydrogen content because of the high solubility of this gas in molten aluminium, which can lead to porosity in the weld bead after solidification has occurred (see Section 3.1.1).

Arc welding The two main processes in use today are TIG, i.e. tungsten inert gas (also known as GTA or gas tungsten arc) and MIG, i.e. metal inert gas (also known as GMA or gas metal arc) in which a shroud of inert gas, commonly argon, replaces the chemical flux. The heat source is an electric arc

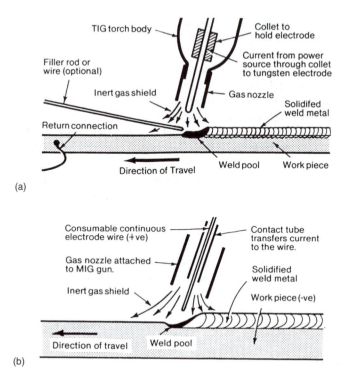

Fig. 3.42 Essential features of (a) TIG or GTA and (b) MIG or GMA welding processes (courtesy W. T. I. A.).

struck between the workpiece and either a non-consumable electrode (TIG), or consumable metal wire (MIG) as shown schematically in Fig. 3.42. Current flows in the arc due to ionization of the inert gas and it is the ionized particles which disrupt the oxide film and clean the surface. TIG welding is mostly carried out manually although the use of mechanized equipment is increasing where higher welding speeds can offset the greater cost of the facility. Where necessary, filler metal is introduced as bare wire. In MIG welding the consumable electrode is fed into the weld pool automatically through a water-cooled gun and the whole process is normally mechanized. It is favoured for volume production work, particularly on material thicker than 4 mm. MIG welding of thin-gauge aluminium presents problems because of the relatively thin wire that would need to be used at normal welding speed which is difficult to handle, or because welding speeds become too high with the thicker wires. For this purpose the pulsed MIG process has been devised in which transfer of metal from the wire tip to the weld pool occurs only at the period of the pulse or peak in welding current. During the intervals between pulses, a background current maintains the arc without metal transfer taking place.

The weldability of the various wrought alloys was mentioned when the individual classes were discussed. Essentially, all are weldable with the exception of most of the 2xxx series and the high-strength alloys of the 7xxx series. An example of a welded structure is shown in Fig. 3.25. The strengths of non-heat-treatable alloys after welding are similar to their corresponding strengths in the annealed condition irrespective of the degree of cold-working prior to welding. This loss of strength is due to annealing of the zone adjacent to the weld and can only be offset by selecting a weldable alloy such as 5083 which has comparatively high strength in the annealed condition. This accounts for the popularity of this series of alloys for welded constructions. When a heat-treatable alloy is welded there is a drop in the original strength to that approximating the T4 condition. The metallurgical condition of this softened zone is complex and may include regions that are partly annealed, solution treated and over-aged, depending upon the actual alloy, speed of welding, material thickness and joint configuration. As in the case of the medium-strength 7xxx alloys (Section 3.4.4), some natural ageing can occur in parts after welding but the overall strength of the welded joint is not usually restored to that of the unwelded alloy. The presence of the range of metallurgical structures adjacent to welds can lead to localized corrosive attack of some alloys in severe conditions (e.g. Fig. 3.27b).

Filler alloys must be selected with due regard to the composition of the parent alloys and common fillers are listed in Table 3.10. Generally speaking selection is based primarily on ease of producing crack-free welds of the highest strength possible. However, in some cases maximum resistance to corrosion or stress-corrosion, or the ability of the weld to accept a decorative or anodized finish compatible with the parent alloy, may be required.

Softening of the heat-affected zones adjacent to aluminium alloy welds also reduces fatigue properties and this effect is accentuated when weld geometry leads to a concentration of cyclic stress in these softened regions, e.g. at weld toes. One consequence is that the fatigue strengths of welded joints tend to show little difference from one another.

Surface treating by peening in the region of weld toes to induce residual compressive stresses is a recognized method for improving the fatigue performance of welded steel components and structures. However, such a procedure is less used for aluminium alloys, presumably because they are normally softer and the peening process may cause unacceptable surface damage. Shot peening can be effective for components that can be introduced into a suitable chamber. Another technique which has shown promise for application in the field is to use a portable needle gun in which long steel needles are vibrated back and forth by compressed air. The resulting surface suffers only superficial damage and this practice has, for example, been used successfully to reduce the incidence of fatigue cracking in welded aluminium rail wagons.

Another method is to smooth out the contour of weld toes so that cyclic stress concentrations are reduced. This can be achieved by using a portable TIG welding gun to remelt the critical toe regions after the earlier MIG welding has

Table 3.10 Aluminium filler alloys for general-purpose TIG and MIG welding of wrought alloys

Base metal welded to base metal	7005	6061 6063 6351	5454 5154A	5086 5083	5052	5005 5050A	3004 Alclad 3004	1100 3003 1200
1100 1200 3003	5356†	4043	4043†	5356‡	4043†	4043†	4043†	1100‡ 1200‡
3004 Alclad 3004	5356†	4043†	5356†	5356†	5356†‡	4043†	4043†	
5005 5050A	5356†	4043§	5356†	5356†	4043†	4043†§§		
5052	5356†	5356†‡	5356ᶜ	5356†	5356§			
5083 5086	5356†	5356†	5356†	5356†				
5154A	5356†	5356‡§	5356‡§					
5454	5356†	5356‡§	5356‡§					
6061 6063 6351	5356†	4043§						
7005	5039†							

†5356 or 5556 may be used
‡4043 may be used for some applications
§5154A, 5356 and 5556 may be used. In some cases they provide: improved colour match after anodizing treatment; higher weld ductility; higher weld strength; improved stress-corrosion resistance
§§Filler metal with the same analysis as the base metal is sometimes used.
   For alloy compositions see Tables 3.2 and 3.4 except: 4043 Al–5Si; 5356 Al–5Mg–0.1Cr–0.1Mn; 5039 Al–3.8Mg–2.8Zn–0.4Mn–0.15Cr; 5154A Al–3.5Mg–0.3Mn

been completed (Fig. 3.43). Providing the heat associated with this dressing operation does not introduce undesirable levels of residual stress, fatigue performance may be significantly improved. An example of the beneficial effects of TIG dressing in delaying or preventing the onset of fatigue cracking in a fillet welded aluminium alloy specimen is shown in Fig. 3.44. In this case, fatigue lives over a wide cyclic stress range have been increased by three to five times.

Friction stir welding Friction stir welding (FSW) is a solid state process, invented at The Welding Institute in England in the early 1990s, which is particularly well suited to joining aluminium alloys that are difficult to fusion weld. A schematic illustration of the welding process is shown in Fig 3.45.

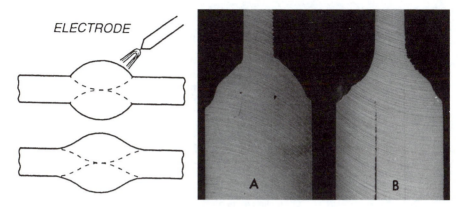

Fig. 3.43 (a) Representation of TIG dressing to smooth the contour of weld toes; (b) cross-section of fillet welds showing (A) undressed welds and (B) the effect of TIG dressing these welds.

Components to be joined (mainly sheet or plate) need to be symmetrical. They are securely clamped to a rigid table and a specially designed rotating tool (probe) is forced into, and traverses down the joint line. The tool, which is made of a hard steel and is non-consumable, consists of a profiled pin with a concentric larger diameter shoulder. The depth of penetration is controlled by the tool shoulder and the length of the pin. Welding speeds may be comparable with those experienced with fusion welding processes.

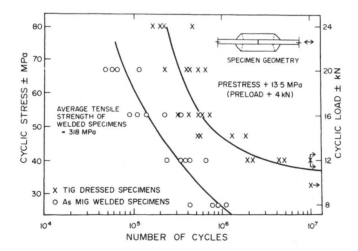

Fig. 3.44 Stress/number of cycles (S/N) curves for TIG dressed and undressed fillet welded specimens (Polmear, I. J. and Wilkinson, D. R. *Welding J.*, **62**(3), 785, 1983).

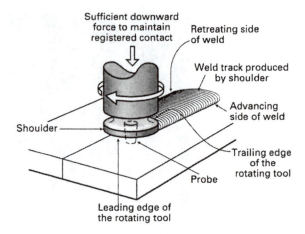

Fig. 3.45 Friction stir welding process (courtesy The Welding Institute, U.K.).

The material being joined is rapidly friction heated within the solid state to a temperature at which it is easily plasticized. The frictional heat is produced from a combination of contact with the rotating shoulder and downward pressure on the tool. As the tool moves along, metal is forced to flow from the leading edge to the trailing face of the pin; it then cools to form the solid state joint. Because of the extreme levels of plastic deformation involved, which are not unlike that experienced in equal channel angular pressing (Section 7.8), a very fine, dynamically recrystallized microstructure (e.g. grain sizes 2–10 μm) is obtained in the weld zones. Welds are stronger, comparatively defect-free and, because the heat generated is much less than for fusion welding, residual stresses are reduced. Furthermore, since there is no melting, undesirable brittle and low melting point eutectic phases are not formed. It is for these reasons that the more highly alloyed compositions, such as the 7000 series alloys, and even complex metal matrix composites, can be welded by FSW. It is also possible to weld dissimilar aluminium alloys. Another advantage over fusion welding is that environmental hazards associated with fume and spatter are eliminated.

FSW can be automated and used to produce symmetrical butt and lap joints including corner configurations. The process is being applied in the automotive, shipping, defence and aerospace industries. One important advance is with the joining of thin sheet that may buckle severely if fusion welded. Armour plate that is FSW has been shown to be more resistant to cracking under ballistic impact. Another new application is for welding of the large external, 8.5 metre diameter, Al–Cu–Li–Mg–Ag alloy "super light weight" fuel tank now used for launching the Space Shuttle (Fig. 3.38). At the other end of the scale, FSW has been successfully adapted to joining 1–2 mm thick sheets by using a modified rotating tool. It is also useful for welding together sheets of differing thicknesses.

Laser welding Resistance spot welding is the most important joining method used for assembling stamped automotive steel sheet sections. With aluminium alloy sheet components however, the presence of the relatively thick surface oxide film causes excessive pitting of the welding electrodes requiring them to be changed frequently. The oxide film may also have an adverse effect on the strength of the spot welded joints. Laser welding is showing promise as an alternative to spot welding when assembling aluminium alloy structures.

Laser welding is a comparatively new method of joining in which the laser beam is focused onto the workpiece producing a metal-vapour filled cavity, or "keyhole". A molten metal layer, that forms in dynamic equilibrium with the metal vapour, establishes the weld pool which is then translated rapidly through the thickness of the workpiece. Melting takes place at the leading edge of the cavity and solidification occurs at the rear. Filler compositions are similar to those used in arc welding. Because of the high power density of what is a narrow beam, the overall heat input during laser welding is less that in conventional fusion welding, so that the widths of the fusion and heat affected zones are both reduced. Welding speeds are high (e.g. 5–8 m/minute for thin sections) and thermal distortion of joints is less. Because of these characteristics, laser welding is particularly well suited to joining sheet materials and it is now being used in the aircraft and automotive industries.

Two types of lasers are used for welding which differ in the way the beam is generated. One involves a gaseous CO_2 device and the other is a solid state system comprising a neodymium yttrium-aluminium garnet (Nd-YAG) crystal. The latter has generally proved more suitable for welding light metals and alloys. Because of the narrow fusion zone, severe thermal cycles of heating and cooling may be involved when laser welding thicker sections and the major concerns are possible hot cracking, loss of alloying elements and porosity.

3.5.2 Brazing

Brazing involves the use of a filler metal having a liquidus above about 425 °C but below the solidus of the base metal. Most brazing of aluminium and its alloys is done with aluminium-base fillers at temperatures in the range 560–610 °C. The non-heat-treatable alloys that have been brazed most successfully are the 1xxx and 3xxx series and the low-magnesium members of the 5xxx series. Of the heat-treatable alloys, only the 6xxx series can be readily brazed because the solidus temperatures of most of the alloys in the 2xxx and 7xxx series are below 560 °C. These alloys can, however, be brazed if they are clad with aluminium or the A1–1Zn alloy 7072. Most commercial filler metals are based on the Al–Si system with silicon contents in the range 7–12%. Sheet alloys to be joined by brazing may first be clad or roll-bonded to the filler metal which is particularly convenient for mass producing complex assemblies such as automative heat exchangers. For other situations, the filler metals can be applied separately in the form of wire, thin sheet, or as dry or wet powders.

Brazing of aluminium was first made possible through the development of fluxes that dissolve the surface oxide film and protect the underlying metal until the joining operation is completed. Fluxes are compounded so that they melt just before the brazing alloy and they commonly comprise a mixture of alkaline earth chlorides and fluorides. Such fluxes may leave a corrosive residue which must be removed and this can present difficulties with assemblies containing narrow passage ways, pockets, etc. More recently, non-corrosive fluxes based on the compound potassium aluminofluoride have been developed.

The easiest method of heating and fluxing aluminium joints simultaneously is to immerse the whole assembly in a bath of molten flux that is maintained at the brazing temperature (e.g. 540 °C). The assemblies are usually preheated to avoid cooling the flux bath and to shorten the processing time, which usually involves several minutes. This procedure is known as dip or flux-bath brazing. It suffers from the disadvantage that corrosion problems may arise in service if the flux is not completely removed.

Furnace brazing of aluminium alloy components has now become more common because of the high volume that can be handled on a continuous production line. This method is used for the assembly of automotive radiators and other heat exchangers in accessories such as air conditioners and oil coolers that were formerly made from copper. Two processes are employed: controlled atmosphere brazing (CAB) that uses a non-corrosive flux, and vacuum brazing which is flux-free. The latter has the advantages that no costs are involved for the flux and cleaning is less important, although the surfaces to be joined must be prepared with more care. CAB is conducted in an atmosphere that usually contains the inert gas nitrogen. The flux, which melts and dissolves the aluminium oxide films, becomes inactive in the presence of magnesium so that the filler must be magnesium-free and it is desirable for the aluminium alloys being joined to have low contents of magnesium. On the contrary, vacuum brazing requires the presence of magnesium, at least in the filler, in order to disrupt the oxide films and promote wetting in the absence of a flux. In both processes, the components are assembled and placed in the furnace that is already at the brazing temperature of around 595 °C.

Limitations on the magnesium contents of alloys to be CAB brazed has led to the development of compositions for use for the tubes, fins and header tanks of automotive heat exchangers. Examples are as follows:

tubes - alloy 3005 : Al–1.2Mn–0.40Mg–0.35Fe–0.30Si–0.15Cu
fins - alloy FA6815 : Al–1.6Mn–1.5Zn–0.85Si–0.25Fe–0.15Zr
headers - alloy FA7827–Al–0.7Si–0.45Mg–0.3Cu–0.25Fe–0.15Ti.

The alloys for fins and headers were developed in Sweden. Like all aluminium alloys for heat exchangers, strengthening relies on a combination of solid solution and dispersion hardening, supplemented in some instances by precipitation hardening.

Aluminium may be successfully brazed to many other metals including carbon steels, stainless steels, copper, nickel and titanium. Special filler alloys and melting conditions may be required and care must be taken to apply protective coatings to the final assembly for protection against galvanic corrosion.

3.5.3 Soldering

Soldering is distinguished from brazing in that the metals or alloys being joined are not themselves melted. For aluminium alloys the filler metal, or solder, melts below a temperature of 450 °C. Thus, aluminium-base fillers are not used and most are alloys of zinc, tin, cadmium and lead; examples are Zn–10Cd (melting range 265–400 °C), Sn–30Zn (199–311 °C) and Pb-34Sn–3Zn (195–256 °C). Fluxing must again be used to remove surface oxide films. Whereas it is possible to solder all aluminium alloys, this method of joining is not commonly used. The quality of joints varies with composition and, for the wrought series of alloys, generally decreases in the following order: 1xxx, 3xxx, 6xxx, 2xxx and 7xxx.

3.5.4 Diffusion bonding

Diffusion bonding is an attractive means of joining because there is minimal disturbance to the microstructures of the parent alloys. Another advantage is the prospect of joining different alloys, or even dissimilar materials. With aluminium alloys, however, the presence of the chemically stable oxide film makes it difficult to achieve such a bond.

It is possible to disrupt a surface oxide film by imposing plastic deformation, but early work on diffusion bonding of aluminium alloys revealed that the minimum amount needed was around 40% before bonds of reasonable strength could be obtained. Another approach has been to use alloys containing active elements such as lithium or magnesium which serve to weaken or decompose the oxide films during the bonding process. More recently a new method has been developed which is known as transient liquid phase (TLP) diffusion bonding. This process involves interposing a thin (e.g. 5–10 μm) layer of copper between the surfaces to be joined which are lightly clamped in an evacuated chamber and heated close to the Al–Cu eutectic temperature of 548 °C for times of around 10 minutes. Local melting and solute diffusion occurs and, by imposing a temperature gradient across the reaction zone, a sinusoidal/cellular interface is formed which both increases the surface area of the interface and disperses the oxide particles. Bonds with alloys such as 6082 have shown shear strengths equal to that of the parent material. Successful bonds have also been prepared between this alloy and a metal matrix composite containing SiC particles.

3.6 SPECIAL PRODUCTS

3.6.1 Aircraft alloys

Designers of aircraft require materials that will allow them to produce light-weight, cost-effective structures which are durable and damage tolerant at ambient, sub-zero and occasionally elevated temperatures. As mentioned earlier, strong aluminium alloys date from the accidental discovery of the phenomenon of age-hardening by Alfred Wilm in Berlin in 1906. His work led to the development of the wrought alloy known as Duralumin (Al–3.5Cu–0.5Mg–0.5Mn) which was quickly adopted in Germany for structural sections of Zeppelin airships, and for the Junkers F13 aircraft that first flew in 1919. Since that time, wrought aluminium alloys have been the major materials for aircraft construction which, in turn, has provided much stimulus for alloy development.

Duralumin was the forerunner of a number of 2xxx series alloys including 2014 and 2024 that are still used today. The other major aircraft group of alloys is the 7xxx series. Both classes were considered in Section 3.4 together with the lithium-containing alloys that have been developed as possible substitute light-weight materials.

Materials selection for structural applications in aircraft depends mainly on a daunting variety of performance requirements which are summarised for a typical passenger aircraft in Fig. 3.46. Examples of older and newer alloys that have been used are given below:

Fuselage skin	:	2024–T3, 7075–T6, 7475–T6, 6013–T78
Fuselage stringers	:	7075–T6, 7075–T73, 7475–T76, 7150–T77
Fuselage frames/bulkheads	:	2024–T3, 7075–T6, 7050–T6
Wing upper skin	:	7075–T6, 7150–T6, 7055–T77
Wing upper stringers	:	7075–T6, 7150–T6, 7055–T77, 7150–T77
Wing lower skin	:	2024–T3, 7475–T73
Wing lower stringers	:	2024–T3, 7075–T6, 2224–T39
Wing lower panels	:	2024–T3, 7075–T6, 7175–T73
Ribs and spars	:	2024–T3, 7010–T76, 7150–T77, 7085–T7651
Empennage (tail)	:	2024–73, 7075–T6, 7050–T76

The fuselage is a semi-monocoque (continuous skin) structure made up of the outer skin to sustain cabin pressure (tension) and shear loads, longitudinal stringers (longerons) to carry longitudinal tension and compression loads caused by bending, circumferential frames to maintain fuselage shape and redistribute loads into the skin, and bulkheads to carry concentrated loads. Strength, stiffness, resistance to fatigue crack initiation and propagation, corrosion resistance and fracture toughness are all important. Traditionally, the alloy 2024 in the cold worked and naturally aged (T3) temper has been favoured for the skin of passenger aircraft because of its high damage tolerance. This property has been progressively improved with the introduction of the compositions 2224, 2324 and 2524 in which the impurity levels have been reduced, and tighter controls have been placed on the

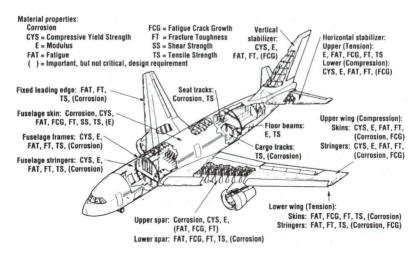

Material properties:
Corrosion
CYS = Compressive Yield Strength
E = Modulus
FAT = Fatigue
FCG = Fatigue Crack Growth
FT = Fracture Toughness
SS = Shear Strength
TS = Tensile Strength
( ) = Important, but not critical, design requirement

Vertical stabilizer: CYS, E, FAT, FT, (FCG)

Horizontal stabilizer:
Upper (Tension): E, FAT, FCG, FT, TS
Lower (Compression): CYS, E, FAT, FT, (FCG)

Fixed leading edge: FAT, FT, TS, (Corrosion)

Seal tracks: Corrosion, TS

Fuselage skin: Corrosion, CYS, FAT, FCG, FT, SS, TS, (E)

Fuselage frames: CYS, E, FAT, FT, TS, (Corrosion)

Fuselage stringers: CYS, E, FAT, FT, TS, (Corrosion)

Floor beams: E, TS

Cargo tracks: TS, (Corrosion)

Upper wing (Compression):
Skins: CYS, E, FAT, FT, (Corrosion, FCG)
Stringers: CYS, E, FAT, FT, (Corrosion, FCG)

Upper spar: Corrosion, CYS, E, (FAT, FCG, FT)

Lower spar: FAT, FCG, FT, TS, (Corrosion)

Lower wing (Tension):
Skins: FAT, FCG, FT, TS, (Corrosion)
Stringers: FAT, FT, TS, (Corrosion, FCG)

Fig. 3.46 Property requirements for structural members of a typical passenger aircraft (from Staley, J. T. and Lege, D. J., *Journal de Physique IV*, **3**, 179, 1993).

contents of the major alloying elements copper, magnesium and manganese. All Al–Cu–Mg sheet alloys all have the disadvantage that they must be clad to prevent corrosion and most cannot be fusion welded. This has led to limited introduction of unclad 6xxx series alloys that are weldable, cheaper and which also provide a 3% weight reduction. The copper-containing Al–Mg–Si alloys 6013 and 6056 have been favoured because of their higher strengths, although both have shown some susceptibility to intergranular corrosion because of the presence of anodic precipitate-free zones. This problem has been minimized by the introduction of a proprietary T78 temper and these alloys have been introduced into parts of the fuselages of the Boeing 777 and larger European Airbus 380 aircraft. Fuselage stringers are usually made from formed sheet or extrusions using 7xxx series alloys such as 7075–T6 and 7150–T77.

Wings are essentially beams that are loaded in bending during flight. The upper surfaces are loaded primarily in compression and compressive yield strength is the key property. Here the focus has been to develop wing skins with higher strengths and Fig. 3.47 shows the incremental improvements that have occurred since the introduction of aluminium alloys to aircraft in 1919. The lower wing skins are in tension during flight and more emphasis must be placed on properties in addition to yield strength such as fatigue resistance and toughness. Alloys that have been used include 2024–T3 and more recently 2324–T39, both in the form of rolled plate. The selection of alloys for internal members, such as wing spars, is dependent upon their positions with respect to compressive and tensile loading regimes.

The empennage (tail) consists of a horizontal stabiliser, a vertical stabiliser, elevators and rudders. The horizontal stabiliser is like an upside down wing

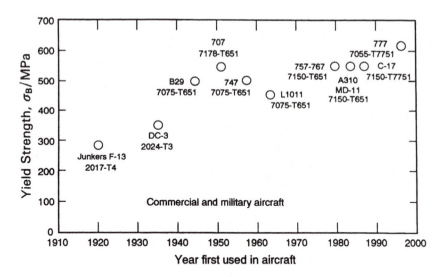

Fig. 3.47 Historical record of wing upper skin alloys and tempers used for passenger aircraft (from Lukasak, D.A. and Hart, R.M. *Light Metal Age*, **49**, Oct, 11, 1991).

whose span is often about half that of the wing. Both the upper and lower surfaces are often in compression due to up and down bending so that modulus of elasticity in compression is the most important property. Sheet, plate and extrusions of 7075–T6 and 2024–T3 have traditionally been used in the empennage although high-stiffness, fibre-reinforced materials are now being adopted.

The relative contribution that improved structural materials have made to reducing costs of operating passenger aircraft through improved fuel efficiency is shown schematically in Fig. 3.48. Reducing the weight of materials is a constant aim, but the designer must first ensure that a proposed new material meets the property requirements for a particular structural component. Furthermore, any additional material costs should not exceed the savings expected from using less fuel, and from other possible economies such as reduced maintenance. In this regard, the Boeing Airplane Company has developed a formula that is useful when evaluating projected weight savings for recurring materials costs:

$$\Delta Q = \frac{1}{u}\left[\frac{\Delta P}{\left(\dfrac{W_o}{W_c} - 1\right)} - P_o\right]$$

P_o = Price per kg of baseline material
ΔP = Difference in price per kg between candidate and baseline material
W_o = Weight of baseline part
W_c = Weight of candidate part
u = Material utilization, i.e., ratio between part and purchased material weight
ΔQ = Cost difference per kg of weight saved

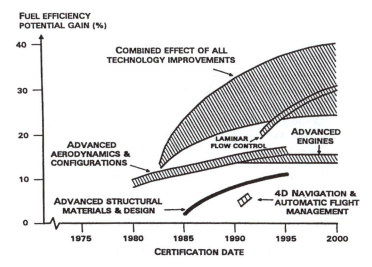

Fig. 3.48 Relative contributions of different technological advances to improved fuel efficiency of aircraft (courtesy of C. J. Peel, QinetiQ).

In recent years, the aluminium industry has faced strong competition from alternative lightweight materials, notably fibre-reinforced organic composites. Predictions made in the 1980's suggested that the aluminium content of civil passenger aircraft could fall from around a traditional figure of around 80% to less than 50% within a decade. Even more drastic reductions were expected for military aircraft. This latter prediction has proved to be correct for combat aircraft (Fig. 3.49) which fly at supersonic speeds that cause significant aerodynamic

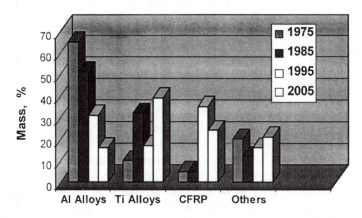

Fig. 3.49 Use of different materials in military combat aircraft (courtesy of C.J. Peel, QinetiQ).

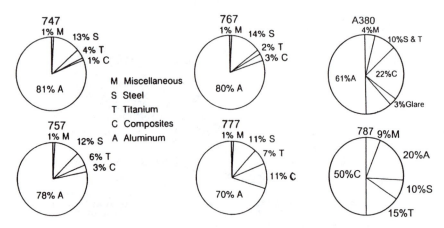

Fig. 3.50 Structural materials used in selected passenger aircraft (courtesy M.V. Hyatt and A.S. Warren, Boeing Airplane Company, and E. Grosjean, Airbus Industrie).

heating during flight, and require the use of titanium alloys or other materials for much of the skin and structure. However, aluminium alloys have retained their dominant position with recent passenger aircraft with the exception of the Boeing 787 which is scheduled to commence service in 2007 (Fig. 3.50). The Boeing 777 has 70% aluminium alloys in its structure and the new large Airbus A380 has 61%, although the wing still has retained more than 80%. A major change with the design of this latter aircraft is use carbon fibre reinforced plastics for much of the critical centre wing box, which is said to allow a weight saving of one and a half tonnes compared to using the most advanced aluminium alloys. Parts of the fuselage and much of the empennage (tail) are made from fibre-reinforced materials that include the GLARE laminate (Section 7.1.1). The actual division of materials in the A380 is: aluminium alloys 61%, fibre reinforced polymer composites 22%, titanium and steel 10%, GLARE laminates 3%, surface protection compounds 2%, and miscellaneous 2%. For the Boeing 787, the radical decision has been made to reduce weight by using to use graphite epoxy composites for the primary structure and the overall aluminium content will be reduced to only 20%. Here the aim is to achieve 20% better fuel efficiency and improve corrosion resistance. Manufacturing costs and cabin noise in flight are also expected to be reduced.

A new process called creep age forming (CAF) has been developed which offers substantial cost benefits for the production of curved aluminium alloy components, such as large wing panels for aircraft. In CAF, the forming operation is combined with artificial ageing by elastically loading the component onto a former using a combination of mechanical clamping and vacuum bagging, and then holding for a fixed time at or more ageing temperatures (typically in the range 150–190 °C). Stress relaxation occurs by creep so that the component undergoes permanent deformation. Allowance must be made for significant springback that may occur when the component is released because the ageing conditions needed to achieve the required mechanical properties are insufficient to relax fully the elastic stresses. Advantages of CAF include accuracy, reproducibility, and the ability to produce multiple curvatures in complex components close to near net shape in a single procedure. CAF has been applied successfully to alloys of the 2xxx and 7xxx series, as well as to the lithium-containing alloy 8090. It is being used for the production of upper wing panels of several civil and military aircraft, including is the new Airbus 380, for which the wing panels are up to 33 m long, 2.8 m wide and have thicknesses varying from 3 to 28 mm. Motivation to use CAF for the production of the more complex lower wing skins is also strong and this requires matching the properties offered by alloys such as 2024 which, in the T351 condition, is stretched and naturally aged to provide high damage tolerance.

3.6.2 Automotive sheet and structural alloys

Serious attention to weight savings in motor vehicles first arose in the 1970s following steep increases in oil prices imposed by Middle Eastern countries. More recently impetus has come from legislation in some countries to reduce levels of exhaust emissions through improved fuel economies. In this regard, each 10% reduction in weight is said to correspond to a decrease of 5.5% in fuel consumption. Moreover, each kg of weight saved is estimated to lower CO_2 emissions by some 20 kg for a vehicle covering 170 000 km.

The two strategies to save weight have been to build smaller vehicles (so-called downsizing) and to use lighter materials (lightweighting). The extent to which each has been adopted has depended on factors such as customer preference and levels of fuel tax in various countries. More recently, the use of lighter materials has been further stimulated by the need to offset weight penalties associated with the demand for accessories such as air conditioners and emission control equipment, as well as additional safety features to provide more protection to occupants.

Particular attention has focused on the replacement of steel and cast iron by aluminium alloys which usually results in weight savings of 40–50%. In this regard, the average aluminium content of motor vehicles built in North America has risen from 3 kg in 1947 to 20 kg in 1960, 113 kg in 1999, and has a projected increase to 156 kg in 2009 (Fig. 3.51). A similar trend has been followed in Western Europe and in Japan. In the year 2000, this amounted to some 5.8 million tonnes of aluminium worldwide which is expected to rise to 8 million tonnes by the year 2010. Also significant is the rise in the content of magnesium which is considered in Chapter 5, and in the use of plastics.

A controlling factor in materials selection is cost, and an additional premium of up to $US3 for each kilogram of weight saved is currently considered to be acceptable for mass-produced cars. This restriction is accommodated more readily with lower-cost aluminium castings for engine components and wheels which currently account for about two thirds of the aluminium used world-wide in cars. For wrought components, steel generally costs less than $US1 per kg as compared

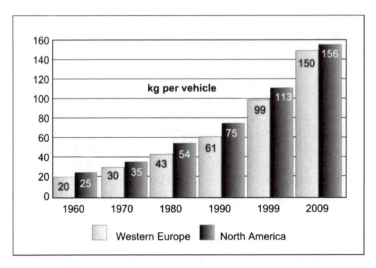

Fig. 3.51 Past average and predicted future use of aluminium alloys per vehicle in North America and Western Europe (from Hall A., *Light Metal Age* April, **60**, 49 April 2002).

with $US3-4 for aluminium alloy sheet and extrusions. Luxury cars such as Rolls-Royce, Rover and Porsche have for many years made use of aluminium alloys for bonnets and other body panels. Now competition with steel has become intense for both body sheet and structural members in mass-produced cars.

In 2004, the major automotive applications of aluminium alloys fell into the following broad categories, each of which is at a different stage of market penetration:

- engines 37%
- transmission 17%
- wheels 13%
- heat exchangers 11.5%
- chassis and suspension 5%
- body and closures 3%
- steering 3%

Predictions for the potential growth in the use of aluminium alloys by components during the current decade are as follows:

- engines: +27% to 47.3 kg per vehicle
- wheels: +33% to 20 kg per vehicle
- chassis and suspension: +142% to 6.3 kg per vehicle
- autobody sheet: +250% to 6.7 kg per vehicle.

Growth in the use of aluminium alloys for automotive body sheet to an estimated modest average level of 6.7 kg per vehicle by the year 2010 would consume some 525 000 tonnes. Material and fabricating costs need to be minimized and the main property requirements for the sheet are:

- sufficient strength for structural stability, fatigue resistance, dent resistance and crash worthiness
- good formability for stretching, bending and deep drawing, as well as the ability to control anisotropy and spring back
- good surface appearance, e.g. freedom from Lüders bands.
- easy joining by welding, clinching, riveting, brazing and adhesive bonding
- high corrosion resistance
- ease of recycling

Formability, notably the capacity for sheet to be drawn, is a critical requirement. This property is controlled by the crystallographic texture developed on rolling, as well as the work hardening exponent n and the R-value (ratio of width to thickness strain), both of which should be as large as possible (Section 2.1.3). An optimum combination of strength and formability requires careful control of alloy composition, working and heat treatment.

The first alloys selected for automotive sheet were 3004, 5052 and 6061. However, the low strength of 3004, problems with Lüders band formation drawing of some 5xxx series alloys and the limited formability of 6061 led to the development of new compositions such as the copper-containing alloys 2008 and 2036, and other 6xxx series alloys including 6009 and 6010 (Table 3.11). Now focus for producing the so-called "body in white" vehicle is on the use of the non-heat treatable Al–Mg alloys or several Al–Mg–Si alloys of the 6xxx series that respond to age hardening

Table 3.11 Typical mechanical properties and formability indices for aluminium automotive sheet alloys (from Hirsch, J., *Mater. Sci. Forum*, **242**, 33, 1997)

Alloy	Nominal Composition	0.2% proof stress MPa	Elongation %	η	R
2008–T4	0.9Cu–0.65Si–0.4Mg	126	28	0.25	0.70
2036–T4	2.6Cu–0.45Mg–0.25Mn	195	24	0.30	0.65
5754–0	3.1Mg	100	28	0.30	0.75
5052–0	2.5Mg–0.25Cr	90	25	0.30	0.75
5182–0	4.5Mg–0.35Mn	140	30	0.31	0.75
5022–T4	3.7Mg–0.35Cu	135	30	0.30	0.65
6009–T4	1.1Si–0.6Mg–0.35Mn	125	27	0.22	0.64
6010–T4	1Si–0.8Mg–0.5Mn–0.4Cu	165	24	0.22	0.70
6016–T4	1.25Si–0.4Mg	120	28	0.27	0.60
6111–T4	0.85Si–0.75Mg–0.7Cu–0.3Mn	160	26	0.26	0.56

η = strain hardening exponent
R = ratio of width to thickness strain

The main advantages of Al–Mg alloys are their relatively high strength combined with good formability which arise from solution strengthening and capacity for extensive strain hardening. In this regard, these alloys have a high stain hardening exponent n ($\sim$0.30) which is important in drawing operations because it serves to reduce metal flow in locally strained regions. One disadvantage that remains is the tendency to serrated yielding during stretching and drawing that may cause Lüders bands to form which disfigure the surface of a sheet. This phenomenon arises because of the pinning of dislocations by magnesium atoms and can often be overcome during sheet production by giving a final light cold rolling pass that increases dislocation density. It is also less of a problem with more dilute alloys such as 5754 (Al–3.1Mg). A second disadvantage is some loss of some strength that may occur due to partial annealing during the paint bake cycle (e.g. 30 minutes at 180 °C). This problem can be minimized by copper additions which induce clustering so that some rapid hardening occurs during the first few minutes at such temperatures (Section 2.3.2). An example of an Al–Mg–Cu alloy is 5022 (Al–3.7Mg–0.35Cu).

Neither of the above problems occurs in 6xxx alloys and paint baking may in fact lead to some increase in strength due to age hardening. Yield strengths may exceed those of the 5xxx series alloys (Table 3.11) but formability is somewhat less (n values 0.22–0.27). Levels of magnesium, silicon and, more recently, copper must be optimised to achieve a required combination of properties. Surplus amounts of silicon have been found to have positive effects on response to age hardening as well as to formability. For these reasons, the alloy 6016 has gained favour with European manufacturers. In North America, attention has been directed more to the addition of copper which increases age hardening and improves the strain hardening characteristic of Al–Mg–Si alloys, although there are concerns about adverse effects on corrosion properties. One such alloy is 6111. Pre-ageing some 6xxx series alloys for short times at temperatures such as 100 °C has been shown to increase the so-called paint bake response (PBR), i.e. ageing during this brief cycle is accelerated which results in additional hardening.

Stronger alloys are regarded as being more suitable for outer panels whereas the lower strength but more formable alloys, such as the earlier composition 5182 (Al–4.5Mg–0.35Mn), are preferred for more complex inner panels. At the same time, there is an advantage to be gained if the combinations of alloys in the same assembly have compositions that are fairly similar in order to facilitate scrap separation during recycling.

Aluminium body sheet alloys usually have yield strengths equal to or better than most plain carbon steel sheet alloys and the respective work-hardening exponents are similar (typically $n = 0.22$–0.25). However, steel sheet displays much better formability because ductilities (e.g. 40%) and R-values (1.6 compared to 0.75) are significantly higher than those of the aluminium sheet alloys. Another disadvantage with aluminium alloy sheet has been the difficulty of

Fig. 3.52 Self-pierce riveting insertion process for sheet joining (courtesy HENROB Corporation).

using traditional spot-welding techniques during assembly. Frequent changes of the tips in the welding electrodes has been necessary and, even then, the strength and fatigue resistance of the spot welds do not compare well with those in steel sheets. This has led to the development of techniques for continuous adhesive bonding of flanges, often combined with less frequent spot welds to improve peel strength and hold the structure together while the adhesive cures. This practice has become known as weld-bonding and has the additional advantage of increasing the overall stiffness. Another advance has been the use of the technique of self-piercing riveting which has been improved so the leak-proof joints can be made because the innermost sheet is not perforated (Fig. 3.52). Joints are claimed to be 30% stronger than those produced by spot welding and fatigue resistance can be up to 10 times better. Furthermore, sheets of different materials can be easily joined. When combined with adhesive bonding the joining system has become known as rivbonding.

Although aluminium alloy sheet has been used for many years for body panels in some automobiles such the Rolls-Royce and the Land Rover in Britain, the first production vehicle to utilize aluminium for the body in a broad and systematic way was the Honda Acura NSX sports car that was introduced in 1990. The primary structure weighed 210 kg which represented a saving of an estimated 140 kg over an equivalent steel construction. The Alcan and Ford Companies have made extensive use of bonded sheet for load-bearing members in the conventional monocoque design of their aluminium intensive vehicle (AIV). Behind this approach has been the belief that existing manufacturing and assembly facilities used for fabricating automotive panels from steel sheet can be retained and easily adapted to handle aluminium alloy sheet. The body and exterior panels are 200 kg lighter than would be the case had steel sheet been used. So far as assembly is concerned, Fig. 3.53 shows one production route developed for a prototype manufacturing line which uses the low-strength, corrosion-resistant and highly formable Al–Mg alloy 5052–0. It is necessary to pretreat the surface of sheet to improve its adhesive properties and to use a lubricant during pressing of panels that is compatible with the

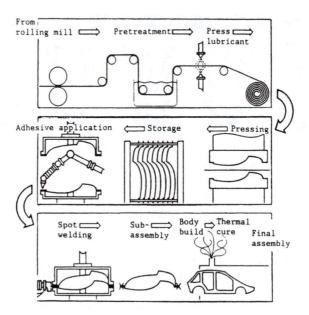

Fig. 3.53 Prototype assembly line for producing adhesively bonded aluminium alloy motor car panels (Shearby, P. G. *et al.*, *Aluminium '86*, Sheppard, T. (Ed.), Institute of Metals, London, 543, 1986).

adhesive. Furthermore, the adhesive itself has to provide a necessary combination of shear, peel and impact strength and to retain these properties over a temperature range of −40 to 120 °C so as to cope with climatic variations and heat generated in the engine compartment.

The Alcoa Company decided to follow a more radical approach to design which has involved the concept of a space frame comprised mainly of hollow aluminium alloy extrusions interconnected at die cast nodes by robotic welding techniques (Fig. 3.54). The extrusions are made from 6xxx series alloys and the nodes are produced by pressure die casting an Al–Si alloy into evacuated dies to minimize porosity and improve fracture toughness to levels well above those normally found for castings. A plant to exploit this technology was built in Soest in Germany to produce components and sub-assemblies for the Audi Company whose A8 vehicle was released in 1994. Fewer than 100 extrusions and nodes were needed for the primary body structure compared with as many as 300 stampings that would be required for a comparable structure made from steel. Several methods have been used to attach the external aluminium alloy body panel to the space frame including shielded argon arc welding, resistance spot welding, punch riveting, clinching and adhesive bonding. Arc welding has been considered in Section 3.5.1 and clinching is a technique for joining two sheet panels by bending over the edges. The all-aluminium alloy body uses

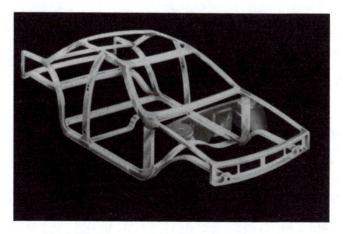

Fig. 3.54 Alcoa extruded aluminium alloy space frame.

55% sheet, 25% extrusions and 20% castings which reduces the overall weight by 40%. More recently, another model, the Audi AL2, has been produced in which cost savings have been made by simpler design in which the extruded members are kept as straight as possible, the number of cast joints is reduced, and the several hundred spot welded and clinched joints have been completely removed.

Economic analysis suggests that space frame technology has a cost advantage for low volume production (up to 60 000–80 000 vehicles per year) whereas the use of stamped sheet components becomes less expensive for higher volumes. Several other companies have been developing vehicles with aluminium alloy bodies made from stamped sheet components. As examples Chrysler in the United States, in collaboration with Reynolds Metals, has built a version of its Neon model in which the use of stamped aluminium alloys for the body has resulted in a weight reduction of 270 kg, and General Motors have adopted the same strategy to reduce the weight of their EV1 electric vehicle.

Another use of wrought 6xxx alloys is for drive shafts which are gaining acceptance and commonly result in weight savings of around 6 kg. Higher-strength alloys are required for bumper bars and bumper reinforcements for which use has been made of Al–Zn–Mg and Al–Zn–Mg–Cu alloys of the 7xxx series. An example is 7029–T5 (Al–4.7Zn–1.6Mg–0.7Cu) which was derived from compositions traditionally used for aircraft construction. This alloy has the following mechanical properties: 0.2% proof stress 380 MPa, tensile strength 430 MPa and elongation 15%. Impurity levels are kept low to improve toughness and bright finishing characteristics.

Two other factors that influence the selection of materials for car bodies are safety (or crashworthiness) and capacity for recycling. Stiffness is a prime concern when considering the deflection of panels or structures and this parameter is directly related to elastic modulus which places aluminium alloys at an apparent disadvantage with respect to steels. What is more significant, however, is that stiffness of a component also varies with the cube power of thickness. As a general rule, when an aluminium structure is half the weight of an equivalent steel structure, the gauge of the sheet or the wall thickness of components is some 50% greater which provides for enhanced rigidity and crush resistance, particularly for tubular sections. With respect to recycling, aluminium scrap has the advantage of relatively high intrinsic value, particularly if similar alloys can be contained and remelted within a closed loop. In this regard, aluminium currently comprises 6% of the weight of an average car but can represent as much as 30% of the recycled value (Section 1.1.4).

A manufacturing innovation, which was developed in the steel industry in the 1990s, is the use of welded "tailored blanks" to produce complete sections of automobiles. This method has become possible because of advances in laser welding which enable panels of varying thicknesses, and sometimes of different alloys, to be joined together. Aluminium alloys of the 5xxx and 6xxx series are being investigated for this purpose and consideration is being given to using different joining methods so that combinations of aluminium and steel can be employed. The aim with tailored blanks is to enable designers to reduce weight and to improve rigidity in precise locations where this is needed.

3.6.3 Shipping

The position of the ship's centre of buoyancy above the centre of mass determines its stability. For passenger vessels, the use of aluminium alloys makes possible an increase in the volume and height of the superstructure without loss of stability which, in turn, allows for the inclusion of more passenger decks than is possible with an equivalent design built in steel.

The introduction of significant tonnages of aluminium alloys for ship building dates back to the end of World War II. An early major advance was made with the passenger liner SS United States of 54,200 tonnes displacement which was launched in 1952 and contained around 2,000 tonnes of aluminium alloy plate and extrusions in its superstructure. This change in design also gave savings of 15–20% in weight and 8% in fuel so that this vessel quickly captured the speed record for crossing the Atlantic Ocean. Although the overall use of aluminium has generally been less, this trend has continued with liners such as the Queen Elizabeth II, and with modern cruise ships. During recent years, there has been a worldwide demand for high speed, ocean going ferries and other coastal vessels that has led to construction of all aluminium designs commonly 50 to 80 metres long and capable of speeds of exceeding 80 kmh. One example of a vehicular ferry is shown in Fig. 3.55.

Although there is no pressing need to exploit the weight advantages of aluminium alloys in cargo ships, their good welding characteristics and cryogenic properties (no ductile/brittle transition) have led to their use for the large

Fig. 3.55 Ocean-going ferry constructed from 5xxx series aluminium alloys (courtesy INCAT, Tasmania, Australia).

spherical tanks that are distinctive features of vessels transporting liquefied natural gases. The other use of aluminium alloys in shipping is with naval vessels because of the need to reduce weight high up, especially for masts and structures carrying radar aerials. Some parts of the superstructures of many of these naval vessels have also been constructed from aluminium alloys leading to accusations that this makes them prone to fire damage through the "burning" of aluminium. This is incorrect because bulk aluminium does not burn in air, nor does it support combustion.

The alloys most commonly used for plate are the 5xxx series based on the composition 5083 (Al–4.5%Mg–0.7%Mn) in the as rolled condition (H temper) that are readily weldable and show a good combination of strength and corrosion resistance. A more recent alloy is 5383 (Table 3.2) for which the nominal levels of magnesium and manganese have been slightly raised, as have the allowable amounts of the impurities zinc and copper, whereas the iron and silicon impurities are reduced. The advantages claimed for 5383 over 5083 are that the proof stress and tensile strength are 12–14% higher both before and after welding, and that the corrosion resistance is improved. Further increases in pre- and post-weld tensile properties have been reported for a European alloy known as ALUSTAR in which the 5083 composition has been adjusted to allow a zinc content as high as 1.3%. The minimum level for the proof stress in the welded condition is set at 160 MPa which compares with 140 and 125 MPa, respectively, for the alloys 5383 and 5083. Friction stir welding (see 3.5.1) is proving to be a useful method for distortion-free joining of long symmetrical plate sections in all these alloys.

Further weight savings are possible with the development of composite panels involving a combination of flat and corrugated sheets, the latter being manufactured by standard roll-forming methods. CORALDEC™ is the trademark of one such assembly in which the flat top and bottom sheets are laser welded to the central corrugated sheets. These panels are being introduced for construction of accommodation and vehicle decks in fast ferry catamarans. Advantages of up to 25% in weight savings are claimed over extruded panels made from 6xxx series alloys such as 6005 and 6082.

3.6.4 Building and Construction

Significant use of aluminium and its alloys for building materials commenced some 50 years ago after the end of World War II. Since then consumption has increased steadily in most countries and this segment is now the major market in both China and Japan which are two of three largest consumers of aluminium (e.g. Fig. 1.5). Advantages of aluminium include its good decorative appearance, high corrosion resistance in most environments, lightness, ease of fabrication, and the fact that extruded sections can be easily prepared for the provision of double glazing or the insertion of insulation and blinds. Competitive disadvantages of aluminium when compared with steel are cost, relatively low elastic modulus which means that beams and other supports must be thicker, the need for greater protection from fires due to the lower melting point of aluminium, and the higher coefficient of thermal expansion which requires more allowance to be made for movement, particularly in extreme climates. Most of these disadvantages are more apparent with plastics which are another major competitor for many building products.

Applications include facades, roofing, guttering, window frames, sun shades, curtain walls and balustrades. Aluminium alloys are also often used as external cladding to retain spalled fragments and disguise discoloration in old stone and concrete buildings. More limited use is being made to construct small bridges. The alloys in most common use for rolled products are those based on the 5xxx series (e.g. 5083) and 3xxx series (e.g. 3003), whereas extrusions are usually made from the 6xxx series (e.g. 6063). Where these alloys are exposed to the environment, their corrosion resistance may be increased by anodising. In addition, a wide range of colours are available that are durable and do not fade.

3.6.5 Packaging

During the last two decades, the use of aluminium in packaging has increased to an extent that, at times, it has been the largest market for this metal in some countries including the United States. In 2004 packaging was placed second is the United States (Fig. 1.5), and also in Japan which is one of the other two leading consumers of aluminium. This situation has arisen because aluminium is an attractive container for food and beverages since it has generally high corrosion resistance, offers good thermal conditions, and is impenetrable by light, oxygen, moisture and microorganisms. Furthermore, simple alloy compositions can be used that are easy both to fabricate and recycle. The largest production items are beverage cans and foil.

Canstock The entry of aluminium into the can market occurred in 1962 with the introduction of the tear-top tab, or so-called easy-open end, which was used for steel beverage cans. This was followed by the two-piece, all-aluminium can with a seamless body that is produced from sheet by cupping and then by drawing and ironing the side walls. Since then growth in use of the all-aluminium can has been phenomenal and now accounts for more than three-quarters of all the aluminium used for packaging.

In 2004, an estimated 3.7 million tonnes of aluminium was used worldwide to produce just over 200 billion beverage cans, half of these in North America. In this region, aluminium has captured more than 95% of this market and the average throughout the world is 80%. This represents the largest single use of aluminium for the one product. As mentioned in Section 1.1.4, efficient recycling has been a key economic factor in guaranteeing the competitive success of aluminium for canstock.

Canstock was first made using the established alloy 3003 (Al–1.2Mn) but increasing demands for thinner gauges led to the development of a new composition 3004 which also contains 1% magnesium to provide solid solution strengthening. Can tops are normally made from the alloy 5182 (Al–4.5Mg–0.4Mn).

Sheet for can bodies is mostly produced by rolling large DC cast ingots (Fig. 3.3) and the following is a typical processing cycle.:

1. Homogenization of the ingots, usually at two temperatures (e.g. 570 °C followed by slow cooling to 510 °C) to obtain a desirable distribution of primary particles (Al_6(Fe,Mn) and α-Al(Fe,MnSi)), and the finer dispersoids e.g. Al_6(Fe,Mn). This helps to control recrystallization in (4) below.
2. Hot-rolling at 450 °C to reduce the thickness from around 600 mm to 25 mm.
3. Warm-rolling at 275 °C to a so-called hot-band thickness of 2.5 mm, during which the grain structure develops a preferred orientation as shown in Fig. 3.56.

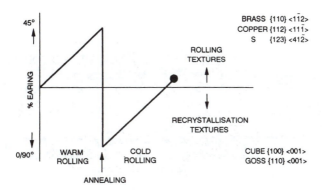

Fig. 3.56 Progressive development of textures and earing during the processing of 3004-H19 canstock.

4. Annealing for 2 h after slow heating to 350 °C to promote recrystallization.
5. Cooling to room temperature and cold rolling in several stages to the final gauge size of around 0.30 mm. At this stage the sheet is said to be in the H19 condition and typical mechanical properties are: 0.2% P.S. 290 MPa, T.S. 310 MPa and elongation 10%.

What is critical is control of sheet texture to minimize earing (Section 2.1.4) during the final deep drawing and ironing of the can. This is achieved by promoting formation of a cube (0–90°) texture in recrystallized grains during the annealing stage of the processing cycle, which then serves to balance the 45° texture that develops during subsequent cold rolling (Fig. 3.56) What is finally desired is consistent slight positive earing of less than 2.5% in each can.

One stage in the demanding process of drawing and ironing of the can body is shown schematically in Fig. 3.57. It can be seen that ironing involves the progressive squeezing of the metal between an inner punch and outer dies that have decreasing internal diameters. Machines are now available that can punch out as many as 500 cans each minute. Lubrication is important, as is the presence of primary particles of the relatively hard intermetallic compound α-Al(Fe,MnSi) in the sheet that serves to wipe away aluminium debris which may be transferred to the dies during the ironing process. This latter phase, the composition of which may vary between α-Al$_{12}$(Fe,Mn)$_3$Si and α-Al$_{15}$(Fe,Mn)$_3$Si$_2$, forms from the interaction of silicon in solid solution in the matrix with primary Al$_6$(Fe,Mn) particles during homogenization.

The thickness of final gauge sheet for can making has been reduced from 0.50 mm in the 1960s to 0.44 mm in the early 1970s, 0.30 mm in the early 1990s, and to an average of 0.27 mm in 2005 in North America. This progressive thinning, or lightweighting, has been crucial because of the relatively high cost of aluminium. Careful attention must be given to the design of cans

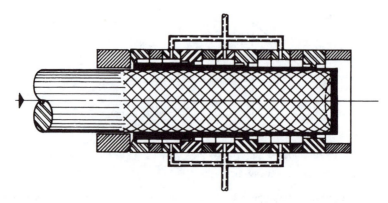

Fig. 3.57 One stage in the drawing and ironing of a seamless aluminium alloy can.

which, in effect, are small pressure vessels that must withstand internal pressures of some 600 kPa when they contain gasified beverages. Wall thicknesses vary depending on location in the can and may be as thin as 0.10 mm in the central regions of the side walls. It is necessary, therefore, to filter the molten aluminium during casting to remove inclusions and to control the size of primary intermetallic particles which might otherwise cause pinholing or perforation of the sheet during can making. These requirements also apply to the production of foil.

Thickness reductions and cost savings have also been achieved with can tops (end stock) through improved designs and the use of stronger alloys. Typical thicknesses were around 0.335 mm in the 1970s and now approach 0.208 mm in some advanced cans. In the mid 1980s, a typical yield strength for the alloy 5182 were 320 MPa which has been raised to 370 MPa by gradually increasing the level of magnesium above the specified limit of 5%, increasing the manganese content, and allowing small increases in minor elements such as copper.

Foil Due to its capability of being cold rolled to very thin gauges, aluminium foil is widely used as an effective barrier to light, water vapour and gases. Foil is commonly defined as rolled sheet having a thickness less than 0.15 mm. Products range from light gauge foil, e.g. 0.01 mm thick, for domestic and other uses to thicker foil for semi-rigid containers. In laminated products with plastic or paper films, aluminium is also an ideal packaging material for applications such as pharmaceuticals and food containers. In North America the foil market consumes some 500 000 tonnes of aluminium annually.

Historically, 1xxx aluminium alloys such as 1145, 1100 and 1200 have been widely used. However, the need for alloys with greater levels of strength and ductility has led to the development of alloys with higher levels of iron, manganese and silicon which contain higher volume fractions of intermetallic compounds. These compounds can control the microstructure during recovery and recrystallisation during annealing resulting in an attractive combination of properties after final cold rolling. Examples are the alloys 8006 and 8011.

The major advance in foil production has been the development of methods for rolling at high speeds whilst maintaining close control of thickness. Rolls capable of handling coil stock weighing as much as 10 tonnes and operating at speeds in excess of 40 m s^{-1} are now common. The coil stock is produced by rolling semi-continuously cast ingots (Fig. 3.3) or, more economically, continuously cast strip (Fig. 3.4b and 3.5) and then cold-rolling to a strip thickness of about 0.5 mm. Foil is produced by further cold-rolling the strip in several stages in a continuous mill, the material being pack-rolled double in the last stage if very light gauges, e.g. <0.025 mm, are needed. Most foil is then softened by annealing. One of a number of coating treatments such as laminating with paper, lacquering, printing and embossing may then be applied for purposes such as advertising.

A common problem associated with foil production from twin roll cast strip is the tendency for alloying elements to segregate to the centre of the strip during casting. This centreline segregation occurs because, during rapid solidification, the shearing action of the rolls squeezes the interdendritic liquid towards the centre of the strip. The result is that this zone contains higher amounts of the relatively hard Al–Fe–Si intermetallic compounds, which if not reduced in size by further thermal treatments and cold working, may perforate the final foil and produce an unacceptable number of pinholes. For most applications, a pinhole limit of less than 500 per m^2 and sizes below 20 μm is required. Centreline segregation may also reduce foil strength and cause higher foil tearing at the mill.

3.6.6 Powder metallurgy products

Powder metallurgy technology provides a useful means for fabricating near net shape components, thereby enabling the costs associated with machining to be minimized. This aspect is discussed further in Section 6.5.8. Another advantage is that microstructures are generally finer than those present in alloys prepared by normal ingot metallurgy, which may result in improved mechanical properties and corrosion resistance.

Normal processing involves the production of powders which are then cold pressed and sintered at elevated temperatures (e.g. 500 °C). Recently several novel techniques have been developed which have enabled new, and sometimes unconventional, alloys to be produced, thereby extending the range of properties available from powder compacts. One method is spray forming which is discussed later in this section; others include rapid solidification processing and mechanical alloying, which are described in Chapter 7. In all cases much of the developmental work was carried out in aluminium-based materials.

Aluminium powders were first produced as flakes by ball milling and resulted in a number of fatal fires and explosions because of the highly exothermic nature of finely dispersed aluminium. Safer milling methods are now available but most powders are currently produced by 'atomization' in which the molten aluminium (alloy) issuing from an orifice is broken up into very small droplets by a stream of high-velocity air or an inert gas. The droplets solidify as a fine powder which is collected in the chamber and graded for size. A problem is that aluminium is always covered by a tenacious oxide film and the thickness on atomised powders can vary from 5 to 15 nm. Although this film cannot be removed, it may be disrupted by sintering in the presence of magnesium. Magnesia has a lower free energy of formation than alumina and magnesium metal can partially reduce alumina to form a spinel $MgAl_2O_4$. This reaction serves to rupture the oxide which exposes the underlying aluminium metal and facilitates sintering. Less than 0.2% magnesium is required.

As mentioned above, the direct production of components from powders normally involves cold compaction in dies followed by sintering. Allowance must be made for shrinkage. Alternatively, if powder alloy billet is required, then the following stages are usually involved:

1. Compacting the powder in an aluminium can.
2. Vacuum degassing and hot pressing. This step may require sealing the can which is later removed by machining.
3. Fabricating the powder alloy billet by the normal processes of extrusion, forging or rolling.
4. Heat treatment.

Although they are more expensive to produce, powder metallury billets are now available weighing several hundred tonnes.

Conventional processing One of the early products produced by powder metallurgy techniques was the material known as SAP that was developed in Switzerland. SAP is an Al–Al$_2$O$_3$ alloy prepared by pressing and sintering finely ground flakes of aluminium, and the final microstructure may contain an oxide content as high as 20% in the form of Al$_2$O$_3$ particles dispersed in the aluminium matrix. The sintered compacts can be hot-worked by extrusion, forging and rolling to give a product which, using Al–15Al$_2$O$_3$ as an example, may have room temperature properties of 0.2% P.S. 240 MPa, T.S. 345 MPa, with 7% elongation. These properties are relatively low but the fact that the oxide is stable up to the melting point means that creep strength is superior to conventional aluminium alloys at temperatures above about 200–250 °C. Although SAP was once considered as a possible skin material for high-speed aircraft which would suffer aerodynamic heating, it has found relatively few applications and its major use has been for fuel element cans in certain nuclear fission reactors which use organic coolants. For this purpose its creep strength, corrosion resistance, low capture cross-section for neutrons and absence of radioactive isotopes are special advantages. The material is also a candidate for part of the internal structure of experimental nuclear fusion reactors which may employ the deuterium–tritium fuel cycle.

A wider range of powders is now available including pre-mixes which give compositions that tend to mimic several wrought alloys normally produced from cast ingots. Compositions of common commercial powder metallurgy alloys are shown in Table 3.12. It will be noted that most are based on the wrought Al–Mg–Si and Al–Cu–Mg alloys 6061 (Al–1Mg–0.6Si–0.28Cu) and 2014 (Al–4.4Cu–0.6Mg–0.85Si–0.8Mn). A common feature is that all but one contain magnesium which reacts with the surface Al$_2$O$_3$ film during sintering to form spinel, MgAl$_2$O$_4$. This disrupts the film and exposes underlying metal which facilitates the sintering process. Another common feature is that they all respond to age hardening.

Table 3.12 Compositions of commercial powder metallurgy aluminium alloys

Alloy	Cu	Mg	Si
602	–	0.6	0.4
601	0.25	1.0	0.6
6711	0.25	1.0	0.8
321	0.2	1.0	0.5
202	4.0	–	–
2712	3.8	1.0	0.75
201	4.4	0.5	0.6
2014	4.4	0.5	0.8
13	4.5	0.5	0.2
123	4.5	0.5	0.7

Since these alloys were not specifically designed to be sintered, it is to be expected that opportunities will exist to improve processing conditions through compositional changes. For example, it has been shown the sintering response of the Al–Cu–Mg alloys can be enhanced by microalloying additions of the low melting point elements tin, lead, bismuth or antimony. Each of these elements has a high diffusivity in aluminium and a high vacancy binding energy. They therefore diffuse into aluminium ahead of the alloying element copper and will bind preferentially with vacancies. This has the effect of reducing the rate at which alloying elements in the pre-mix powders diffuse into aluminium thereby allowing the transient liquid phase to persist for longer times which improves densification. Alloys based on the Al–Zn–Mg–Cu system (7xxx series) do not have good sintering characteristics and have been little used as powder metallurgy materials. However, for the composition Al–8Zn–2.5Cu–1Cu, it has been demonstrated that a very small addition ($<$0.01 at.%) of lead will greatly improve the sintering process and reduce porosity (Fig. 3.58).

Improvements have also been made with lubricants used during pressing of the compacts which have reduced the earlier problems of excessive wear of punches and dies. Better combinations of strength and resistance to both corrosion and stress-corrosion cracking have been obtained for small forgings and extrusions made from pre-alloyed, atomized powders than the corresponding products made from ingot metal. In addition, it has been possible to blend into the powders elements that could not be readily added to conventional alloys. One example has been the addition of cobalt to an Al–Zn–Mg–Cu alloy which produces a fine dispersion of the phase Al_9Co_2 in the powder metallurgy product. Commercial alloys such as 7090 and 7091 (Table 3.4) are now available and, as shown in Fig. 3.59 grain size is much reduced as compared with that for a similar type of alloy produced from an ingot. Tensile properties are increased (Table 3.5) and the presence of fine dispersions of the phase Al_9Co_2 along grain boundaries is claimed to reduce the rate of crack propagation in conditions that favour fatigue or stress-corrosion cracking.

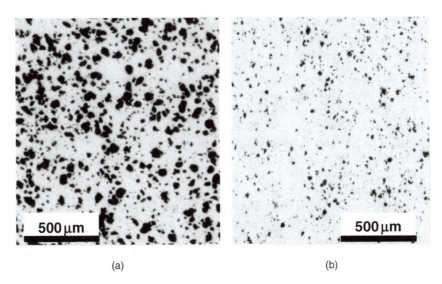

(a)   (b)

Fig. 3.58 Optical micrographs of unetched sections of sintered alloys (a) Al–8.5Zn–2.5Mg–1Cu and (b) Al–8.5Zn–2.5–1Cu–0.07Pb showing the effect of a microadditon of lead in reducing porosity (courtesy G. B. Schaffer).

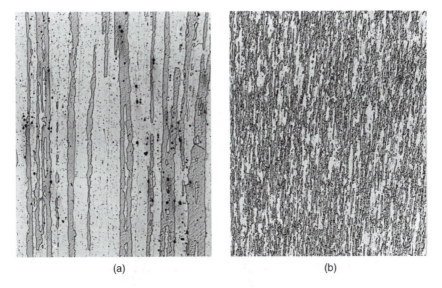

(a)   (b)

Fig. 3.59 Microsections of extrusions produced from (a) 7475 ingot and (b) 7091 powder metallurgy compacts × 100 (courtesy J. T. Staley, Alcoa Research Laboratories).

Powder metallurgy techniques are also enabling the creation of materials with improved tribological properties. Wear resistance can be improved by blending pre-alloyed aluminium powder with hard particles, such as silicon carbide, whereas the incorporation of solid lubricants reduces the friction. In this regard, experiments have shown that the inclusion of up to 4 wt% of graphite reduces the friction coefficient of powder metallurgy aluminium alloys to values below that of grey cast iron.

Spray forming This method, which is known more generally as the Osprey Process, involves atomization of molten alloys by a stream of high velocity gas such as nitrogen. However, in this case, the resulting spray of droplets is deposited on a rotating collector plate to produce a compact preform (Fig. 3.60). Cooling of the molten droplets in flight and on impact is fast and solidification rates may be as high as 10^3 to 10^4 °C s^{-1}. Preforms, which may be 98–99% of the theoretical density, have refined microstructures with respect to the sizes of grains and primary intermetallic compounds.

The fast rate of solidification also provides the opportunity for increasing the supersaturation of alloying elements and a number of experimental compositions have been produced. Examples are Al–11Zn–Mg–Cu–X and Al–20Si–Cu–X, where X is one or more of the transition metals such as iron, chromium, nickel, manganese and zirconium. In the T6 condition, one of the former alloys has shown the following tensile properties: 0.2% P.S. 760 MPa, T.S. 775 MPa, elongation 8% which are significantly greater than those for the normal 7xxx series alloys. Higher values of fracture toughness and fatigue strength are also possible. The other composition is a hypereutectic Al–Si–Cu alloy and mechanical properties (e.g. T.S. 480 MPa) are again much greater than those of the equivalent conventionally cast alloy (Section 4.2.3) due

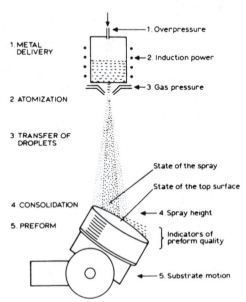

Fig. 3.60 Schematic illustration showing spray forming by deposition of molten droplets on to a rotating collector plate (courtesy D. Apelian).

mainly to the finer dispersion of the primary silicon phase. Wear resistance, which is a feature of these particular alloys, is also increased.

Spray forming offers a number of advantages:

1. Some of the steps involved in normal powder processing are eliminated.
2. Preforms may be fabricated by forging, extrusion, etc.
3. It should be possible to produce alloys with controlled variation in composition (graded materials).
4. Particulates (e.g. SiC) may be injected into the gas stream to produce metal matrix composites directly (Section 7.1.3).
5. The process may be modified to allow direct production of sheet and tube.

Several spray forming facilities have been constructed and preforms are available weighing a few hundred kilograms. However, it is necessary to appreciate that the alloys carry a significant cost premium so that potential applications are likely to be confined to specialized products. Loss of powder due to overspray beyond the confines of the collector plate is also difficult to control and can have a significant effect on the efficiency of alloy production.

3.6.7 Aluminium alloy bearings

The development of aluminium alloys for bearings dates back to the 1930s when the high thermal conductivity, corrosion resistance and fatigue resistance of aluminium were recognized. Compositions containing 2–15% copper were then the most successful and the structure of the alloys comprised hard intermetallic compounds in a softer aluminium matrix. Although they were adopted for a number of applications, their relatively high hardness was a disadvantage with regard to certain property requirements, e.g. conformability to rotating shafts. Al–Sn alloys offered the alternative prospect of a softer bearing alloy and compositions with up to 7% tin were introduced. These are still used. They were first produced by casting solid bearings but the high coefficient of thermal expansion of aluminium made retention of fit with steel shafts virtually impossible at the operating temperatures of modern engines. Consequently, most current aluminium alloy bearings are backed with steel.

Early work revealed that the seizure resistance of Al–Sn bearings continued to improve as the tin content was raised to levels of 20% or more. However, the full potential of these alloys could not then be realized because, regardless of the method of casting, tin contents in excess of 7% resulted in the formation of low-strength, grain boundary films of tin (Fig. 3.61a). Eventually it was discovered that these films could be fragmented and dispersed along the boundaries if the alloys were cold-worked and recrystallized during processing. For alloys with tin contents below 15%, the tin then appears as discrete particles in the grain corners. However, if the alloy contains more than 16% tin this phase remains continuous in the boundaries where three grains meet producing what is known as a reticular structure.

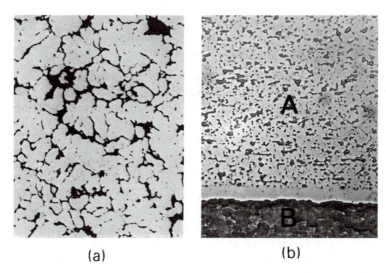

(a)   (b)

Fig. 3.61 (a) Continuous grain boundary film of tin in a cast Al–30Sn–3Cu alloy (from Liddiard, E. A. G., *The Engineers Digest*, 1955) × 120; (b) micro-structure of Al–20Sn–1Cu (A) roll bonded to steel (B) via an aluminium interlayer (courtesy G. C. Pratt. The Glacier Metal Co Ltd) (× 150).

Bonding the bearing alloy to a steel backing presented further problems because intimate contact was prevented by the tenacious Al_2O_3 film, until it was found that this film could be dispersed by severe cold rolling. Moreover, if the tin content exceeded 7%, smearing of the tin occurred during machining. This meant that pure aluminium cladding had to be incorporated during rolling to form an interface with the steel (Fig. 3.61b).

Current practice for producing steel-backed, Al–Sn bearings commonly involves the following procedures:

1. Either chill casting the alloy on a copper plate to promote directional solidi-fication of a slab approximately 300 mm thick, or semi-continuously casting a similar slab which is cut into lengths in order to machine top and bottom.
2. Cladding top and bottom with pure aluminium by heavily cold rolling in stages to produce a bonded composite having a thickness between 4 and 10% of the original dimension. This treatment also elongates the tin particles.
3. Cold roll-bonding this composite to the steel backing, contact being made via the pure aluminium interface. The configuration of the rolls is arranged so that the bearing alloy is more heavily worked than the steel backing so as to prevent the latter being unduly hardened.
4. Annealing at 350 °C for 1 h which recrystallizes the bearing alloy and coa-lesces the tin either into discrete globules in the grain corners, or to form the reticular structure.

A content of 20% tin is often regarded as the optimal level and, in Europe, the most common automotive plain bearing has the composition Al–20Sn–1Cu. An alloy with 30Sn–1Cu is widely used in bushes, and in conditions where lubrication is marginal. These particular alloys offer the added advantage of not requiring a Pb–Sn or Pb–Sn–Cu overlay that is needed to protect Cu–Pb, and some other bearings, from corrosion by chemicals present in modern lubricating oils.

In Japan, especially, the Al–20Sn alloy has been modified by reducing the tin content to 15–17% and adding 2–4% silicon. Hard silicon particles are formed within the grains which have the particular advantage of polishing the widely used spheroidal graphite (SG) cast iron crankshafts.

Other aluminium alloys have been developed as bearing materials and particular interest has been shown in substituting cheaper lead for tin. Lead is also claimed to promote development of a better film of lubricant than tin so that Al–Pb alloys exhibit lower frictional characteristics and have a higher resistance to seizure. One example is an alloy Al–9Pb–3Si–1Cu. However, the preparation of alloys based on the Al–Pb system by conventional casting methods presents special difficulties because of gravity segregation of heavy lead, and the fact that this element is immiscible in both liquid and solid aluminium. Unconventional methods have been used including stir casting, rheocasting, rapid solidification, powder metallurgy and spray deposition, all of which are discussed elsewhere. Each is capable of improving the homogeneity of the dispersion particles of lead although some methods may introduce unacceptable levels of porosity. Hot extrusion has been found to improve homogeneity and reduce porosity in Al–Pb alloys with increased lead contents up to 20%. Another group of alloys that are based on the Al–Si system have been shown to offer greater fatigue strength than Al–Sn alloys and are used in some high-speed diesel engines.

3.6.8 Superplastic alloys

Superplasticity is the ability of certain materials to undergo abnormally large extensions (commonly 1000% or more) without necking or fracturing. After many years as a laboratory curiosity, the phenomenon has now attracted commercial interest because of the possibility of forming complex shapes in a small number of operations using relatively inexpensive tooling.

The usual requirement for a material to show superplastic characteristics is the presence of roughly equal proportions of two stable phases in a very fine dispersion, e.g. grain size 1–2 μm, which must be maintained at the working temperature. Such materials are commonly eutectics or eutectoids and the eutectic alloys Al–33Cu and Al–35Mg exhibit the phenomenon although neither has commercial potential. One eutectic composition that did show early promise was Al–5Ca–5Zn, which was developed in Canada. However, tensile properties at room temperature were relatively low and all the aluminium alloys currently in commercial use have their grain structures stabilized by fine particles.

One example is a modified version of the corrosion-resistant alloy 5083 (Al–4.5Mg–0.7Mn–0.15Cr), known as 5083 SPF, in which the fine grain structure is controlled by submicron, manganese-bearing particles. This alloy has relatively low tensile properties (0.2% P.S. 150 MPa, T.S. 300 MPa) and more attention has been given to compositions that respond to age-hardening after forming. In this regard, it was known that Al–5Cu can recrystallize to give a fine grain size. However, this structure is not stable at temperatures of 400–500 °C where superplastic flow might be expected, because most of the Al₂Cu particles that might restrict grain growth have redissolved. The possibility of adding a third element that might form small stable particles was then considered, and zirconium was found suitable for this purpose. A composition Al–6Cu–0.5Zr was subsequently developed which was found to be superplastic in the temperature range 420–480 °C and, under the name of Supral 100, this material is now being used to produce products such as those shown in Fig. 3.62.

The production route for the alloy is basically similar to that used for manufacturing conventional aluminium alloy sheet, although some changes are necessary in order to achieve adequate supersaturation of zirconium. Higher casting temperatures (~800 °C) are used and the semi-continuous, direct-chill casting process is modified so that the rate of solidification is increased. The cast slabs are sometimes clad with pure aluminium during the hot-rolling stage for subsequent corrosion protection, and reduction to sheet is not greatly different from standard practice. Zirconium precipitates as small particles of a metastable, cubic phase Al₃Zr and the fine-grained structure develops during superplastic deformation at temperatures between 420 and 480 °C. Strain rates are of the order of 5×10^{-3} s^{-1} and, although the flow stresses are greater than

Fig. 3.62 Examples of products produced by superplastic forming of the alloy Al–6Cu–0.5Zr (courtesy T. I. Superform Ltd).

those needed to form thermoplastics, similar methods of fabrication can be used. Once a component has been formed, it can be strengthened by precipitation hardening using standard heat treatment operations. Mechanical tests on clad 1.6 mm sheet of the alloy Al–6Cu–0.5Zr after solution treating, quenching and ageing 16 h at 165 °C have given values of 0.2% P.S. 300 MPa, T.S. 420 MPa, with 5% elongation. Similar figures for the alloy in the as-formed condition are 125 MPa, 200 MPa and 7% respectively.

Attention has also been directed at achieving superplastic behaviour in high-strength aluminium alloys of the 7xxx series in which it has been shown that thermomechanical processing by the ITMT route (Section 3.1.6) can produce very fine grain sizes (e.g. 10 μm) in rolled components. As shown in Fig. 3.63, the treatment requires the alloy (e.g. 7475) to be over-aged at a relatively high temperature after homogenization so that coarse (e.g. 1 μm) η phase precipitates are formed. The alloy is then warm-worked (e.g. 80% reduction at 200 °C) which causes intense deformation zones to form around the larger precipitates. These regions then provide sites at which recrystallization occurs during subsequent solution treatment leading to a fine grain size. This treatment also causes the large precipitates to redissolve. The alloy is then water quenched and artificially aged in the usual way to develop the normal high-strength properties.

The major benefits of such a fine-grained 7475 alloy are improved formability and resistance to stress-corrosion cracking. The microstructure is amenable to a degree of superplasticity at appropriate strain rates and temperatures so that elongations exceeding 500% have been achieved. The process presents some handling problems but it does make possible the superplastic forming of complex airframe shapes in conventional high-strength sheet alloys.

In each of the examples mentioned above, the strain rates needed to achieve superplasticity are quite slow (e.g. 5×10^{-3} s^{-1} to 10^{-4} s^{-1}) which limits productive output. Moreover, although tooling is usually inexpensive, the superplastic aluminium alloys themselves usually carry a significant cost penalty

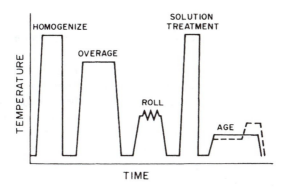

Fig. 3.63 Schematic diagram of a thermomechanical treatment (ITMT) used to achieve very fine grain size in alloy 7475 (from Paton, N. E. *et al., Met. Trans.* A, **12A**, 1267, 1981).

when compared with conventional sheet alloys, many of which possess relatively good cold forming characteristics. To date, these factors have limited applications to niche markets, notably in the aircraft industry. What is required is the development of less expensive alloys that exhibit superplastic deformation at higher strain rates of 10^{-2} s^{-1} or less. If such alloys become available, then there are opportunities to expand to mass produce components that are normally difficult to form, particularly in the automotive industry.

3.6.9 Electrical conductor alloys

The use of aluminium and its alloys as electrical conductors has increased significantly in recent decades, due mainly to fluctuations in the price and supply of copper. The conductivity of electrical conductor (EC) grades of aluminium and its alloys average about 62% that of the International Annealed Copper Standard (IACS) but, because of its lower density, aluminium will conduct more than twice as much electricity for an equivalent weight of copper. As a consequence, aluminium is now the least expensive metal with a conductivity high enough for use as an electrical conductor and this situation is unlikely to change in the future.

Aluminium has virtually replaced copper for high voltage overhead transmission lines although the relatively low strength of the EC grades requires that they be reinforced by including a galvanized or aluminium-coated high-tensile steel core with each cable. Aluminium is also widely used for insulated power cable, especially in underground systems. In this case, instead of the substitution of copper wires by aluminium, each strand of wire is usually replaced by a solid aluminium conductor which is continuously cast by the Properzi process (Section 3.1.1) and sector shaped by rolling (Fig. 3.64). Cable manufacture is thus simplified and economies are also achieved with insulating materials. For other applications, e.g. wiring for electric motors, communication cables or power supply to buildings, properties such as tensile strength and ductility also become critical requirements and growth in the use of aluminium has been slower.

Stronger alloys such as some heat-treatable Al–Mg–Si compositions have been used as electrical conductors for a number of applications but their conductivity is relatively low (55% IACS or less). In general, alloying elements which are either in solid solution or present as finely dispersed precipitates cause significant increases in resistivity so that these methods of strengthening tend to be unacceptable for aluminium conductors. Work-hardening is less deleterious in this regard, but conductors strengthened in this way tend to exhibit poor thermal stability and may be susceptible to mechanical failure in service. For these reasons, attention has been directed to alternative methods by which adequate strengthening can be achieved whilst retaining a relatively high electrical conductivity (>60% IACS).

As mentioned in Section 2.1.2, aluminium is amenable to substructure strengthening and much of the developmental work has been directed at

Fig. 3.64 An underground cable consisting of four-core solid aluminium conductors, insulated with PVC and enclosed in lead, then in steel to protect against mechanical damage, and lastly in PVC as an outer layer.

stabilizing the substructure with low volume fractions of finely dispersed intermetallic compounds. These compounds also assist in improving ductility in the final product by increasing strain-hardening which delays localized deformation and necking. It is necessary for the compounds to be uniformly distributed and this presents some difficulties because they precipitate in the interdendritic regions during casting. Casting methods must be used which ensure a rapid rate of solidification as this reduces the dendrite arm spacing and refines the microstructure. Extensive deformation during rod production and wire drawing further assists in distributing the compounds. Rod production involves hot-working at a temperature at which dynamic recovery occurs during processing, thereby ensuring the formation of subgrains rather than cells, and the work-hardening exponent n approaches 0.5.

The general trend to increase the level of electrical currents carried by transmission lines leads to higher operating temperatures and may cause softening through changes to alloy microstructure. Zirconium in solid solution is known to have a particularly strong effect in stabilizing grain structure by raising recrystallization temperatures but this element causes a marked decrease in electrical conductivity (1% reduction for each 0.03% present). Work in Japan has shown that this effect may be partly offset by adding trace amounts of rare earth elements, such as yttrium, and electrical conductors are available with a rating of 60% IACS which are capable of continuous operation at 150 °C.

Table 3.13 Compositions and typical properties of some aluminium alloy electrical conductor wires (from Starke, E. A. Jr., *Mater. Sci. Eng.*, **29**, 99, 1977)

Alloy		Yield strength (MPa)	Tensile strength (MPa)	Elongation in 250 mm (%)	Electrical conductivity (% IACS)
Old alloys	EC (99.6 Al)	28	83	23	63.4
	5005–H19 (Al–0.8Mg)	193	200	2	53.5
	6201–T81(Al–0.75Mg–0.7Si)	303	317	3	53.3
New alloys	Triple E (Al–0.55Fe)	68	95	33	62.5
	Super T (Al–0.5Fe–0.5Co)	109	129	25	61.1
	8076 (Al–0.75Fe–0.15Mg)	61	109	22	61.5
	Stabiloy (Al–0.6Fe–0.22Cu)	54	114	20	61.8
	Nico (Al–0.5Ni–0.3Co)	68	109	26	61.3
	8130 (Al–0.6Fe–0.08Cu)	61	102	31	62.1

Note: All alloys except 5005 and 6201 are in annealed condition (O-temper)

The compositions and properties of some newer conductor alloys are compared with those of the older EC materials in Table 3.13. Additions such as iron and nickel have been selected because they have a very low solubility in aluminium, they form stable compounds and they cause relatively small increases in resistivity when out of solution. The presence of magnesium in the alloy 8076 leads to improved creep resistance. The success of the new products has been reflected in their increased usage for electrical wire.

3.6.10 Electric storage batteries

As mentioned in Chapter 1, a large amount of electrical energy is required to reduce alumina to aluminium. Aluminium has therefore been regarded as an energy bank, providing this energy can be released by electrochemical conversion to aluminium hydroxide. A particularly attractive way of making this conversion is to couple an aluminium anode through an aqueous electrolyte to an air electrode which is supplied with an inexhaustible supply of the cathode reactant (oxygen) from air. In such an arrangement the theoretical specific energy (W h g^{-1}) of aluminium is 8.1. Only lithium has a higher value but lithium anodes are difficult to fabricate and use safely.

The concept of the aluminium/air battery is simple and is shown in Fig. 3.65. The electrolyte may be either a neutral chloride (e.g. NaCl) or an alkali (e.g. NaOH) and the net reaction is: $4Al + 6H_2O + 3O_2 \rightarrow 4Al(OH)_3$. Refuelling is also conceptually easy and quick since it involves only the regular addition of water and removal of the solid reaction products, together with the occasional replacement of the aluminium anodes. Major reasons for delays in the commercial exploitation of the system have been the development of cheap and efficient electrodes as well as a convenient system for separating and removing the reaction products.

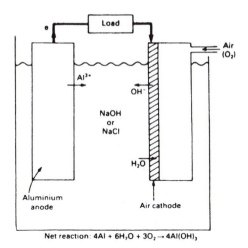

Fig. 3.65 Concept of aluminium/air electric storage battery (Fitzpatrick, N. and Scamans, G. M., *New Scientist*, No. 1517, 17 July 1986).

The fact that the aluminium anode is protected by an adherent, insulating oxide film has been a key barrier to overcome before controlled dissolution can occur in an electrolyte. Initially it was necessary to use super-purity aluminium (99.995%) to which was added small amounts (e.g. 200 p.p.m.) of so-called activating elements such as tin, indium or gallium. Experiments have revealed that, when an aluminium/air battery is operating, the current actually comes from very small areas on the surface of the anode. Modification of aluminium electrochemistry occurs when these activating elements are in solid solution, and dissolution at appropriate anode potentials results in the rejection and deposition of the more noble alloying element on to the anode surface. Once this has occurred, rapid surface diffusion and agglomeration of these atoms appears to take place under the aluminium oxide film. Localized defilming results allowing rapid dissolution of aluminium by galvanic action rather than by the much slower process of ionic transport through the oxide layer. If this so-called 'superactive' state can be sustained, the anodes show high coulombic efficiencies over a wide range of current density (15 to >1000 mA cm^{-2}). What needs to be avoided is the 'hyperactive' state in which current densities may reach 4 A cm^{-2} and undesirable copious amounts of hydrogen are evolved.

An Al–Sn–Mg anode exhibiting the desired characteristics was patented by Alcan in 1988 and development since then has concentrated on strategies that would allow the use of lower-purity aluminium. Essentially this has involved coping with iron impurities and up to 40 p.p.m. can now be tolerated which has more than halved the cost of the aluminium.

The air cathodes have also presented significant manufacturing problems. One form consists of a current collector composed of a nickel wire mesh on to which is pressed a mixture of carbon and fluorocarbon resin (Fig. 3.66).

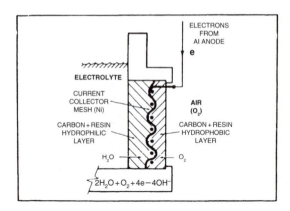

Fig. 3.66 Schematic design of an air cathode for an aluminium/air electric storage battery (Fitzpatrick, N. and Scamans, G. M., *New Scientist*, No. 1517, 17 July 1986).

Oxygen from the air can diffuse into the carbon so that it comes into contact with the electrolyte. The coatings are formulated so that the electrolyte penetrates the cathode through the hydrophilic layer to the mesh, but cannot leak through the hydrophobic layer to the air. Catalysts are incorporated in the cathode to improve performance. It has also been necessary to develop a method for continuous production of the cathodes from low-cost materials. This has been achieved by rolling the various layers together as a laminate and cathodes 280 mm wide and only 0.5 mm thick have been produced in rolls up to 90 m in length.

Individual cells generate approximately 1.4 V and they are connected in series to give the desired power output. Efficiency is limited by the electrolyte mainly because the dissolved aluminium forms soluble complexes with either chloride or hydroxyl ions, thereby showing down precipitation of the aluminium hydroxide. Precipitation can be encouraged by additives and by seeding in a crystallizer through which the electrotype must be pumped in order to be regenerated.

Interest in the aluminium/air battery was first stimulated by its potential as a power unit for electric vehicles. Using an alkaline electrolyte, this battery was seen as a viable alternative to the internal combustion engine so far as acceleration, refuelling time and range were concerned. Specific energy yields of around 4 W h g^{-1} were obtained and it has been claimed that such a battery would be capable of providing the power needed to drive a conventional sized motor car some 400 km between stops for water to replenish the alkaline electrolyte, and 2000 km before more aluminium was needed. To date, however, the system has not proved to be competitive with engines powered by petroleum fuels. One limited application has been the use of aluminium/air batteries as reliable and compact reserve units to back up d.c. electrical systems. In this regard, they provide a quiet and clean alternative to more costly diesel generators.

FURTHER READING

Altenpohl, D., *Aluminium : Technology, Applications and Environment* 6th Ed., TMS, Warrendale, Pa, USA, 1998

Davis, J.R. and Associates (Eds.), *ASM Specialty Handbook on Aluminium Alloys*, ASM International, Materials Park, Ohio, USA, 1993

Hatch, J.E. (Ed), *Aluminium : Properties and Physical Metallurgy*, ASM, Cleveland, Ohio, USA, 1984

Mondolfo, L.F., *Aluminium Alloys: Structure and Properties*, Butterworths, London, 1976

Van Horn, K.R. (Ed.), *Aluminium*, Vols. **1–3**, ASM, Cleveland, Ohio, USA, 1967

Kim. N.J., Designing with aluminium alloys, *Handbook of Mechanical Alloy Design*, Eds. Totten, G.E., Xie, L. and Funatani, K., Marcel Dekker Inc., New York, USA, 441, 2004

Greer, A.L., Grain refinement of alloys by inoculation of melts, *Phil. Trans. R. Soc. Lond. A*, **361**, 479, 2003

Quested, T.E. Understanding mechanisms of grain refinement of aluminium alloys by inoculation, *Mater. Sci. and Tech.*, **20**, 1357, 2004

International Alloy Designations and Chemical Composition Limits for Wrought Aluminium and Wrought Aluminium Alloys, The Aluminium Association Inc., New York, USA, 2004

Tempers for Aluminium Alloy Products, The Aluminium Association, New York, USA, 2004

Nie, J.-F., Morton, A.J. and Muddle, B.C., (Eds.), *Proc. 9th Inter. Conf. on Aluminium Alloys*, Brisbane, Australia, Inst. Materials. Eng. Australasia, 2004

Gregson, P.J. and Harris, S.J., *Proc. 8th Inter. Conf. on Aluminium Alloys*, Cambridge, UK, Trans Tech Publications, 2002. *Mater. Sci. Forum* Vols. **396–402**, 2002

Starke Jr, E.A, Sanders Jr, T.H. and Cassada, W.A., *Proc. 7th Inter. Conf. on Aluminium Alloys*, Charlottesville, Va, USA, Trans Tech Publications, 2000. *Mater. Sci. Forum*, Vols. **331–337**, 2000

Sato, T., Kumai, S., Kobayashi, T. and Murakami, Y., *Proc. 6th Inter. Conf. on Aluminium Alloys*, Toyohashi, Japan, Japan Inst. of Light Metals, 1998

Driver, J.H., Dubost, B., Durand, F., Fougeres, R., Guyot, P., Sainfort, P. and Suery, M., *Proc. 5th Inter. Conf. on Aluminium Alloys*, Grenoble, France, Trans Tech Publications, 1996. *Mater. Sci. Forum*, Vols. **217–222**, 1996

Wanhill, R.J.H., The status of Al–Li alloys, *Inter. J. Fatigue*, **16**, 3, 1994

Mishra, R.S., Friction stir processing technologies, *Adv. Mater.* and *Structures*, **152**, October, 43, 2003

Zhao, H., White D.R. and DebRoy, T., Laser welding of automotive aluminium alloys, *Inter. Mater., Rev.*, **44**, 238, 1999

Karlsson A., Braze clad aluminium materials for automotive heat exchangers – some recent developments, *Inter. Conf. on Aluminium, INCAL '98*, New Dehli, India, 1998

Staley, J.T., History of wrought aluminium alloy development, *Treatise on Materials Science and Technology* Vol. **31**, *Aluminium Alloys – Contemporary Research and Applications*, Vasudevan, A.K. and Doherty, R.D. (Eds.), Academic Press, 3, 1989

Sanders, R.E., Baumann, S.F. and Stumpf, H.C., *ibid.*, p. 65

Starke Jr, E.A. and Staley, J.T., Application of modern aluminium alloys to aircraft, *Prog. Aerospace Sci.*, **32**, 131, 1996

Williams, J.C. and Starke Jr, E.A., Progress in structural materials for aerospace systems, *Acta Mater.*, **51**, 5775, 2003

Davies, G., *Materials for Automobile Bodies*, Elsevier, London, 2003

Bottema J. and Miller, W.S., Aluminium in transport replacing weight with intelligence, *Materials Australasia*, **37**, No. 2, 15, 2004

Bryant, A.J., Aluminium at sea, *Light Metal Age*, **59**, April, 48, 2001

Hosford, W.F. and Duncan, J.L., The aluminium beverage can, *Scientific American*, **271**, (3), 48, 1994

Pratt, G.C. Materials for plain bearings, Inter, Met. Rev., **18**, 62, 1973

Grimes, R., Superplastic forming : evolution from metallurgical curiosity to major manufacturing tool, *Mater. Sci. and Tech.*, **19**, 3, 2003

Tuck, C.D.S.(Ed.) *Modern Battery Technology*, Ellis Horwood, New York, USA, 487, 1991

4

CAST ALUMINIUM ALLOYS

Aluminium is one of the most versatile of the common foundry metals and the ratio of cast to wrought aluminium alloy products is increasing primarily because of the larger amounts of castings being used for automotive applications. This ratio varies from country to country and in 2004 it was approximately 1:2 in North America. A wide range of cast aluminium alloys is available for commercial use and, for example, nearly 300 compositions were registered with the United States Aluminium Association in 2005. The most widely used are those based on the Al–Si, Al–Si–Mg and Al–Si–Cu systems. In general, alloys are classed as "primary" if prepared from new metals and "secondary" if recycled materials are used. Secondary alloys usually contain more undesirable impurity elements that complicate their metallurgy and often lead to properties inferior to those of the equivalent primary alloys.

The most commonly used processes are sand casting, permanent mould (gravity die) casting, cold chamber and hot chamber pressure die casting. Sand moulds are fed with molten metal by gravity. The metal moulds used in permanent mould casting are fed either by gravity or by using low-pressure air or other gas to force the metal up the sprue and into the mould. In pressure diecasting, molten aluminium is forced into a steel die through a narrow orifice (or gate) at high speeds ranging from 20 to 100m/s (Fig. 5.7). This is achieved by means of a piston and cylinder (or hot sleeve) where the piston is driven by a hydraulic ram capable of exerting a pressure of up to 100MPa on the metal. The aim is to continue feeding the casting as it solidifies rapidly in the die. General features about the melting and alloying of aluminium as well as grain refinement, where needed, are similar to those described in Chapter 3.

Apart from light weight, the special advantages of aluminium alloys for castings are the relatively low melting temperatures, negligible solubility for all gases except hydrogen, and the good surface finish that is usually achieved with final products. Most alloys also display good fluidity and compositions can be selected with solidification ranges appropriate to particular applications. The

205

major problem with aluminium castings is the relatively high shrinkage that occurs in most aluminium alloys during solidification. Allowance for this must be made in the design of moulds in order to achieve dimensional accuracy in castings, and to avoid or minimize hot tearing, residual stresses and shrinkage porosity. These problems are considered in Section 4.1.3.

As with wrought materials, there are cast alloys which respond to heat treatment and these are discussed below. It should be noted, however, that pressure diecastings are not normally solution treated because blistering may occur due to the expansion of air entrapped during the casting process. Moreover, there is the possibility of distortion as residual stresses are relieved.

In all areas, except creep, castings normally have mechanical properties that are inferior to wrought products and these properties also tend to be much more variable throughout a given component. As it is common practice to check tensile properties by casting separate test bars, it is necessary to bear in mind that the results so obtained should be taken only as a guide. The actual properties of even a simple casting may be 20–25% lower than the figures given by the test bar.

The demand for a greater assurance in meeting a specified level of mechanical properties within actual castings has led to the concept of 'premium quality' castings and represents a major advance in foundry technology. Specifications for such castings require that guaranteed minimum property levels are met in any part of the casting and match more closely those obtained in test bars. Mechanical properties previously thought unattainable have been achieved through strict control of factors such as melting and pouring practices, impurity levels, grain size and, in the case of sand castings, the use of metal chills to increase solidification rates. Certain radiographic requirements may also need to be met. Moreover, castings are submitted in lots and if one casting selected at random fails to meet specification requirements, all are rejected. Premium quality castings are more expensive to produce although they may be cost-effective if wrought components can be replaced. Some of the improved procedures have been translated into more general foundry practices.

Because of its relatively low melting point and general ease of handling, aluminium has been a model material for developing several novel casting processes. Examples are rheocasting and squeeze casting which both promote improved mechanical properties in products. These are discussed in Section 4.6. Another important development in which special attention has been paid to aluminium alloys has been the computer simulation of solidification processes. Applied initially to predicting heat transfer and freezing patterns, this technique is now being used to model most aspects of the casting process including the computer-aided design of moulds, evolution of microstructures and the estimation of thermal stresses generated during solidification. Software packages are available to assist with gating and feeder design, and to describe fluid flow during mould filling. It is also possible to assess the performance of a casting process relative to the geometry of an engineered cast component before the first melt is poured.

4.1 DESIGNATION, TEMPER AND CHARACTERISTICS OF CAST ALUMINIUM ALLOYS

No internationally accepted system of nomenclature has so far been adopted for identifying aluminium casting alloys. However, the Aluminium Association of the United States has introduced a revised system which has some similarity to that adopted for wrought alloys and this is described below. Details are also given of the traditional British system.

4.1.1 United States Aluminium Association system

This Association now uses a four-digit numerical system to identify aluminium and aluminium alloys in the form of castings and foundry ingot. The first digit indicates the alloy group in Table 4.1

In the 1xx.x group, the second two digits indicate the minimum percentage of aluminium, e.g. 150.x indicates a composition containing a minimum of 99.50% aluminium. The last digit, which is to the right of the decimal point, indicates the product form with 0 and 1 being used to denote castings and ingot respectively.

In the 2xx.x to 9xx.x alloy groups, the second two digits have no individual significance but serve as a number to identify the different aluminium alloys in the group. The last digit, which is to the right of the decimal point, again indicates product form. A modification to the above grouping of alloys is used in Australia in which the 6xx.x series is allocated to Al–Si–Mg alloys.

When there is a modification to an original alloy, or to the normal impurity limits, a serial letter is included before the numerical designation. These letters are assigned in alphabetical sequence starting with A but omitting I, O, Q and X, the X being reserved for experimental alloys. The temper designations for castings are the same as those used for wrought products (see Fig. 3.16). This does not apply for ingots.

Table 4.1 Four-digit system for aluminium and its alloys

	Current designation	Former designation
Aluminium, 99.00% or greater	1xx.x	
Aluminium alloys grouped by major alloying elements:		
Copper	2xx.x	1xx
Silicon with added copper and/or magnesium	3xx.x	3xx
Silicon	4xx.x	1 to 99
Magnesium	5xx.x	2xx
Zinc	7xx.x	6xx
Tin	8xx.x	7xx
Other element	9xx.x	7xx
Unused series	6xx.x	

4.1.2 British system

Most alloys are covered by the British Standard 1490 and compositions for ingots and castings are numbered in no special sequence and have the prefix LM. The condition of castings is indicated by the following suffixes:

M   as-cast
TB   solution treated and naturally aged (formerly designated W)
TB7   solution treated and stabilized
TE   artificially aged after casting (formerly P)
TF   solution treated and artificially aged (formerly WP)
TF7   solution treated, artificially aged and stabilized (formerly WP-special)
TS   thermally stress-relieved.

The absence of a suffix indicates that the alloy is in ingot form.

There are also some aerospace alloys which are covered separately by the L series of British Standards and by the DTD specifications. These specifications, which are not systematically grouped or numbered, are concerned with specific compositions and often with particular applications. In both cases, the condition of the casting is not indicated by a suffix and is identified only by the number of the alloy.

4.1.3 General characteristics

In general terms, an alloy can be described as being castable if it consistently and reliably produces castings that have sound microstructures and are dimensionally accurate for a wide range of products, processes and plants. The property of castability of an alloy is the sum of several factors of which the most notable are fluidity, mould-filling ability, volume shrinkage, susceptibility to hot tearing, porosity-forming characteristics and surface quality. These factors are discussed below. Some depend strongly on chemical composition although mould design and process conditions are frequently more important.

Aluminium foundries occasionally experience intermittent outbreaks of high reject rates because of porosity and shrinkage defects even though the chemical composition of an alloy appears to meet the require specification. In some instances, the batch of aluminium appears to be the culprit and it should be noted that several sources are used by foundries including:

1. Primary metal ingots supplied from a smelter as ingots that are provided in grades usually ranging from 99.5% to 99.85% aluminium, or as pre-alloyed ingots.
2. Secondary metal ingots prepared by melting mixtures of recycled aluminium alloy products adjusted in composition to meet particular specifications.
3. Molten metal delivered directly from a smelter or secondary producer to the foundry.
4. In-house scrap returns.

The use of primary aluminium minimizes remelting and therefore reduces the formation of surface dross which is a mixture of aluminium metal, aluminium oxide and other melt reaction products. Secondary metal is cheaper to purchase because remelting of aluminium scrap consumes only about 5% of the energy needed to extract aluminium from its ore. However, impurity levels inevitably are higher and there is also a danger that undesirable elements may accumulate. The additional remelting associated with the use of secondary metal increases both the loss of aluminium to dross and the opportunity for the entrainment of oxide. Because of this many foundries are adopting the practices of degassing and filtering melts prior to casting.

Fluidity The fluidity of an alloy is commonly measured by pouring into a sand-moulded spiral and alloys are compared by measuring the respective "running lengths" of travel before solidification occurs. Composition influences fluidity due to the effects of alloying elements on viscosity, surface tension, freezing range and mode of solidification. Fluidity is reduced as the purity of aluminium decreases and alloying elements are added. This is due mainly to a widening of the freezing range, the formation of intermetallic compounds, and to changes in the solidification pattern from a planar front to a mushy mode. This situation reverses close to a eutectic composition when fluidity is usually at a maximum.

Another aspect of fluidity is the ability to feed the interdendritic spaces remaining once the mould has been filled, but solidification is incomplete. As these spaces between the dendrites continue to narrow, melt viscosity and surface tension become more important and the pressure head needed to maintain flow increases. Volumetric shrinkage assists in this regard as normally there is an overall contraction during the change from liquid to solid.

Volumetric shrinkage Total volumetric shrinkage during solidification also depends on alloy composition since each phase present has its own density characteristics. Shrinkage amounts to 6% for pure aluminium to as little as 1–2% for hypereutectic Al–Si alloys. Most alloying elements cause less change. Shrinkage may be evident as an overall contraction, as localised effects at surfaces, and as internal defects such as large isolated voids, interconnected porosity, or microporosity. The actual type that may occur in a casting depends on alloy composition, mould design, cooling rate and mode of solidification. In pressure die castings, shrinkage will also occur if metal in the gate freezes before feeding of the casting has been completed.

Porosity Most casting processes result in some internal porosity in a casting. This problem is usually most apparent in pressure die castings because of the speed of the operation, turbulent metal flow and the opportunity to entrap gases. As mentioned earlier, it is because of the presence of porosity that die castings cannot normally be solution treated since this leads to surface blistering. In general, porosity in castings must be controlled and minimized so as to avoid detrimental effects on mechanical properties, pressure tightness, and surface appearance, particularly if machining is necessary. By attention to mould design and control of casting conditions, every effort is made to concentrate porosity in risers, or to redistribute it uniformly throughout a casting as less harmful micropores. Porosity arises primarily from shrinkage, the formation of internal oxide films, and from the entrapment of gases (air, steam, dissolved hydrogen, or products derived from the burning of organic lubricants). Porosity is usually worst in pressure die castings because of the highly turbulent metal flow and rapid solidification rates, although the entrapment of gases can be overcome by evacuating the mould prior to the entry of the molten metal. With regard to alloy composition, porosity formation is closely related to the extent of the freezing range. Long ranges tend to result in the formation of dispersed, interdendritic or intergranular micropores, whereas short freezing ranges may promote formation of more localised regions macro-shrinkage or larger intergranular pores. Porosity is also increased by the presence of intermetallic compounds with morphologies that may impede interdendritic feeding during casting. Examples are needles of the compound Al_5FeSi and $Al_{15}(MnFe)_3Si_2$ which has the appearance of Chinese script.

Hot tearing Hot tearing (or hot shortness) arises when the tensile stresses generated within a solidifying casting exceed the fracture stress of the partially solidified metal. The result is usually evident as surface tearing and cracking which tend to occur at the sites of local hot spots where the casting is physically restrained. Generally alloys are more susceptible if they have wide solidification ranges and low volumes of eutectics that can fill and repair hot tears. Alloys based on the Al–Cu and Al–Mg system are the most prone to hot tearing whereas those based on Al–Si show the highest resistance.

Die soldering A particular problem with the pressure die casting of aluminium alloys is the tendency for the casting to stick (solder) to ferrous dies during solidification because of the high affinity iron and molten aluminium have for each other. Rapid interatomic diffusion can occur when they come into contact resulting in the formation of a series of intermetallic compounds at the die surface. Die soldering may prevent automatic ejection of a casting which then has to be removed manually. This can cause delays and serious cost penalties, particularly if die surfaces are damaged and require repair. Soldering is influenced by several factors:

(i) Die casting alloy composition. Iron has the greatest beneficial effect and increasing amounts up to saturation level progressively alleviate soldering. For the widely used Al–Si casting alloys, the critical level of iron that is required can be estimated by the empirical relationship:

$$Fe_{crit} \approx 0.075 \ [Si\%] - 0.05$$

Manganese and small amounts of titanium are also beneficial, whereas the presence of nickel has been found to be detrimental. Magnesium, silicon and strontium have no significant effect.

(ii) Die steel composition. The plain carbon steel H–13 (Fe–0.35C–0.35Mn) is commonly used. More highly alloyed compositions can better resist attack by molten aluminium but are often not used because of cost of the steels.

(iii) Melt injection temperature. This should as low as possible and as upper limit of around 670 °C before the molten metal enters the shot sleeve has been found to be critical minimizing sticking to some die steels.

(iv) Use of die surface coatings. Such coatings can serve as diffusion barriers to the iron-aluminium reaction and thereby retard soldering. A number of coatings are used by the die casting industry and boron-based ceramics have proved particularly effective. Boron nitride deposited by physical vapour depositon (Section 7.7) is also provide a barrier providing it is compatible with the steel substrate. An alternative method is to aluminize the surface by immersing the die in molten aluminium at 760 to 800 °C after which it is removed, held for 12 to 24 h at 300 °C to form a controlled layer of iron aluminide, and then the surface is highly polished.

The features of the different types of aluminium casting alloys are discussed in succeeding sections and are classified according to the major alloying element that is present, e.g. Al–Si, Al–Mg etc. Because of the problems of identification in different countries, individual alloys are represented by their basic compositions. It will be noted that castings in commercial-purity aluminium are not considered separately because their only major use is for certain electrical applications which were discussed in Chapter 3.

Although many different casting alloys are available, the number used in large quantities is much smaller. In Britain, for example, most cast aluminium alloy components are made from only four alloys designated LM2, LM4, LM6, and LM21. A representative list of alloy compositions is given in Table 4.2 together with the type of casting process for which they are used. Some mechanical properties and the relative ratings of the various casting characteristics are given in Table 4.3.

Selection of alloys for castings which are produced by the various casting processes depends primarily upon composition which, in turn, controls characteristics such as solidification range, fluidity, susceptibility to hot-cracking, etc. Sand castings impose the least limitation on choice of alloy and commonly used alloys are 208 (Al–4Cu–3Si), 413 (Al–11.5Si), 213 (Al–7 Cu–2Si–2.5Zn) and 356 (Al–7Si–0.3Mg). Alloys 332 (Al–9Si–3Cu–1Mg) and 319 (Al–6Si–4Cu) are favoured for permanent mould castings whereas 380 (Al–8.5Si–3.5Cu) and 413 (Al–11.5Si) are most commonly used for pressure

Table 4.2 Compositions of selected aluminium casting alloys

Association number	BS 1490 LM number	Casting process	Si	Fe	Cu	Mn	Mg	Cr	Ni	Zn	Ti	Other
150.1	LM 1	Ingot	†	†	0.10							99.5 Al min
201.0		S	0.10	0.15	4.0–5.2	0.20–0.50	0.15–0.55			0.05	0.15–0.35	Ag 0.40–1.0
208.0		S	2.5–3.5	1.2	3.5–4.5	0.50	0.10		0.35	1.0	0.25	
213.0		PM	1.0–3.0	1.2	6.0–8.0	0.6	0.10		0.35	2.5	0.25	
	LM 4	S and PM	4.0–6.0	0.8	2.0–4.0	0.20–0.6	0.15		0.30	0.50	0.20	
238.0		PM	3.5–4.5	1.5	9.0–11.0	0.6	0.15–3.5		1.0	1.5	0.25	
242.0	LM 14	S and PM	0.7	1.0	3.5–4.5	0.35	0.15–0.35		1.7–2.3	0.35	0.35	
295.0		S	0.7–1.5	1.0	4.0–5.0	0.35	0.03	0.25		0.35	0.25	
308.0		PM	5.0–6.0	1.0	4.0–5.0	0.50	0.10			1.0	0.25	
319.0	LM 21	S and PM	5.5–6.5	1.0	3.0–4.0	0.50	0.10		0.35	1.0	0.25	
328.0		S	7.5–8.5	1.0	1.0–2.0	0.20–0.6	0.20–0.6	0.35	0.25	1.5	0.25	
A332.0	LM 13	PM	11.0–13.0	1.2	0.50–1.5	0.35	0.7–1.3		2.0–3.0	0.35	0.25	
355.0	LM 16	S and PM	4.5–5.5	0.6‡	1.0–1.5	0.50‡	0.40–0.6	0.25		0.35	0.25	
356.0	LM 29	S and PM	6.5–7.5	0.6	0.25	0.35	0.20–0.40			0.35	0.25	
A356.0		S and PM	6.5–7.5	0.20	0.20	0.10	0.20–0.40			0.10	0.20	
357.0	LM 25	S and PM	6.5–7.5	0.15	0.05	0.03	0.45–0.60			0.05	0.20	Be 0.04–0.07

Alloy	BS (LM)	Process										Sn
360.0	LM 9	D	9.0–10.0	2.0	0.6	0.35	0.40–0.6		0.50	0.50		
380.0	LM 24	D	7.5–9.5	2.0	3.0–4.0	0.50	0.50	0.10	0.50	3.0		
A380.0	LM 24	D	7.5–9.5	1.3	3.0–4.0	0.10	0.10		0.50	3.0		
390.0	LM 30	D	16.0–18.0	1.3	4.0–5.0	0.10	0.45–0.65			0.10		
	LM 6	S, PM and D	10.0–13.0	0.6	0.10	0.50	0.10		0.10	0.10	0.20	
413.0	LM 20	D	11.0–13.0	2.0	1.0	0.35	0.10		0.50	0.50		
	LM 2	D	9.0–11.5	1.0	0.7–2.5	0.50	0.30		0.50	2.0	0.20	
443.0	LM 18	S	4.5–6.5	0.8	0.6	0.50	0.05	0.25		0.50	0.25	
514.0	LM 5	S	0.35	0.50	0.15	0.35	3.5–4.5		0.15	0.15	0.25	
518.0		D	0.35	1.8	0.25	0.35	7.5–8.5		0.15	0.15		
520.0	LM 10	S	0.25	0.30	0.25	0.15	9.5–10.6		0.15	0.15	0.25	
535.0		S	0.15	0.15	0.05	0.1–0.25	6.2–7.5				0.10–0.35	
705.0		S and PM	0.20	0.8	0.20	0.40–0.6	1.4–1.8	0.2–0.4		2.7–3.3	0.25	
707.0		S and PM	0.20	0.6	0.20	0.40–0.6	1.4–1.8	0.2–0.4		4.0–4.5	0.25	
712.0		PM	0.15	0.50	0.25	0.10	0.50–0.65	0.4–0.6	0.15	5.0–6.5	0.10–0.25	
713.0		S and PM	0.25	1.18	0.40–1.0	0.6	0.20–0.50	0.35	0.15	7.0–8.0	0.25	
850.0		S and P	0.7	0.7	0.7–1.3	0.10	0.10		0.7–1.3	0.20		5.5–7.0

Notes: Compositions are in % maximum by weight unless shown as a range

S = sand casting; PM = permanent mould (gravity die) casting; D = pressure diecasting

†Ratio Fe: Si minimum of 2:1

‡If iron exceeds 0.45% manganese content must be less than one-half the iron content

Table 4.3 Mechanical properties and foundry characteristics of selected casting alloys

Aluminium Association number	BS 1490 LM number	Casting process	Temper	0.2% proof stress (MPa)	Tensile strength (MPa)	Elongation (% in 50 mm)	Casting characteristics					
							Fluidity	Pressure tightness	Resistance to hot tearing	Corrosion resistance	Weldability	Machining
201.0		S†	T6	345	415	5.0	C	D	D	D	C	B
208.0		S	T533	105	185	1.5	B	B	B	D	B	C
213.0		PM	T533	185	220	0.5	B	C	C	E	C	B
	LM4	S and PM	T21	95	175	3.0	B	B	B	C	C	B
			T6	230	295	2.0						
242.0	LM14	PM	T6	230	295	1.0	C	C	D	D	D	B
295.0		PM	T61	195	260	4.0	C	D	D	D	C	B
319.0	LM21	S	T21	125	185	1.0	B	B	B	C	B	C
		PM	T21	125	200	2.0						
332.0	LM13	S and PM	T61	240	260	0.5	B	C	A	C	B	B
			T65	295	325	0.5						
355.0	LM16	S	T4	125	210	3.0	B	B	B	C	B	B
		PM	T4	140	245	6.0						
		PM	T6	235	280	1.0						
356.0	LM25	S	T6	205	230	4.0	B	A	A	B	B	C
		PM	T6	225	240	6.0						
357.0		S	T6	275	345	3.0	B	A	A	B	B	C
		PM	T6	295	360	6.0						

Alloy No.	Alloy	Process	Condition	0.2% PS	Tensile	Elong. %						
360.0	LM9	S	T5	110	185	2.0	A	B	A	B	B	C
		S	T6	215	255	—						
		PM	T5	130	245	2.5						
		PM	T6	265	310	1.0						
	LM6	S	F1	65	185	8.0	A	B	A	A	B	C
		PM	F1	90	205	9.0						
		D	F1	130	250	2.5						
413.0	LM20	D	F1	140	265	2.0	A	A	A	B	A	C
443.0	LM18	S	F1	65	130	5.0	B	A	A	A	B	C
		PM	F1	70	160	6.0						
514.0	LM5	S	F1	80	170	5.0	C	D	C	A	C	B
		PM	F1	80	230	10.0						
518.0		D	F1	130	260	10.0	D	E	C	A	A	A
520.0	LM10	S	T1	175	320	15.0	D	E	B	A	E	B
535.0		S	F	145	275	13.0	D	E	C	A	A	A
705.0		S	T1	130	240	9.0	D	C	E	B	D	A
707.0		S	T1	185	255	3.0	D	C	E	B	D	A
713.0		S	T5	175	235	4.0	D	C	E	B	D	A

Notes: S = sand casting; PM permanent mould (gravity die) casting; D = pressure diecasting

†Results for sand cast alloys obtained from separately cast test bars

Tensile properties for all alloys generally represent "best practice" in casting procedures

Ratings for casting characteristics A through to E in decreasing order of merit diecastings. In the latter case, the prime consideration is a low melting point which increases production rates and minimizes die wear.

In order of decreasing castability, the groups of alloys can be classified in the order 3xx, 4xx, 5xx, 2xx, and 7xx. Corrosion resistance is also a function of composition and the copper-free alloys are generally regarded as having greater resistance than those containing copper. The 8xx series is confined to Al–Sn bearing alloys which were discussed in Section 3.6.7.

4.2 ALLOYS BASED ON THE ALUMINIUM–SILICON SYSTEM

Alloys with silicon as the major alloying addition are the most important of the aluminium casting alloys mainly because of the high fluidity imparted by the presence of relatively large volumes of the Al–Si eutectic. Fluidity is also promoted because of the high heat of fusion of silicon ($\sim$1810 kJ/kg compared with $\sim$395 kJ/kg for aluminium) which increases "fluid life" (i.e. the distance the molten alloy can flow in a mould before being too cold to flow further), particularly in hypereutectic compositions. Other advantages of castings based on the Al–Si system are high resistance to corrosion, good weldability and the fact that the silicon phase reduces both shrinkage during solidification and the coefficient of thermal expansion of the cast products. However, machining may present difficulties because of the presence of hard silicon particles in the microstructure. Commercial alloys are available with hypoeutectic and, less commonly, hypereutectic compositions.

The eutectic is formed between an aluminium solid solution containing just over 1% silicon and virtually pure silicon as the second phase. The eutectic composition has been a matter of debate but recent experiments with high-purity binary alloys has shown it to be Al–12.6Si, with the transformation occurring at 577.6 °C. Slow solidification of a pure Al–Si alloy produces a very coarse microstructure in which the eutectic comprises large plates or needles of silicon in a continuous aluminium matrix (Figs. 4.1a and b). The eutectic itself is composed of individual cells within which the silicon particles appear to be interconnected. Alloys having this coarse eutectic exhibit low ductility because of the brittle nature of the large silicon plates. Rapid cooling, as occurs during permanent mould casting, greatly refines the microstructure and the silicon phase assumes a fibrous form with the result that both ductility and tensile strength are much improved. The eutectic may also be refined by the process known as modification.

4.2.1 Modification

The widespread use of Al–Si alloys for other types of castings derives from the discovery by Pacz in 1920 that a refinement or modification of microstructure, similar to that achieved by rapid cooling, occurred when certain alkali fluorides were added to the melt prior to pouring (Fig. 4.1c).

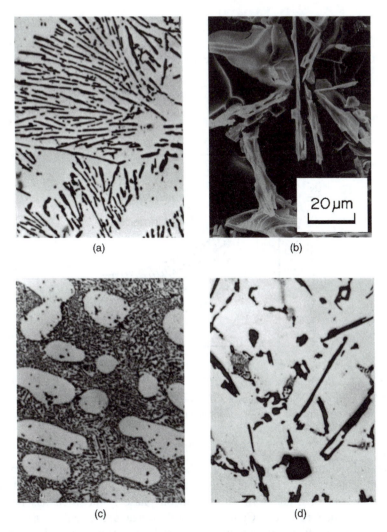

Fig. 4.1 As-cast binary Al–Si alloys in the following conditions: (a) 12% Si unmodified; (b) 7% Si unmodified (scanning electron micrograph); (c) 12% Si modified with sodium; (d) 12% Si with excess phosphorus [(a), (c), (d) × 400 (courtesy Alcoa of Australia Ltd), (b) × 600 (courtesy J. A. Cheng)]

As shown in Table 4.4, mechanical properties may be substantially improved due to refinement of the microstructure and to a change to a planar interface during solidification which minimizes porosity in the casting. Fracture toughness is also significantly raised (Fig. 4.2).

For many years modification of alloys based on the Al–Si system was achieved only by the addition of sodium salts or small quantities (0.005–0.015%)

Table 4.4 Mechanical properties of Al–13% Si alloys (from Thall, B. M. and Chalmers, B., J. Inst. Met., **77**, 79, 1950)

Condition	Tensile strength (MPa)	Elongation (%)	Hardness (Rockwell B)
Normal sand cast	125	2	50
Modified sand cast	195	13	58
Normal chill cast	195	3.5	63
Modified chill cast	220	8	72

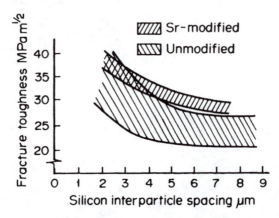

Fig. 4.2 Fracture toughness of alloy A357 (Al–7Si–0.5 Mg). Castings solidified over a range of cooling rates, with and without strontium modification (from Chadwick, G. A., *Metals and Materials*, **2**, 693, 1986).

of metallic sodium to the melt although the actual amount of sodium needed may be as little as 0.001%. The mechanism by which the microstructure and, more particularly, the size and form of the silicon phase is modified has been the subject of much research. Controversy still remains although most theories involve possible effects of sodium on the nucleation and/or growth of eutectic silicon during solidification.

Sodium may depress the eutectic temperature by as much as 12 °C and a finer microstructure is therefore to be expected because the rate of nucleation will be greater in the under-cooled condition. Depression of the eutectic temperature implies that sodium reduces the potency of nucleating sites for the eutectic phases, notably silicon. It is known that silicon itself is readily nucleated at the surface of particles of the compound AIP which is formed by reaction of aluminium with impurity amounts of phosphorus. In fact, an excessive level of phosphorus can lead to the formation of a third, granular type of microstructure containing large particles of silicon which results in poor mechanical properties (Fig. 4.1d). Accordingly, a possible explanation for the behaviour of sodium is that it neutralizes the effect of phosphorus, probably by the preferential formation of the compound NaP. In a similar way, the phenomenon known as over-modification, whereby coarse silicon particles may reappear when an excess of sodium is present, has been attributed to formation of another compound, AlNaSi, which once again provides sites for the easy nucleation of silicon.

Strong support for the alternative concept that sodium exerts its effect by restricting growth of silicon particles comes from the observation that these particles appear to be interconnected within each cell, suggesting that there is no need for repeated nucleation to occur after each cell has formed. Several theories have been proposed to account for the possible effect of sodium on the growth of silicon. For example, it has been suggested that sodium segregates at the periphery of growing silicon plates and prevents, or poisons, further growth. Another theory involves twinning of the silicon plates. Normally, crystal growth in diamond cubic systems, such as silicon, tends to be highly anisotropic, leading to the plate or flake form. If the plates are twinned, then the so-called twin plane re-entrant angle (TPRE) mechanism of growth may operate in which the grooves between the planes act as preferred sites for the attachment of silicon atoms. Growth is then promoted in other crystallographic directions. Twin density is known to be much greater in modified alloys and this is thought to produce numerous alternative growth directions thereby leading to the desirable fibrous form of the silicon. What is uncertain is the mechanism by which sodium promotes twinning. One suggestion is that the greater under-cooling known to occur in the modified alloys may induce significant stresses in the silicon plates due to the large (6:1) difference in the coefficients of thermal expansion between this phase and aluminium. Restricted growth theories do not, however, account for over-modification when coarsening of silicon occurs in the presence of an excess of sodium.

Taken overall, it does seem likely that modification of Al–Si alloys results in changes to both the processes of nucleation and growth of silicon in the eutectic and this conclusion is supported by results of a recent study of the effects of iron on its formation. Commercial alloys based on the Al–Si system invariably contain iron as an impurity and quite small amounts (0.0015% or more) will cause formation of particles of an intermetallic compound during solidification that has been identified as β–$Al_9Si_2Fe_2$. These particles can also serve as sites on which eutectic silicon can nucleate. If the subsequent growth of silicon into the residual eutectic liquid is uninhibited, it will adopt a coarse, platelike morphology. The addition of modifying agents such as sodium is known to increase the viscosity of the eutectic liquid, with the result that the wetting angle with the solid particles of β–$Al_9Si_2Fe_2$ is altered thereby preventing nucleation of silicon. Undercooling of the eutectic liquid then occurs leading to formation of arrays of solid α-aluminium grains through which silicon is forced to grow causing it adopt a the more desirable fibrous morphology.

The use of sodium as the modifying agent does present founding problems because the fluidity of the melt is reduced, but the major difficulty is its rapid and uncertain loss through evaporation or oxidation. It is therefore necessary to add an excess amount and difficulties in controlling the content in the melt can lead to the under- or over-modification of the final castings. For the same reason, the effects of modification are lost if Al–Si castings are remelted which prevents foundries being supplied with pre-modified ingots. Attention has therefore been directed at alternative methods and modification today is carried out mostly by using additions of strontium. The amounts needed vary with the silicon content of the alloy and range from 0.015 to 0.02% for hypoeutectic permanent mould castings and slightly more for sand castings that solidify more slowly, or for alloys with higher silicon contents. Strontium is added as an Al–Sr or Al–Si–Sr master alloy which also refines the Al–Si eutectic and results in castings having tensile properties comparable with those obtained when using sodium. Loss of strontium through volatilisation during melting is much less and the modified microstructure can be retained if alloys are remelted. Over-modification is also less of a problem with strontium because excess amounts are taken into compounds such as Al_3SrSi_3, Al_2SiSr_2 and Al_4Si_2Sr. Yet another advantage of strontium additions is that they suppress formation of primary silicon in hyper-eutectic compositions which may improve their ductility and toughness. This effect is not observed when sodium is used. Additions of antimony (e.g. 0.2%) also cause modification but result in a lamellar rather than a fibrous eutectic.

4.2.2 Binary Al–Si alloys

Binary Al–Si alloys up to the eutectic composition retain good levels of ductility, providing the iron content is controlled to minimize formation of large, brittle plates of the compound β–AlFeSi. In this regard, additions of manganese have been found to be beneficial because this element favours formation of the finer α–AlFeSi phase which has the so-called Chinese script morphology. An accepted rule in industry is that the Mn:Fe ratio needs to be at least 0.5:1 for β–AlFeSi to be suppressed, although the latter phase has been observed in alloys with Mn:Fe ratios as high as 1:1. However, such high levels of manganese can be a disadvantage since they may promote formation of greater volume fractions of inter-metallic compounds than are present for the same levels of iron. This follows because α–AlFeSi contains a lower atomic percentage of iron so that it is possible for larger colonies of this phase to form than the β–AlFeSi it replaces. Thus the preferred strategy is to keep the iron levels as low as possible in the first place.

If the silicon content is below 8%, modification is not necessary to achieve acceptable levels of ductility because the primary aluminium phase is present in relatively large amounts. The eutectic composition, which has a high degree of fluidity and low shrinkage on solidification, has particular application for thin-walled castings, e.g. Fig. 4.3. As a class, the alloys are used for sand and permanent mould castings for which strength is not a prime consideration,

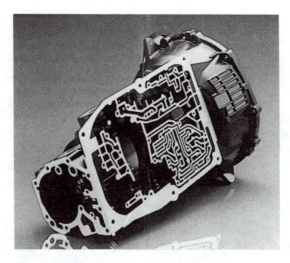

Fig. 4.3 Thin-walled cast Al–Si alloy automotive transmission casing (courtesy Vereinigte Aluminium-Werke, A. G., Bonn).

e.g. domestic cookware, pump casings and certain automobile castings, including water-cooled manifolds.

When as-cast alloys containing substantial amounts of silicon are subjected to elevated temperatures they suffer growth due to precipitation of silicon from solid solution. Dimensional stability can be achieved by heating for several hours in the temperature range 200–500 °C prior to subsequent machining or use, and tempers of the T5 or T7 types should be given to castings which are to be used at temperatures of 150 °C or above.

4.2.3 Al–Si–Mg alloys

Large quantities of sand and permanent mould castings are made from the Al–Si–Mg alloys 356 or LM25 (Al–7Si–0.3Mg) and 357 (Al–7Si–0.5Mg) in which the small additions of magnesium induce significant age hardening through precipitation of phases that form in the Al–Mg–Si system (Table 2.3). For example, the yield strengths of these alloys in the T6 condition are more than double that of the binary alloy containing the same amount of silicon. The alloys also respond well to secondary precipitation (Section 2.3.4). Moreover they display excellent corrosion resistance. Both alloys are used for critical castings for aircraft such as the engine support pylon shown in Fig. 4.4, and precision cast wing flap tracks for models of the European Airbus series. Automotive components include cylinder heads and wheels.

The critical nature of some of these applications has required more attention to be paid to relationships between microstructure and fracture characteristics. Tensile fracture of Al–Si–Mg castings has been shown to initiate mainly by

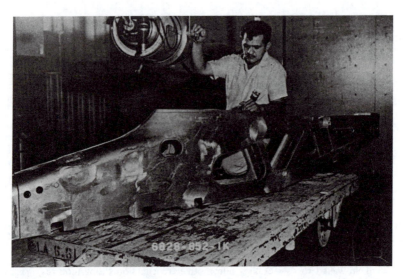

Fig. 4.4 Pylon for a fighter aircraft prepared by premium casting techniques (courtesy Defense Metals Information Center, Columbus, USA).

cracking of eutectic silicon particles as a result of stresses imposed during plastic deformation of the softer aluminium matrix. While solution treatment has the beneficial effect of fragmenting and spheroidizing the silicon particles in alloys that are well modified, it has little effect on the coarser particles in the unmodified condition. The higher magnesium content of the alloy 357 increases the response to age hardening and results in higher tensile properties but there is the disadvantage in that relatively large particles of an intermetallic compound π ($Al_9FeMg_3Si_5$) may form (Fig. 4.5). These particles tend to crack preferentially when the alloy is strained which may reduce the ductility of 357 when compared with the lower magnesium alloy 356. The π phase also removes magnesium so that the yield stress is less than expected. Another result is that, as shown for alloy 357 in Table 3.7, secondary precipitation can be used to increase the tensile properties and fracture toughness of both cast and wrought Al–Si–Mg alloys.

Studies of fatigue properties of the alloy 356 have revealed that, unlike wrought aluminium alloys, the relationship between alternating stress and time to failure is essentially the same for the as cast and heat treated (age hardened) conditions. This behaviour is illustrated in Fig. 4.6 in which specimens machined from continuously cast billets were tested in the as cast (AC) condition, and after solution treatment at 540 °C for times ranging from 1 to 200 h to change the size of the eutectic silicon phase. These alloys were then quenched into water at 80 °C, pre-aged at room temperature for 48 h, and either underaged 4 h at 155 °C or to peak strength after 6 h at 165 °C. Tensile properties varied greatly from 0.2% P.S. of 120 MPa and T.S. 250 MPa for the as cast state, to a 0.2% P.S. of 250 MPa and T.S. 320 MPa when aged to peak strength. Nevertheless, the fatigue results all fell on the same curve. As shown in the figure, this contrasts with

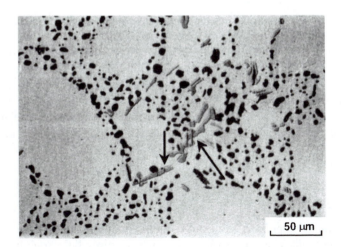

Fig. 4.5 Microstructure of alloy 357 showing relatively large π-phase particles together with smaller eutectic silicon particles. As shown by the arrows, the π particles have cracked preferentially when the cast alloy was deformed (courtesy Q.G. Wang).

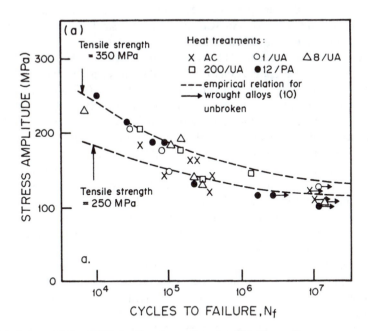

Fig. 4.6 Fatigue lifetime (S/N) data for the casting alloy 356 (R = −1).

AC = as cast condition. 1/UA, 8/UA, 200/UA = alloys solution treated for 1, 8 or 200 h at 540 °C, pre-aged 48 h at room temperature, and underaged 4 h at 155 °C. 12/PA = alloy solution treated 8 h at 540 °C, pre-aged 48 h at room temperature and aged 6 h at 165 °C (T6 temper). Wrought alloy curves shown as dotted lines. (Couper, M. J., Neeson, A. E., and Griffiths, J. R., *Fatigue Fract. Engng. Mater. Struct.*, **13**, 213, 1990).

results for a wrought alloy for which an increase in tensile strength from 250 to 350 MPa was accompanied by improved fatigue performance.

The insensitivity of fatigue strength of cast alloys to heat treatment arises because of fatigue cracks are always initiated at casting defects, notably shrinkage porosity and oxide films. Thus the fatigue strength of castings tends to be defined by the maximum size of defects whereas it is the volume fraction of defects that controls yield strength and ductility. Reducing the size of casting defects will increase fatigue life up to the stage that the initiation of cracks occurs normally in persistent slip bands at some surface (Section 2.5.3). This situation is considered to hold for all aluminium alloys cast under industrial conditions.

4.2.4 Al–Si–Cu alloys

The addition of copper to cast Al–Si alloys also promotes age hardening and increases strength. Copper also improves machinablitiy but castability, ductility and corrosion resistance are all decreased. Commercial Al–Si–Cu alloys have been available for many years and compromises have been reached between these various properties (Table 4.3). Compositions lie mostly within the ranges 3–10.5% silicon and 1.5–4.5% copper. The higher silicon alloys (e.g. Al–10Si–2Cu) are used for pressure diecastings, whereas alloys with lower silicon and higher copper (e.g. Al–3Si–4Cu) are used for sand and permanent mould castings. The strength and machinability of some of these castings is often improved by artificial ageing (T5 temper). In general the Al–Si–Cu alloys are used for many of the applications listed for the binary alloys but where higher strength is needed. One example is the use of alloy 319 (Al–6Si–3.5Cu) for die cast (permanent mould) automative engine blocks and cylinder heads in place of cast iron. As with the wrought alloys, some compositions contain minor additions of elements such as bismuth and lead which improve machining characteristics.

More complex compositions are available where special properties are required. One example is the piston alloys for internal combustion engines, e.g. A332(Al–12Si–1Cu–1Mg–2Ni) in which nickel, in particular, improves elevated temperature properties by forming stable intermetallic compounds that cause dispersion hardening. Another example is the range of hypereutectic compositions such as A390 (Al–17Si–4Cu–0.55Mg) which have been used for sand and permanent mould castings of all-aluminium alloy automotive cylinder blocks. Here the main direction of the developmental programmes has been the desire to eliminate using cast iron sleeves as cylinder liners which is the case in several production engines. Alloy 390 is also used for parts in automotive automatic transmissions and for casting numerous small engines for lawn mowers and chain saws. In a recent development, BMW in Germany has designed a light weight automobile engine that has cylinders cast in alloy A390 that serve as a core around which the engine block is sand cast in a magnesium alloy (Fig. 4.7).

In all these applications it is necessary to incorporate, in the eutectic matrix, sufficient quantities of hard primary silicon particles to provide high wear resistance in the cylinders during service, and yet keep the dispersion low

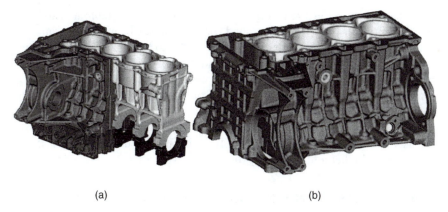

(a)   (b)

Fig. 4.7 Lightweight BMW magnesium alloy automotive engine block containing cast aluminium alloy A390 cylinders. (a) cutaway section, (b) engine block with cylinders in place (courtesy J.-M. Ségaud, BMW Group).

enough to avoid serious problems with machining. It is also necessary to ensure that the primary silicon is well refined. This may be achieved by adding a small amount of phosphorus to the melt which reacts with aluminium to form small, insoluble particles of AlP. This compound has lattice spacings similar to those of silicon, and the particles serve as nuclei on which the primary particles can form (Fig. 4.8). The final microstructure is not unlike that of some modern

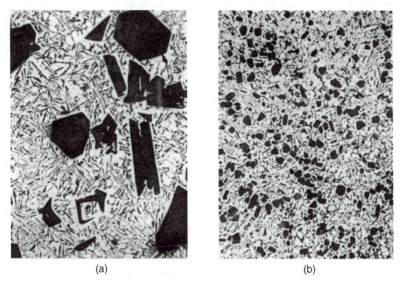

(a)   (b)

Fig. 4.8 Effect of minor additions of phosphorus in refining the size of the primary silicon plates in the hypereutectic alloy A390: (a) no phosphorus; (b) small additions of phosphorus (from Jorstad, J. L., *Met. Soc. AIME*, **242**, 1219, 1968) ($\times$ 100).

metal matrix composites (e.g. Fig. 7.7). Phosphorus is available as Cu–P pellets, a proprietary Al–Cu–P compound, or phosphorus-bearing salts, and it is usual to aim for a retained level of between 0.002 to 0.004%. An excess should be avoided because the AlP particles tend to agglomerate and may become entrapped as undesirable inclusions in a casting if they are not removed by fluxing or filtering. Another characteristic of alloy 390 is its particularly high heat of fusion. Although this has the desirable effect of increasing fluidity during casting, cycle times for permanent mould and pressure die castings are increased which reduces productivity. Tool life is also decreased as a consequence of the higher temperatures experienced.

4.3 ALLOYS BASED ON THE ALUMINIUM–COPPER SYSTEM

Alloys with copper as the major alloying addition were the first to be widely used for aluminium castings although many have now been superseded. Most existing compositions contain additional alloying elements. As a group, these alloys may present casting problems, e.g. hot-tearing, and it is also essential to provide generous feeding during solidification to ensure soundness in the final product. The alloys respond well to age-hardening heat treatments.

Several compositions are available which have elevated temperature properties that are superior to all other classes of aluminium casting alloys. Examples are 238 (Al–10Cu–3Si–0.3Mg), which is used for permanent mould casting of the soleplates of domestic hand irons, and 242 (Al–4Cu–2Ni–1.5Mg), which has been used for many years for diesel engine pistons and air-cooled cylinder heads for aircraft engines. Each alloy relies on a combination of precipitation hardening together with dispersion hardening by intermetallic compounds to provide stability of strength and hardness at temperatures up to around 250 °C.

The highest strength casting alloy is 201 which has the nominal composition Al–4.7Cu–0.7Ag–0.3Mg (Table 4.2). A similar European alloy, known as Avoir, also contains 1.3%Zn. These alloys show a high response to age hardening due to precipitation of the finely dispersed Ω phase (Fig. 3.18) that is described in Section 3.4.1. Using premium quality casting techniques, guaranteed properties of 345 MPa P.S. and 415 MPa T.S. with a minimum elongation of 5% have been obtained from a variety of castings heat treated to the T6 temper, and values as high as 480 MPa P.S. and 550 MPa T.S. with 10% elongation have been recorded. These tensile properties are much higher than can be obtained with any other aluminium casting alloys and compare well with the high-strength wrought alloys. The alloys may be susceptible to stress-corrosion cracking in the T6 condition but resistance is greatly improved by heat treating to a T73 temper. Because of the expense of adding silver, alloys 201 and Avoir are used only for military and other specialised applications.

4.4 ALUMINIUM–MAGNESIUM ALLOYS

This group contain several essentially binary alloys together with some more complex compositions based on Al–4Mg. Their special features are a high resistance to corrosion, good machinability and attractive appearance when anodized. Most show little or no response to heat treatment.

Casting characteristics are again less favourable than Al–Si alloys and more control must be exercised during melting and pouring because magnesium increases oxidation in the molten state. In this regard, special precautions are needed when sand casting as steam generated from moisture in the sand will react to produce MgO and hydrogen, which results in roughening and blackening of the surface of the casting. This mould reaction can be reduced by adding about 1.5% boric acid to the sand which forms a fused, glassy barrier towards steam produced within the mould. An alternative method is to add small amounts (0.03%) of beryllium to the alloy, which results in the formation of an impervious oxide film at the surface. Beryllium also reduces general oxidation during melting and casting although special care must be taken because of the potentially toxic nature of BeO. As a group, Al–Mg alloys also require special care with gating, and large risers and greater chilling are needed to produce sound castings.

The magnesium contents of the binary alloys range from 4 to 10%. Most are sand cast although compositions with 7 and 8% magnesium have limited application for permanent mould and pressure diecastings. Al–10Mg responds to heat treatment and a desirable combination of high strength, ductility and impact resistance may be achieved in the T4 temper. The castings must be slowly quenched from the solution treatment temperature, otherwise residual stresses may lead to stress-corrosion cracking. In addition, the alloy tends to be unstable, particularly in tropical conditions, leading to precipitation of Mg_5Al_8 (perhaps Mg_2Al_3) in grain boundaries which both lowers ductility and may cause stress-corrosion cracking after a period of time. Consequently other alloys such as Al–Si–Mg tend to be preferred unless the higher strength and ductility of Al–10Mg are mandatory.

Casting characteristics are somewhat improved by ternary additions of zinc and silicon and alloys such as Al–4Mg–1.8Zn and Al–4Mg–1.8Si can be diecast for parts of simple design.

4.5 ALUMINIUM–ZINC–MAGNESIUM ALLOYS

Several binary Al–Zn alloys have been used in the past but all are now obsolete except for compositions which are used as sacrificial anodes to protect steel structures in contact with sea water. Engineering alloys currently in use contain both zinc and magnesium, together with minor additions of one or more of the elements copper, chromium, iron, and manganese.

The as-cast alloys respond to ageing at room temperature and harden over a period of weeks. In this condition, or when artificially aged or stabilized after casting, the P.S. values range from 115 to 260 MPa and the T.S. range from 210 to 310 MPa, depending upon composition. Casting characteristics are relatively poor and the alloys are normally sand cast because the use of permanent moulds tends to cause hot-cracking.

One advantage of the alloys is that eutectic melting points are relatively high which makes them suitable for castings that are to be assembled by brazing. Other characteristics are good machinability, dimensional stability and resistance to corrosion. The alloys are not recommended for use at elevated temperatures because over-ageing causes rapid softening.

4.6 NEW CASTING PROCESSES

4.6.1 Semi-solid processing

As metals and alloys freeze, the primary dendrites normally grow and intermesh with each other so that, as soon as a relatively small fraction (e.g. 20%) of the melt has frozen, viscosity increases sharply and flow practically ceases. If, however, the dendrites are broken up by vigorously agitating the melt during solidification, reasonable fluidity persists until the solid content reaches as much as 60%. Each broken piece of dendrite becomes a separate crystal and very fine grain size can be achieved without recourse to the use of grain refiners (Fig. 4.9). Moreover, the semi-solid slurry has thixotropic characteristics, in that viscosity decreases on stirring, and this has interesting implications for casting processes.

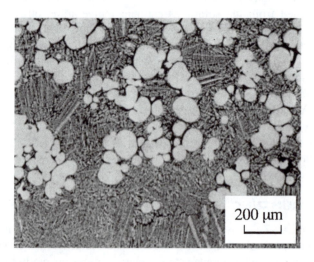

Fig. 4.9 Microstructure of the alloy Al–6.5Si sheared at a rate of 180 s^{-1} showing rounded α–Al dendrite fragments of the primary α–Al phase in a matrix of finely dispersed eutectic (after Ho, Y., Flemings, M. C. and Cornie, J. A., *Nature and Properties of Semi-Solid Materials*, Sekhar, J. A. and Dantzig, J. (Eds), The Minerals, Metals and Materials Society, Warrendale, Pa, U.S.A., p. 3, 1991).

In early laboratory studies of semi-solid processing, a slurry was produced by agitating an alloy in a narrow, annular region between the furnace wall and a central cylindrical stirrer. A temperature difference was maintained so that the alloy within this region was in the two-phase semi-liquid state, while above it was fully molten. The slurry was then transferred and cast in a pressure diecasting machine. The process was known as rheocasting. Slugs of the slurry could also be removed and forged in a die (thixoforging).

Commercialization of the slurry production process presented two major problems. One was the design of a furnace-stirrer system that would provide adequate quantities of slurry. The other was rapid chemical attack and erosion of materials used for the stirrer. These limitations appear to have been overcome with the development of the thixomoulding process by the Dow Chemical Company which was first applied to semi-solid magnesium alloys. Thixomoulding is a one-step process which combines plastic injection moulding technology with metal pressure die casting (see Figs. 5.9 and 5.10). The feedstock in the form of solid alloy pellets is heated to the semi-solid temperature and sheared into a thixotropic state in a screw feeder before passing directly into the diecasting machine. Another advantage of this process is that oxidation of the molten alloy is easily prevented by maintaining an inert atmosphere in the relatively small entry chamber (Fig. 5.9).

Semi-solid processing offers the following advantages:

1. Significantly lower capital investment and operating costs when compared with conventional casting methods. The whole process can be contained within one machine so that the need for melting and holding furnaces as well as melt treatment are all avoided. Foundry cleanliness is easy to maintain and energy requirements are less because complete melting is not required, cycle times are reduced and scrap is minimized.
2. Shrinkage and cracking within the mould are reduced because the alloy is already partly solidified when cast.
3. Lower operating and pouring temperatures lead to an increase in the lives of metal dies.
4. Composite materials can be readily produced by adding fibres or other solid particulates into the feedstock (compocasting).

Commercialization of semi-solid processing is still at an early stage but examples of automative components in current production are master brake cylinders, and pistons and compressors for air conditioners.

4.6.2 Squeeze casting

This process involves working or compressing the liquid metal in a hydraulic press during solidification which effectively compensates for the natural contraction that occurs as the liquid changes to solid. Pressures around 200 MPa are used which is several orders of magnitude greater than the melt pressures experienced in conventional foundry practice. As a consequence, the flow of the melt into incipient shrinkage pores is facilitated and entrapped gases tend to remain in solution. The high pressure also promotes intimate contact between casting and mould walls or tooling which assists heat extraction thereby leading to refinement of the microstructure.

Squeeze casting may be carried out in what are known as direct and indirect machines. In direct squeeze casting, metered amounts of molten metal are poured into a die similar to that used in permanent mould casting and then pressure is applied to the solidifying metal via the second, moving half of the die. For the indirect method, molten metal is first poured into a shot sleeve and then injected vertically into the die by a piston which sustains the pressure during solidification. In both cases a high casting yield is obtained because runners and feeding systems are not required so that virtually all the molten metal enters the die cavity. Casting mass may vary from 200 g to 35 kg depending upon the capacity of the machine.

Squeeze casting has been used for half a century in Russia and is now being exploited in other countries, notably Japan. For example, the Toyota Motor Company in Japan introduced squeeze-cast aluminium alloy wheels into their product line for passenger cars in 1979. An indication of the potential of this technique is evident from Table 4.5, in which a comparison of the properties of the alloy 357 (Al–7Si–0.5Mg) prepared by sand, gravity die and squeeze casting shows the latter clearly to be superior. Here the properties in each case are influenced mainly by the different levels of porosity, with the improvements in ductility being particularly notable.

Another advantage is that it is possible to squeeze cast some complex, high-strength alloys that are normally produced only as wrought products. For example, the Al–Zn–Mg–Cu aircraft alloy, 7010, has been successfully squeeze cast and, after a T6 ageing treatment, the minimum required properties for the wrought (die forged) condition are easily met (Fig. 4.10). A feature of the squeeze-cast material is that the casting has isotropic properties whereas wrought products always suffer from directionality effects (see Fig. 2.38). What is also interesting is that the S–N curves obtained in fatigue tests on squeeze-cast 7010 lie within the scatter band for wrought materials whereas castings normally have much poorer performance (Fig. 4.11).

Table 4.5 Mechanical properties of 357 aluminium alloy produced by different casting processes (from Lavington, M. H., *Metals and Materials*, **2**, 713, 1986)

Process	0.2% proof stress MPa	Tensile strength MPa	Elongation %
Sand cast	200	226	1.6
Chill cast	248	313	6.9
Squeeze cast	283	347	9.3
Cosworth	242	312	9.8

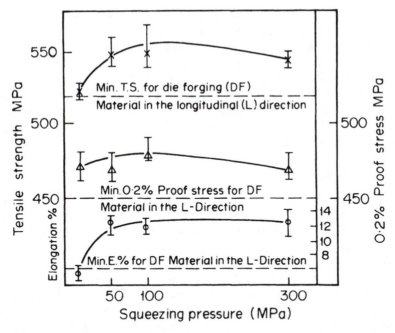

Fig. 4.10 Tensile properties of squeeze-cast alloy 7010 as a function of squeezing pressure (from Chadwick, G. A., *Metals and Materials*, **2**, 693, 1986).

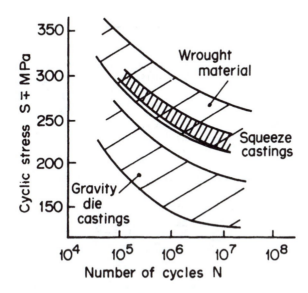

Fig. 4.11 Fatigue (*S/N*) curves for alloy 7010 in wrought, gravity diecast and squeeze-cast conditions (from Chadwick, G. A., *Metals and Materials*, **2**, 693, 1986)).

Properties may be further enhanced by incorporating fibres or particulates into squeeze-cast alloys, thereby producing the metal matrix composites that are discussed in Section 7.1.3. In this regard squeeze casting is superior to other low-pressure techniques for infiltrating liquid metal into a mesh or pad of fibres which has been a major limitation in the past. An advantage of such composites is that the fibres can be confined to strategic locations in castings where there is a need for increased strengthening or wear resistance. One recent example is the fibre-reinforcement of regions of aluminium alloy pistons for cars that are now manufactured commercially.

4.6.3 Cosworth Process

Conventional methods for casting aluminium and its alloys all involve turbulent transfer which has the undesirable effect of dispersing fine particles of oxides and other inclusions through the melt. Such particles are known to act as nuclei for the formation of microporosity in solidified castings.

The Cosworth Process allows quiescent transfer of metal from the stage of melting of the ingots to the final filling of the mould (Fig. 4.12). The melting/ holding furnace has sufficient capacity to ensure that the dwell time for the alloy between melting and casting is long enough to allow oxide particles or inclusions to separate by floating or sinking. No fluxes or chemicals are used and a protective atmosphere is employed to minimize the risk of gas absorption and oxide formation. A unique feature of the process is that the mould is located above the furnace and gently filled with molten metal from below by a programmable

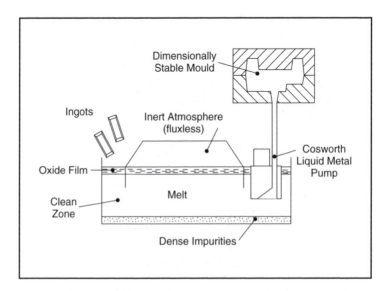

Fig. 4.12 Diagrammatic representation of a casting unit using the Cosworth Process (from Lavington, M. H., *Metals and Materials*, **2**, 213, 1986).

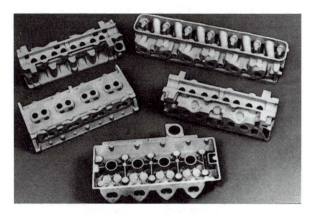

Fig. 4.13 Cylinder head castings produced by the Cosworth Process (courtesy Cosworth Research and Development Ltd).

electromagnetic pump which, itself, has no moving parts. The mould is permeable to allow air to escape from the cavity. A recent development has been to roll over the mould under an applied positive pressure immediately after filling. This practice further improves the quality of castings and allow higher production rates to be achieved. Taken overall, the generation and entrapment of oxide particles is minimized. Porosity is much reduced and this is reflected in improved tensile properties and ductility when compared with conventionally produced sand castings (Table 4.5). Fatigue properties at both room and elevated temperatures are also superior and scatter in the test results is much reduced.

Another feature of the Cosworth Process is the use of reclaimable zircon sand rather than conventional silica for making moulds and cores. This avoids the volume change associated with the phase transition from α to β quartz that occurs in silica at temperatures close to those used for melting aluminium alloys. The stability of both moulds and cores is thus much improved which allows close dimensional tolerances of castings to be obtained and repeated. For example, within one mould piece, it is possible to apply tolerances of ±0.15 mm up to sizes of 100 mm and ±0.25 mm up to 800 mm. This feature, when combined with the greatly reduced porosity, allows thin-walled, pressure-tight castings to be produced. These have proved particularly beneficial in the manufacture of complex cylinder heads and other engine components for high-performance motor cars (e.g. Fig. 4.13). Size ranges of castings have so far been within the range 0.5–55 kg and integrated production lines have been established.

4.6.4 Improved Low Pressure Casting (ILP) Process

Another precision casting method that utilizes transfer of molten metal vertically through a riser tube into the bottom of the mould cavity is the ILP process developed in Australia. Degassed and filtered metal is delivered to the casting furnace,

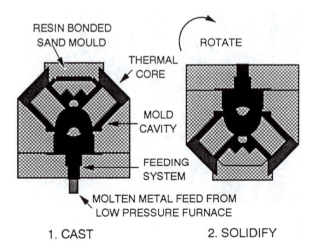

Fig. 4.14 Diagrammatic representation of the ILP process (courtesy Comalco Aluminium Ltd.).

which uses a pressurized atmosphere of nitrogen to enable quiescent, computer-controlled filling of the mould. Once filled, the mould is sealed and immediately removed so that solidification occurs remote from the casting section. This allows another pre-assembled mould to be positioned on the station which facilitates high productivity, and cycle times of around 60 s become possible.

As shown in Fig. 4.14, a special feature of the ILP process is the use of a combination of metal cores and resin-bonded silica sand for the mould which promotes rapid unidirectional solidification in those regions of the casting where optimal properties are required. Furthermore, the mould can be inverted to facilitate this controlled solidification, which may provide dendrite arm spacings less than 20 μm adjacent to the metal cores, and less than 0.5% micro-shrinkage overall.

The ILP process has been adapted to allow robotic handling of the moulds so that movements are precisely repeatable. It is being used in Mexico for the mass production of cylinder heads and blocks for engines of some models of automobiles that are manufactured for the North American market.

4.7 JOINING

Most aluminium casting alloys can be arc welded in a protective atmosphere of an inert gas, e.g. argon, provided they are given the correct edge preparation. Ratings of weldability were included in Table 4.3. In addition some surface defects and service failures in sand and permanent mould castings may be repaired by welding. Filler metals are selected which are appropriate to the compositions of alloy castings with 4043 (Al–5Si) and 5356 (Al–5Mg–0.1Mn–0.1Cr) being commonly used. Special care must be taken when repair welding cast components such as automotive wheels that have previously been heat treated. Unless an appropriate re-heat treatment schedule is available this process should be avoided. With respect to joining of castings by brazing, similar conditions apply to those discussed for wrought aluminium alloys in Section 3.5.2.

FURTHER READING

Van Horn, K. (Ed.), *Aluminium*, Vols. **1–3**, American Society for Metals, Cleveland, Ohio, USA, 1967

Davis, J.R., and Associates (Eds.), *ASM Specialty Handbook on Aluminium Alloys*ASM International, Materials Park, Ohio, USA, 1993

Chadwick, G.A., Casting technology, in *Future Developments in Metals and Ceramics*, Charles, J.A., Greenwood, G.W. and Smith, G.C. (Eds.), Institute of Materials, London, 179, 1992

Taylor, J.A., Metal-related castibility effects in aluminium foundry alloys, *Cast Metals*, **8**, 225, 1995

Pehlke, R.D., Computer Simulation of Solidification Processes – The Evolution of a Technology, *Metall. Mater. Trans. A*, **33A**, 2251, 2002

Wang, W., Makhlouf, M., and Apelian, D., Aluminium die casting alloys: alloy composition, microstructure, and properties-performance relationships, *Inter. Mater. Rev.*, **40**, 221, 1995

Makhlouf, M. and Guthy, H.V., The aluminium-silicon eutectic reaction: mechanisms and crystallography, *Journal of Light Metals*, **1**, 199, 2001

Shankar, S. and Apelian, D., Mechanism and Preventitative measures for die-soldering during Al casting in a ferrous mould, *JOM*, **58**, no. 8, 47, 2002

Couper, M.J., Neeson, A.E., and Griffiths, J.R., Casting defects and the fatigue behaviour of an aluminium casting alloy, *Fatigue Fract. Engng. Mater. Struct.*, **13**, 213, 1990

Ye, H., An overview of the development of Al–Si–alloy based material for engine applications, *Journal of Materials Engineering Performance*, **12**, 288, 2003

Jorstad, J.L. Hypereutectic Al–Si alloy parts manufacturing: practical processing techniques, *Proc. 1st Inter. Light Metals Technology Conf.*, A. Darle, Ed., 93, 2003

Lavington, M.H., The Cosworth Process – a new concept in aluminium alloy casting production, *Metals and Materials*, **2**, 213, 1986

Kirkwood, D.H., Semisolid processing, *Inter. Mater. Rev.* **39**, 173, 1993

Sekhar, J.A. and Dantzig, J., (Eds), *Nature and Properties of Semi-Solid Materials*, TMS, Warrendale, Pa, USA, 1991

Bosworth, J., Repair welding of aluminium castings, *Modern Casting*, **74**, No. 3., p. 32; No. 4, p. 19, 1984

5

MAGNESIUM ALLOYS

5.1 INTRODUCTION TO ALLOYING BEHAVIOUR

Magnesium is readily available commercially with purities exceeding 99.8% but it is rarely used for engineering applications without being alloyed with other metals. Key features that dominate the physical metallurgy of the alloys are the hexagonal lattice structure of magnesium (Fig. 5.1) and the fact that its atomic diameter (0.320 nm) is such that it enjoys favourable size factors with a diverse range of solute elements. Aluminium, zinc, cerium, silver, thorium, yttrium and zirconium are examples of widely different elements that are present in commercial alloys.

Solubility data for binary magnesium alloys are shown in Table 5.1 with the first section containing elements used in commercial compositions. Apart from magnesium and cadmium which form a continuous series of solid solutions, the magnesium-rich sections of the phase diagrams show peritectic or, more commonly, eutectic systems. A wide range of intermetallic compounds may form, with the three most frequent types of structures being as follows:

1. AB. simple cubic CsCl structure. Examples are MgTl, MgAg, CeMg, SnMg and it will be seen that magnesium can be either the electropositive or the electronegative component.
2. AB_2. Laves phases with ratio $R_A/R_B = 1.23$ preferred. Three types exist namely:

    $MgCu_2$ (f.c.c., stacking sequence abcabc)

    $MgZn_2$ (hex., stacking sequence ababab)

    $MgNi_2$ (hex., stacking sequence abacaba)
3. CaF_2. f.c.c. This group contains Group IV elements and examples are Mg_2Si and Mg_2Sn.

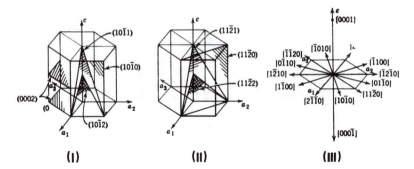

Fig. 5.1 Principal planes and directions in the magnesium unit cell.

Table 5.1 Solubility data for binary magnesium alloys (from Massalski, T. B., *Binary Phase Diagrams*, 2nd edn, Vols 1–4, ASM International, Metals Park, Ohio, 1990; Nayeb-Hashemi, A. A. and Clark, J. B. *Phase Diagrams of Binary Magnesium Alloys*, ASM International, Metals Park, Ohio, 1988)

Element	At. %	Wt. %	System
Lithium	17.0	5.5	Eutectic
Aluminium	11.8	12.7	Eutectic
Silver	3.8	15.0	Eutectic
Yttrium	3.75	12.5	Eutectic
Zinc	2.4	6.2	Eutectic
Neodymium	~1	~3	Eutectic
Zirconium	1.0	3.8	Peritectic
Manganese	1.0	2.2	Peritectic
Thorium	0.52	4.75	Eutectic
Cerium	0.1	0.5	Eutectic
Cadmium	100	100	Complete S.S.
Indium	19.4	53.2	Peritectic
Thallium	15.4	60.5	Eutectic
Scandium	~15	~24.5	Peritectic
Lead	7.75	41.9	Eutectic
Thulium	6.3	31.8	Eutectic
Terbium	4.6	24.0	Eutectic
Tin	3.35	14.5	Eutectic
Gallium	3.1	8.4	Eutectic
Ytterbium	1.2	8.0	Eutectic
Bismuth	1.1	8.9	Eutectic
Calcium	0.82	1.35	Eutectic
Samarium	~1.0	~6.4	Eutectic
Gold	0.1	0.8	Eutectic
Titanium	0.1	0.2	Peritectic

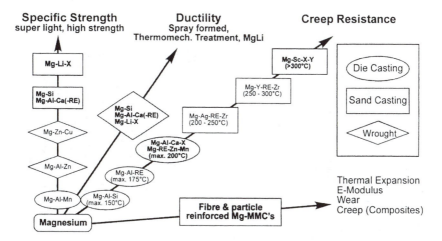

Fig. 5.2 Directions in the development of magnesium alloys (from Mordike, B. L. and Ebert, T., *Mater. Sci. and Eng.* **A302**, 37, 2001).

General directions in the development of die cast, sand cast and wrought magnesium alloys for specific requirements are summarised in Fig. 5.2. A special feature has been the successful introduction of several rare earth (RE) elements, notably cerium, neodymium, and yttrium into a number of commercial magnesium alloys. Most RE elements have high solid solubilities in magnesium because their atomic sizes are favourable, and they have electron negativities (e.g. Ce 1.21, Nd 1.19, and Y 1.20) similar to magnesium (1.20). These elements increase the strength of magnesium and its alloys, particularly at elevated temperatures.

Cast magnesium alloys have always predominated over wrought alloys, particularly in Europe where, traditionally, they have comprised 85–90% of all products. The first alloying elements to be used commercially were aluminium, zinc and manganese and the Mg–Al–Zn system is still the one most widely used for castings. The first wrought alloy was Mg.1.5Mn that was produced as sheet, extrusions and forgings, but this material has largely been superseded. General effects of alloying elements that may be present in commercial magnesium alloys are summarized in Table 5.2

Early Mg–Al–Zn castings suffered severe corrosion in wet or moist conditions which was significantly reduced with the discovery, in 1925, that small additions (0.2 wt %) of manganese gave increased resistance. The role of this element was to remove iron and certain other heavy metal impurities into relatively harmless intermetallic compounds, some of which separate out during melting. In this regard, classic work by Hanawalt and colleagues has shown that the corrosion rate increases abruptly once so-called 'tolerance limits' are exceeded: these are 5, 170 and 1300 p.p.m. for nickel, iron and copper respectively. An example of this behaviour is illustrated in Fig. 5.3.

Table 5.2 General effects of elements used in magnesium alloys. (from Neite, G. et al *Magnesium-Based Alloys* in *Materials Science and Technology : A Comprehensive Treatment* Eds. Cahn, R. W., Haasen, P., and Kramer, E. J., **Vol. 8**, *Structure and properties of Non-Ferrous Alloys*, VCH, Weinheim, Germany, p. 113, 1994)

Alloying element	Melting and casting behavior	Mechanical and technological properties	Corrosion behavior I/M produced
Ag	Improves castability, tendency to microporosity	Improves elevated temperature tensile and creep properties in the presence of rare earths	Detrimental influence on corrosion behavior
Al		Solid solution hardener, precipitation hardening at low temperatures (<120 °C)	Minor influence
Be	Significantly reduces oxidation of melt surface at very low concentrations (<30 ppm), leads to coarse grains		
Ca	Effective grain refining effect, slight suppression of oxidation of the molten metal	Improves creep properties	Detrimental influence on corrosion behavior
Cu	System with easily forming metallic glasses, improves castability		Detrimental influence on corrosion behavior, limitation necessary
Fe	Magnesium hardly reacts with mild steel crucibles		Detrimental influence on corrosion behavior, limitation necessary
Li	Increases evaporation and burning behavior, melting only in protected and sealed furnaces	Solid solution hardener at ambient temperatures, reduces density, enhances ductility	Decreases corrosion properties strongly, coating to protect from humidity is necessary
Mn	Control of Fe content by precipitating Fe-Mn compound, refinement of precipitates	Increases creep resistance	Improves corrosion behavior due to iron control effect

Ni	System with easily forming metallic glasses		Detrimental influence on corrosion behavior, limitation necessary
Rare earths	Improve castability, reduce microporosity	Solid solution and precipitation hardening at ambient and elevated temperatures; improve elevated temperature tensile and creep properties	Improve corrosion behavior
Si	Decreases castability, forms stable silicide compounds with many other alloying elements, compatible with Al, Zn, and Ag, weak grain refiner	Improves creep properties	Detrimental influence
Th	Suppresses microporosity	Improves elevated temperature tensile and creep properties, improves ductility, most efficient alloying element	
Y	Grain refining effect	Improves elevated temperature tensile and creep properties	Improves corrosion behavior
Zn	Increases fluidity of the melt, weak grain refiner, tendency to microscopy	Precipitation hardening, improves strength at ambient temperatures, tendency to brittleness and hot shortness unless Zr refined	Minor influence, sufficient Zn content compensates for the detrimental effect of Cu
Zr	Most effective grain refiner, incompatible with Si, Al, and Mn, removes Fe, Al, and Si from the melt	Improves ambient temperature tensile properties slightly	

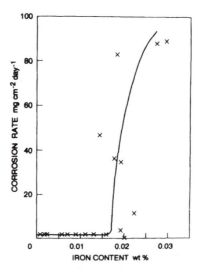

Fig. 5.3 Effect of iron on corrosion of pure magnesium (alternate test in 3%NaCl) (from Hanawalt, J. D., Nelson, C. E. and Peloubet, J. A., *Trans. AIME*, **147**, 273, 1942).

Another problem with early magnesium alloy castings was that grain size tended to be large and variable, which often resulted in poor mechanical properties, microporosity and, in the case of wrought products, excessive directionality of properties. Values of proof stress also tended to be particularly low relative to tensile strength. In 1937 Sauerwald, in Germany, discovered that zirconium had an intense grain-refining effect on magnesium, although several years elapsed before a reliable method was developed to alloy this metal. Paradoxically, zirconium could not be used in most existing alloys because it was removed from solid solution due to the formation of stable compounds with aluminium and manganese. This led to the evolution of a completely new series of cast and wrought zirconium-containing alloys that were found to have much improved mechanical properties at both room and elevated temperatures. Compositions, tensile properties, and the general characteristics of various commercial cast and wrought magnesium alloys are shown in Tables 5.3 and 5.4. These are considered below and divided into two groups depending on whether or not they contain the grain refining element zirconium. Some new and experimental alloys are shown later in Table 5.6.

Another characteristic of alloy systems in which solubility is strongly influenced by atomic size factors is that solid solubility generally decreases with decreasing temperature. Such a feature is a necessary requirement for precipitation hardening and most magnesium alloys are amenable to this phenomenon although the responses are significantly less than is observed in some aluminium alloys. Precipitation processes are usually complex and are not completely understood. Probable precipitation sequences in alloys of commercial interest are shown in Table 5.5. A feature of the ageing process in several alloys is that one stage

Table 5.3 Nominal composition, typical tensile properties and characteristics of selected magnesium casting alloys

ASTM designation	Nominal composition											Condition	Tensile properties			Characteristics
	Al	Zn	Mn	Si	Cu	Zr	RE (MM)	RE (Nd)	Th	Y	Ag		0.2% proof stress (MPa)	Tensile strength (MPa)	Elong-ation (%)	
AZ63	6	3	0.3									As-sand cast	75	180	4	Good room temperature strength and ductility
												T6	110	230	3	
AZ81	8	0.5	0.3									As-sand cast	80	140	3	Tough, leaktight castings
												T4	80	220	5	With 0.0015 Be, used for pressure diecasting
AZ91	9.5	0.5	0.3									As-sand cast	95	135	2	General-purpose alloy used for sand and diecastings
												T4	80	230	4	
												T6	120	200	3	
												As-chill cast	100	170	2	
												T4	80	215	5	
												T6	120	215	2	
AM50	5		0.3									As-diecast	125	200	7†	High-pressure diecastings
AM20	2		0.5									As-diecast	105	135	10†	Good ductility and impact strength
AS41	4		0.3	1								As-diecast	135	225	4.5†	Good creep properties to 150 °C
AS21	2		0.4	1								As-diecast	110	170	4†	Good creep properties to 150 °C
AE42	4		0.3				2					As-diecast	135	225	10.5	Good creep properties to 150 °C
ZK51		4.5				0.7						T5	140	235	5	Sand castings, good room temperature strength and ductility
ZK61		6				0.7						T5	175	275	5	As for ZK51
ZE41		4.2				0.7	1.3					T5	135	180	2	Sand castings, good room temperature strength, improved castability

(*Continued*)

243

Table 5.3 (*Continued*)

ASTM designation	Nominal composition											Condition	Tensile properties			Characteristics
	Al	Zn	Mn	Si	Cu	Zr	RE (MM)	RE (Nd)	Th	Y	Ag		0.2% proof stress (MPa)	Tensile strength (MPa)	Elong-ation (%)	
ZC63		6	0.5		3							T6	145	240	5	Pressure-tight castings, good elevated temperature strength, weldable
EZ33		2.7				0.7	3.2					Sand cast T5 / Chill cast T5	95 / 100	140 / 155	3 / 3	Good castability, pressure-tight, weldable, creep resistant to 250 °C
HK31						0.7			3.2			Sand cast T6	90	185	4	Sand castings, good castability, weldable, creep resistant to 350 °C
HZ32		2.2				0.7			3.2			Sand or chill cast T5	90	185	4	As for HK31
QE22						0.7		2.5			2.5	Sand or chill cast T6	185	240	2	Pressure tight and weldable, high proof stress to 250 °C
QH21						0.7		1	1		2.5	As-sand cast T6	185	240	2	Pressure-tight, weldable, good creep resistance and proof stress to 300 °C
WE54						0.5		3.25‡		5.1		T6	200	285	4†	High strength at room and elevated temperatures.
WE43						0.5		3.25‡		4		T6	190	250	7†	Good corrosion resistance, weldable

†Values quoted for tensile properties are for separately cast test bars and may not be realized in certain parts of castings RE = rare earth element; MM = misch metal; Nd = neodymium

‡Contains some heavy metal rare earth elements

Table 5.4 Nominal composition, typical tensile properties and characteristics of selected wrought magnesium alloys

ASTM desig-nation	Nominal composition								Condition	Tensile properties			Characteristics
	Al	Zn	Mn	Zr	Th	Cu	Li	Y		0.2% proof stress (MPa)	Tensile strength (MPa)	Elongation (%)	
M1			1.5						Sheet, plate F	70	200	4	Low- to medium-strength alloy, weldable, corrosion resistant
									Extrusions F	130	230	4	
									Forgings F	105	200	4	
AZ31	3	1	0.3 (0.20 min.)						Sheet, plate O	120	240	11	Medium-strength alloy, weldable, good formability
									H24	160	250	6	
									Extrusions F	130	230	4	
									Forgings F	105	200	4	
AZ61	6.5	1	0.3 (0.15 min.)						Extrusions F	180	260	7	High-strength alloy, weldable
									Forgings F	160	275	7	
AZ80	8.5	0.5	0.2 (0.12 min.)						Forgings T6	200	290	6	High-strength alloy
ZM21		2	1						Sheet, plate O	120	240	11	Medium-strength alloy, good formability, good damping capacity
									H24	165	250	6	
									Extrusions	155	235	8	
ZK30		3		0.6					Forgings	125	200	9	High-strength alloys
									Forgings T6	215	300	9	
									Extrusions T6	240	290	14	
ZK60		6		0.6					Forgings T6	235	315	8	Good formability
									Extrusions T6	280	320	12	
ZMC711		6.5	0.75			1.25			Extrusions T6	300	325	3	High-strength alloy

(Continued)

Table 5.4 (*Continued*)

ASTM desig-nation	Nominal composition								Condition	Tensile properties			Characteristics
	Al	RE	Mn	Zr	Th	Cu	Li	Y		0.2% proof stress (MPa)	Tensile strength (MPa)	Elongation (%)	
HK31				0.7	3.2				Sheet, plate H24	170	230	4	High creep resistance to 350 °C, weldable
									Extrusions T5	180	255	4	
HM21			0.8		2				Sheet, plate T8	135	215	6	High creep resistance to 350 °C, short time exposure to 425 °C, weldable
									T81	180	255	4	
									Forgings T5	175	225	3	
WE43		3		0.5				4	Extrusions T6	160	260	6	High Temperature creep resistance
WE54		3.5		0.5				5.25	Forgings T6	180	280	6	
LA141	1.2		0.15 min.				14		Sheet, plate T7	95	115	10	Ultra-light weight (S.G. 1.35)

Table 5.5 Probable precipitation processes in magnesium alloys

Alloy system	Precipitation process
Mg–Al	SSSS ⟶ bcc equilibrium precipitate β (Mg$_{17}$Al$_{12}$) (incoherent)? (i) Most precipitates have relationship (011)$_\beta$//(0002)$_{Mg}$, [111]$_\beta$//[2$\bar{1}\bar{1}$0]$_{Mg}$ Faceted lath morphology, habit plane // (0002)$_{Mg}$ (ii) Minor fraction of precipitates (010)$_\beta$//(1$\bar{1}$00)$_{Mg}$, [111]$_\beta$ // [0001]$_{Mg}$ Prismatic rod morphology, long axis // [0001]$_{Mg}$ (iii) Few precipitates have relationship (1$\bar{1}$0)$_\beta$ ~ //(1$\bar{1}$00)$_{Mg}$, [11$\bar{5}$]$_\beta$ ~ //[0001]$_{Mg}$
Mg–Zn(–Cu)	SSSS ⟶ GP zones ⟶ MgZn$_2$ ⟶ MgZn$_2$ ⟶ Mg$_2$Zn$_3$ **GP zones** Discs // {0002}$_{Mg}$ (coherent) **MgZn$_2$** Rods ⊥ {0002}$_{Mg}$ cph, a = 0.52 nm, c = 0.85 nm (coherent) **MgZn$_2$** Discs //{0002}$_{Mg}$ (11$\bar{2}$0)$_{MgZn_2}$//(10$\bar{1}$0)$_{Mg}$ cph, a = 0.52 nm, c = 0.848 nm (semi-coherent) **Mg$_2$Zn$_3$** Trigonal a = 1.724 nm, b = 1.445 nm, c = 0.52 nm, γ = 138° (incoherent)
Mg–RE(Nd)	SSSS ⟶ GP zones ⟶ β″ ⟶ β′ ⟶ β **GP zones** Disks //(00$\bar{1}$0)$_{Mg}$ fcc a = 0.74 nm (coherent) **β″** Mg$_3$Nd? hex. a = 0.64 nm, c = 0.52 nm, DO$_{19}$ superlattice plates (0001)$_{\beta''}$//(0001)$_{Mg}$ [00$\bar{1}$0]$_{\beta''}$//[10$\bar{1}$0]$_{Mg}$ **β′** Mg$_3$Nd hex. a = 0.52 nm, c = 1.30 nm (semi-coherent) **β** Mg$_{12}$Nd bct a = 1.30 nm, c = 0.59 nm or Mg$_{41}$Nd$_5$ (incoherent)
Mg–RE–Zn	SSSS ⟶ GP zones ⟶ γ″ ⟶ γ′ ⟶ γ **GP zones** ordered **γ″** hex. a = 0.556 nm, c = 1.563 nm (0001)$_{\gamma''}$//(0001)$_{Mg}$ [2$\bar{1}\bar{1}$0]$_{\gamma''}$//[10$\bar{1}$0]$_{Mg}$ **γ′** fcc a = 0.72 nm rods// ⟨10$\bar{1}$0⟩$_{Mg}$

(Continued)

Table 5.5 (*Continued*)

Alloy system	Precipitation process

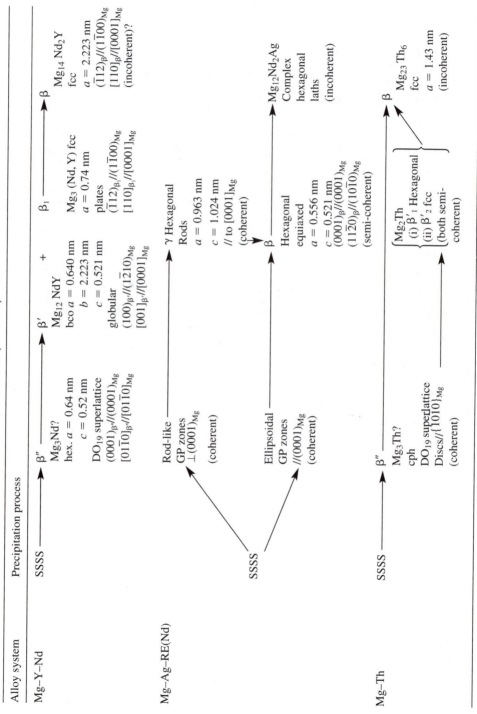

Mg–Y–Nd

SSSS → β″ → β′ + β₁ → β

β″: Mg₃Nd?
hex. $a = 0.64$ nm
$c = 0.52$ nm
DO₁₉ superlattice
$(0001)_{\beta''}//(0001)_{Mg}$
$[01\bar{1}0]_{\beta''}//[01\bar{1}0]_{Mg}$

β′: Mg₁₂NdY
bco $a = 0.640$ nm
$b = 2.223$ nm
$c = 0.521$ nm
globular
$(100)_{\beta'}//(1\bar{2}10)_{Mg}$
$[001]_{\beta'}//[0001]_{Mg}$

β₁: Mg₃(Nd, Y) fcc
$a = 0.74$ nm
plates
$(\bar{1}12)_{\beta_1}//(1\bar{1}00)_{Mg}$
$[110]_{\beta_1}//[0001]_{Mg}$

β: Mg₁₄Nd₂Y
fcc
$a = 2.223$ nm
$(\bar{1}12)_{\beta}//(1\bar{1}00)_{Mg}$
$[110]_{\beta}//[[0001]_{Mg}$?
(incoherent)

Mg–Ag–RE(Nd)

SSSS → Rod-like GP zones ⊥$(0001)_{Mg}$ (coherent) → γ Hexagonal Rods $a = 0.963$ nm, $c = 1.024$ nm // to $[0001]_{Mg}$ (coherent)

SSSS → Ellipsoidal GP zones //$(0001)_{Mg}$ (coherent) → β Hexagonal equiaxed $a = 0.556$ nm, $c = 0.521$ nm $(0001)_{\beta}//(0001)_{Mg}$ $(11\bar{2}0)_{\beta}//(10\bar{1}0)_{Mg}$ (semi-coherent) → Mg₁₂Nd₂Ag Complex hexagonal laths (incoherent)

Mg–Th

SSSS → β″ → β → β

β″: Mg₃Th?
cph
DO₁₉ superlattice
Discs//{1010}_{Mg}
(coherent)

Mg₂Th
(i) β′₁ Hexagonal
(ii) β′₂ fcc
(both semi-coherent)

β: Mg₂₃Th₆
fcc
$a = 1.43$ nm
(incoherent)

248

involves the formation of an ordered, hexagonal precipitate with a DO_{19} (Mg_3Cd) crystal structure that is coherent with the magnesium lattice. This structure is analogous to the well-known θ'' (GP zones 2) phase that may form in aged Al–Cu alloys and it is commonly found in alloys in which there is a large difference in the atomic sizes of the constituents. The phase contributes to hardening in those magnesium alloys in which it forms and it is present at peak hardness over a wide temperature range.

The DO_{19} cell has an *a* axis twice the length of the *a* axis of the magnesium matrix whereas the *c* axes are the same. The precipitate forms as plates or discs parallel to the $\langle 0001 \rangle_{Mg}$ directions which lie along the $\{10\bar{1}0\}_{Mg}$ and $\{11\bar{2}0\}_{Mg}$ planes. In this regard it is significant to note that alternate $(10\bar{1}0)$ and $(11\bar{2}0)$ planes in a structure of composition Mg_3X consist of all magnesium atoms (Fig. 5.4). Thus, the formation of a low-energy interface along these planes is to be expected since only second nearest neighbour bonds need to be altered. This structural feature would account for the fact that the phase is relatively stable over a wide temperature range and it may be the most significant factor in promoting creep resistance in those magnesium alloys in which it occurs.

Various solute species cause solid solution strengthening of magnesium by increasing the critical resolved shear stress for slip along the basal planes, Special attention has been given to Mg–Al and Mg–Zn alloys and, as shown in Fig. 5.5, zinc is approximately three times more effective, on an atomic per cent basis, than aluminium in increasing the yield strength for alloys that are in the solution treated and quenched condition. After correcting for effects arising from changes in grain size for different compositions, the yield strength of Mg–Al alloys has been found to increase linearly with c^n, where c is the atom concentration and n = 1/2 ~ 2/3. Solid solution strengthening is also significant in the aged alloys

Fig. 5.4 Model showing low-energy interfaces along the $\{10\bar{1}0\}$ and $\{11\bar{2}0\}$ planes of precipitates having the DO_{19} structure (Mg_3X). White balls = magnesium atoms, black balls = solute atoms (from Gradwell, K. J., *Precipitation in a High Strength Magnesium Casting Alloy*, PhD Thesis, University of Manchester, 1972).

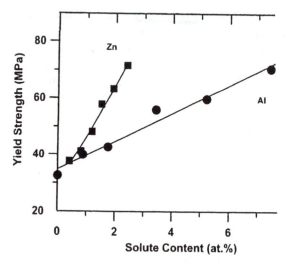

Fig. 5.5 Solid solution strengthening of binary Mg-Al and Mg-Zn alloys (from Abbott. T. B. *et al, Handbook of Mechanical Design*, Eds, Totten, G.E. *et al*, Marcel Dekker Inc,.. New York, 487, 2004).

and, for Mg–Al alloys aged to peak strength, it has been shown that about half the aluminium atoms available for precipitation still remain in solid solution.

5.2 MELTING AND CASTING

5.2.1 Melting

It is usual for magnesium to be melted in mild steel crucibles for both the alloying and refining or cleaning stages before producing cast or wrought components. Unlike aluminium and its alloys, the presence of an oxide film on molten magnesium does not protect the metal from further oxidation. On the contrary, it accelerates this process. Melting is complete at or below 650 °C and the rate of oxidation of the molten metal surface increases rapidly with rise in temperature such that, above 850 °C, a freshly exposed surface spontaneously bursts into flame. Consequently, suitable fluxes or inert atmospheres must be used when handling molten magnesium and its alloys.

For many years, thinly fluid salt fluxes were used to protect molten magnesium which were mixtures of chlorides such as $MgCl_2$ with KCl or NaCl. In Britain, it was usual then to thicken this flux with a mixture of CaF_2, MgF_2 and MgO which formed a coherent, viscous cake that continued to exclude air and could be readily drawn aside when pouring. However, the presence of the chlorides often led to problems with corrosion when the cast alloys were used in service.

In the 1970s, it became usual to replace the salt fluxes with cover gases comprising either a single gas (e.g. SO_2 or argon), or a mixture of an active gas diluted with CO_2, N_2, or air. Sulphur hexafluoride (SF_6) was widely accepted as the active gas because it is non-toxic, odourless, colourless, and effective at low concentrations. Replacing the fluxes and SO_2 also improved occupational health and safety and increased equipment life because corrosion was reduced. SF_6 is, however, relatively expensive, and is now realised to be a particularly potent greenhouse gas with a so-called Global Warming Potential (GWP) of 22 000 to 23 000 on a 100 year time horizon. Efforts are therefore being made to find other active gases containing fluorine and one promising alternative is the organic compound HFC 134a (1,1,1,2-tetrafluoroethane) that is readily available worldwide because of its use as a refrigerant gas. It is also less expensive than SF_6. HFC 134a has a GWP of only 1600, and an estimated atmospheric lifetime of 13.6 years compared with 3 200 years for SF_6. Moreover less is consumed on a daily basis so that the overall potential to reduce greenhouse gas emissions is predicted to be 97%.

Most alloying elements are now added in the form of master alloys or hardeners. Zirconium has presented special problems as early attempts to use either zirconium metal or an Mg–Zr hardener were ineffective. Success was achieved eventually by means of mixtures of reducible zirconium halides, e.g. ones containing fluorozirconate, K_2ZrF_6, together with large amounts of $BaCl_2$ to increase the density of the salt reaction products. These salt mixtures were supplied under licence to foundries. Subsequently it was found possible to prepare hardener alloys from the weighted salt mixtures and proprietary hardeners made in this way were used for adding zirconium to magnesium alloy melts. Prior to the use of $BaCl_2$, severe problems were encountered with persistent flux inclusions which arose through entrainment of salt reaction products in the melt and could not be removed by pre-solidification or any flux-refining step.

Since the late 1960s, Mg–Zr master alloy hardeners have been the preferred method for introducing zirconium as a grain refiner and these alloys usually contain between 10 and 60% of zirconium in weight percentage depending on the supplier. Since the approximate maximum solubility of zirconium in molten magnesium is only 0.6% (Fig. 5.6), almost all the zirconium present in the hardener is undissolved. Because of this, the Mg–Zr master alloys have similar microstructures comprising zirconium or zirconium-rich particles often embedded as clusters in a Mg–0.6Zr matrix. The major difference between the different master alloys lies in the size the particles and how they are distributed in the matrix.

As with aluminium, hydrogen is the only gas that dissolves in molten magnesium although it is less of a problem in this case because of its comparatively high solid solubility (average of about 30 ml 100 g^{-1}). The main source of hydrogen is from water vapour in damp fluxes or corroded scrap/ingot, so pick-up can be minimized by taking adequate precautions with these materials. A low hydrogen content reduces the tendency to gas porosity which is common in Mg–Al and Mg–Al–Zn alloys and these materials should be degassed with chlorine. The optimum temperature for degassing is 725–750 °C. If the

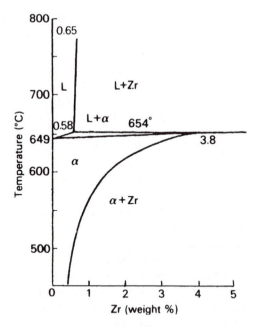

Fig. 5.6 Section of the Mg–Zr phase diagram.

melt is below 713 °C then solid $MgCl_2$ will form which gives little protection from burning, whilst at temperatures much above 750 °C, magnesium losses through reaction with chlorine become excessive. Gas porosity is not normally a problem with zirconium-containing alloys since zirconium will itself remove hydrogen as ZrH_2 and it is generally unnecessary to degas these alloys. However, it should be noted that such a treatment does improve the tensile properties of certain Mg–Zn–Zr alloys, presumably by minimizing the loss of zirconium as the insoluble ZrH_2. In such a case the degassing operation is completed before zirconium is added.

5.2.2 Grain refinement

Although grain refinement is carried out during melting, its importance and complexity in magnesium alloys merits separate consideration. Quite different practices are needed depending upon the presence or absence of zirconium.

The group of alloys based mainly on the Mg–Al system tends to have large and variable grain size. The first method devised to control grain size was to super-heat the melt to a temperature of 850 °C and above for periods of about 30 min, after which the melt was quickly cooled to the normal casting temperature and poured. A comparatively fine grain size was achieved with fair success, although the reasons for the effect are somewhat obscure. The probable explanation is that foreign nuclei with suitable crystal structures such as Al_4C_3 precipitate on cooling to the casting temperature and act as nuclei for the magnesium grains during sub-sequent solidification. The superheating effect is only significant in Mg–Al alloys and presents problems because crucible and furnace lives are reduced and power requirements are increased.

An alternative technique was developed in Germany in which a small quantity of anhydrous FeCl₃ was added to the melt (Elfinal process) and grain refinement was attributed to nucleation by iron-containing compounds. This method also had its disadvantages because the deliquescent nature of $FeCl_3$ made it hazardous, and the presence of as little as 0.005% iron could decrease the corrosion resistance of the alloys. The addition of manganese was made to counter this latter problem but effectively prevented grain refinement by $FeCl_3$.

The method in current use for alloys containing aluminium as a major alloy-ing element is to add volatile carbon-containing compounds to the melt and hexachlorethane (0.025–0.1% by weight) is commonly used in the form of small briquettes which are held at the bottom of the melt whilst they dissociate into carbon and chlorine. Grain refinement has been attributed to inoculation of the melt with particles of Al_4C_3 or $AlN.Al_4C_3$. However, rod-like substances rich in aluminium, carbon and oxygen have now been detected inside the α-Mg grains which appear to serve as nucleation centres. In this regard, it has also been noted that the compound Al_2OC has lattice dimensions a = 0.317 nm and c = 0.5078 nm that are similar to magnesium (a = 0.320 and c = 0.520) and the same hexagonal crystal structure. Release of chlorine causes some degassing of the melt which is a further advantage of the method.

The ability of zirconium to grain refine most other magnesium alloys can also be attributed to the fact that its α-allotrope (stable below 862 °C) has a hexagonal crystal structure and lattice dimensions (a = 0.323 nm, c = 0.514 nm) close to those of magnesium. This implies that zirconium may nucleate magnesium during solidification and microprobe analysis which has revealed the presence of zirconium-rich cores in the centres of magnesium grains. It is presumed that the cores have formed as a consequence of the peritectic reaction between the liquid and the zirconium particles as shown in the Mg–Zr phase diagram (Fig. 5.6). However, since such a reaction would not be expected unless the zirconium content exceeded 0.58%, at least in binary Mg–Zr alloys, it must also be assumed that some grain refinement occurs by direct nucleation on the zirconium particles, or possibly a zirconium compound. Although major grain refinement is attributed to zirconium in solution, it has been proposed that undissolved zirconium parti-cles may contribute up to 30% of this effect.

Undissolved zirconium particles tend to settle to bottom of a crucible and, the longer the time elapsed after alloying with zirconium, the coarser will be the grain size of castings that happen to be poured from the top of a melt. It is therefore desirable to stir a melt before pouring if this is feasible. Because of the settling problem it is also desirable to pour zirconium-containing alloy cast-ings directly from the crucible in which they are melted. If transfer to another crucible is necessary, the alloy should be replenished with zirconium. Since the settling rate of the zirconium particles is dictated mainly by their size, attention is being directed to producing master alloys containing finer dispersions of particles. One possibility is to deform these master alloys, e.g. by rolling, in order to fragment the particles into smaller sizes.

Another potential problem is that zirconium will react preferentially with any iron present in the melt to form an iron-zirconium intermetallic compound. The resultant very low iron content in the melt will then provide a driving force for the rapid uptake of iron from the crucible, which is usually made of mild steel, if the temperature exceeds about 730 °C. Practical solutions are to maintain a considerable excess of zirconium, which increases costs, or to keep the melt temperature as low as possible and delay introducing the magnesium-zirconium master alloy until just before casting.

5.2.3 Casting and working

Most magnesium alloy components are now produced by high-presure diecasting (Fig. 5.7). Cold chamber machines are used for the largest castings and molten shot weights of 10 kg or more can now be injected in less than 100 ms at pressures that may be as high as 1500 bar. Hot chamber machines are used for most applications and are more competitive for smaller sizes due to the shorter cycle times that are obtainable. Magnesium alloys offer particular advantages for both these processes, namely:

1. Most molten alloys show high fluidity which allows casting of intricate and thinwalled parts, (e.g. 2 mm, Fig. 5.8). Magnesium may be used for casings with thinner walls (1–1.5 mm) than is possible with aluminium (2–2.5 mm) or plastics (2–3 mm).
2. Magnesium has a latent heat of fusion per unit volume that is 2/3 lower than that of aluminium. This means that magnesium castings cool more quickly and die wear is reduced.
3. High gate pressures can be achieved at moderate pressures because of the low density of magnesium.
4. Iron from the dies has very low solubility in magnesium alloys, which is beneficial because it reduces any tendency to die soldering.

A disadvantage with high pressure die castings is that they may contain relatively high levels of porosity. This restricts opportunities for using heat treatment to improve their properties because exposure to high temperatures may cause the pores to swell and form surface blisters (e.g. Fig. 3.12). Sand castings and low pressure permanent mould castings generally contain less porosity and are used to produce components having more complicated shapes. They can then be heat treated if the alloys respond to age hardening. With permanent mould casting, turbulence can be minimized by introducing the molten metal into the bottom of the mould cavity, under a controlled pressure, thereby allowing unidirectional filling of the mould. This method has features in common with the Cosworth process for producing high-quality aluminium castings (Fig. 4.12).

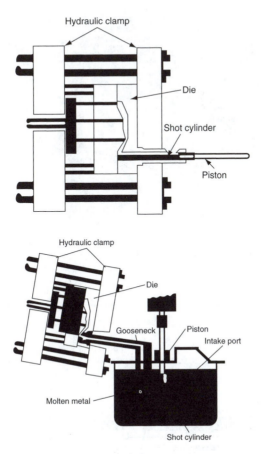

Fig. 5.7 (a) Cold-chamber and (b) hot-chamber pressure diecasting machines (courtesy H. Westengen).

Squeeze casting has also been used to prepare higher-quality castings from existing alloys and to produce castings in alloys that could not be successfully cast by conventional processes. Magnesium alloys are also amenable to thixotropic casting which offers the opportunity to produce high-quality fine-grained products more cheaply than by high-pressure die casting. One promising technique involves heating alloy granules to approximately 20 °C below the liquidus temperature and injection moulding the resulting slurry into a die by means of a high torque screw drive (Fig. 5.9). An example of a thin-walled casting produced by thixomoulding is shown in Fig. 5.10. Mechanical properties of thixomoulded parts are compare well with those obtained with standard high pressure die casting. Thixomoulding has the advantages that microporosity is reduced, die lives are prolonged because heating and cooling cycles are less extreme, and there is no need for a furnace to melt the magnesium alloys prior to casting.

Fig. 5.8 Thin-walled magnesium alloy case and chassis for a hand-portable cellular telephone (courtesy Magnesium Services Ltd, Slough, U.K.).

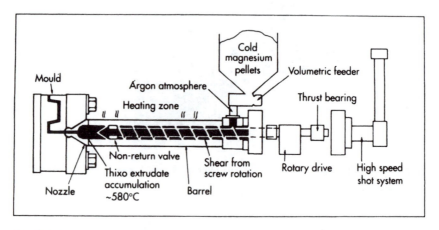

Fig. 5.9 Thixomould for producing diecast magnesium alloys (courtesy Thixomet Inc.).

Magnesium alloys can also be successfully sand cast providing some general principles are followed which are dictated by the particular physical properties and chemical reactivity of magnesium.

1. Suitable inhibitors must be added to the moulding sand in order to avoid reaction between molten magnesium and moisture which would liberate hydrogen. For green sand or sand gassed with carbon dioxide to aid bonding, sulphur is used, whereas for synthetic sands compounds such as KBF_4 and $KSiF_6$ are also added. Boric acid is also used for some sands, both as

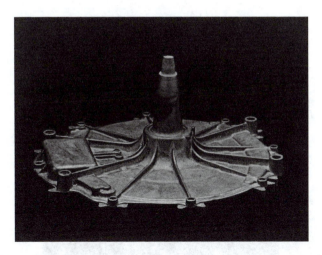

Fig. 5.10 Thin-walled magnesium alloy AZ91D casting produced by thixo-moulding (courtesy Thixomet Inc.).

a moulding aid and as a possible inhibitor through its tendency to coat the sand grains.

2. Metal flow should be as smooth as possible to minimize oxidation.
3. Because of the low density of magnesium, there is a relatively small pressure head in risers and sprues to assist the filling of moulds. Sands need to be permeable and the mould must be well-vented to allow the expulsion of air.
4. Magnesium has a relatively low volumetric heat capacity which necessitates provision of generous risers to maintain a reservoir of hotter metal. Because of this need to make special provision by way of risers and other feeding devices, the volume ratio of poured metal to actual castings may average as much as 4:1 for magnesium alloys.

Ingots for producing wrought products have mainly been cast in permanent metal moulds. However, with a revival in interest in producing magnesium alloy sheet and extrusions, more attention is being given to semi-continuous direct chill methods which are broadly similar to those described for aluminium alloys (Fig. 3.2). Following extensive pilot scale trials, a new vertical direct chill caster for magnesium alloys has been constructed in Austria that is capable of producing extrusion billets with a maximum weight of 4 000 kg. Smaller, horizontal direct chill casters have also been developed.

For the conventional production of magnesium alloy sheet, cast ingots are usually produced with dimensions up to 0.3 m × 1 m × 2 m. These ingots are homogenised (e.g. at 480 °C) for several hours, scalped and hot rolled in stages in a reversing hot mill to a thickness of 5–6 mm. The sheet is then annealed (e.g. at 340 °C) before each subsequent rolling pass which reduces the thickness by only between 5 and 20% because of the hexagonal crystal structure of the

Fig. 5.11 Pilot plant for twin-roll casting of magnesium alloys (courtesy D. Liang, CSIRO, Manufacturing and Infrastructure Technology Division, Melbourne.).

alloys. This latter part of the rolling process is costly and time consuming so that the current useage of magnesium alloy sheet world wide is limited to around 1000 tonnes. This makes the prospect of strip casting an attractive alternative if sheet can be cast to near final thickness and given a final finish roll. Casting of magnesium sheet does, however, present some additional problems apart from the tendency of the molten metal to oxidize with the danger of catching fire. These problems arise because magnesium alloys freeze faster than aluminium alloys and generally have longer freezing ranges which makes as-cast sheet more prone to surface defects and internal segregation. Nevertheless, some success has been achieved through the adaption of twin-roll casting practices developed for strip casting aluminium alloys. As an example, Fig. 5.11 shows a pilot plant that can directly cast a range of magnesium alloys as sheet up to 600 mm wide which has a thickness of 2.5 mm that can be further reduced down to 0.5 mm using a conventional finishing mill.

5.3 ALLOY DESIGNATIONS AND TEMPERS

No international code for designating magnesium alloys exists although there has been a trend towards adopting the method used by the American Society for Testing Materials. In this system, the first two letters indicate the principal alloying elements according to the following code: A—aluminium; B—bismuth;

C—copper; D—cadmium; E—rare earths; F—iron; G—magnesium; H—thorium; K—zirconium; L—lithium; M—manganese; N—nickel; P—lead; Q—silver; R—chromium; S—silicon; T—tin; W—yttrium; Y—antimony; Z—zinc. The letter corresponding to the element present in greater quantity in the alloy is used first, and if they are equal in quantity the letters are listed alphabetically. The two (or one) letters are followed by numbers which represent the nominal compositions of these principal alloying elements in weight %, rounded off to the nearest whole number, e.g. AZ91 indicates the alloy Mg–9Al–1Zn the actual composition ranges being 8.3–9.7 Al and 0.4–1.0 Zn. A limitation is that information concerning other intentionally added elements is not given, and the system may need to be modified on this account. Suffix letters A, B, C etc. refer to variations in composition within the specified range and X indicates that the alloy is experimental. This system will be used when discussing alloys in this book.

The heat-treated or work-hardened conditions, i.e. tempers, of alloys are specified in the same way that has been described for aluminium alloys in Section 3.2. Commonly used tempers are T5 (alloys artificially aged after casting), T6 (alloys solution treated, quenched and artificially aged) or T7 (alloys solution treated and stabilized).

5.4 ZIRCONIUM-FREE CASTING ALLOYS

5.4.1 Alloys based on the magnesium–aluminium system

The Mg–Al system has been the basis of the most widely used magnesium casting alloys since these materials were introduced in Germany during the First World War. Most alloys contain 8–9% aluminium with small amounts of zinc, which gives some increase in tensile properties, and manganese, e.g. 0.3%, which improves corrosion resistance (Section 5.1). The presence of aluminium requires that the alloys be grain refined by superheating or inoculation. A section of the Mg–Al phase diagram together with a summary of the effects of composition and heat treatment on tensile properties of an alloy similar to AZ80A is shown in Fig. 5.12 and other property data are included in Table 5.3.

In the as-cast condition, the β phase $Mg_{17}Al_{12}$ (sometimes referred to as Mg_4Al_3) appears in alloys containing more than 2% aluminium. A network of β forms around grain boundaries as the aluminium content is increased (Fig. 5.13a) and ductility decreases rapidly above 8% (Fig. 5.12). In more slowly cooled castings, discontinuous precipitation of the β phase may occur at grain boundaries with the formation of a cellular or pearlitic structure (Fig. 5.13b). Annealing at temperatures around 420 °C causes the cellular constituent and all or part of the β phase along grain boundaries to redissolve leading to solid solution strengthening. Interdendritic coring (Fig. 5.13a) is also reduced and both tensile strength and ductility are significantly improved. Discontinuous precipitation of lamellae of the β phase is considered to be undesirable in Mg–Al alloys subject to creep conditions and attempts have been made to prevent its formation by adding

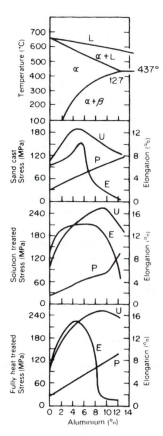

Fig. 5.12 Relation of properties to constitution in the Mg–Al alloys, showing effects of composition and heat treatment on the tensile properties of sand cast AZ80A alloy. P = 0.2% proof stress, U = tensile strength, E = elongation (from Fox, F. A., *J. Inst. Metals,* **71**, 415, 1945).

microalloying elements. So far, the only effective addition has been a trace of gold (~0.1 at.%) which apparently segregates in grain boundaries and suppresses the growth of discontinuous precipitates, perhaps because of the strong inter-action that is known to occur between gold and aluminium atoms. As a conse-quence of this change in microstructure, creep resistance is improved although the use of such an expensive element is not a practical solution to this problem. Discontinuous precipitation may also be eliminated by adding high concentra-tions of zinc to Mg–Al alloys but the Mg–Zn–Al alloys also show only moderate creep resistance.

Because the solid solubility of aluminium in magnesium decreases from a maximum of 12.7 wt.% at 437 °C to around 2% at room temperature (Fig 5.12), it might be expected that ageing treatments would induce significant precipitation

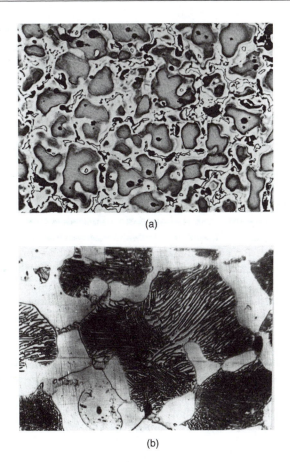

(a)

(b)

Fig. 5.13 Cast structures in alloy AZ80: (a) chill cast alloy with the β-phase ($Mg_{17}Al_{12}$) present in the grain boundaries. Note also the interdendritic aluminium-rich coring (white) around the edges of the α-grains (grey) ($\times$ 200); (b) discontinuous precipitation in more slowly cooled alloys ($\times$ 500) (courtesy Magnesium Elektron Ltd).

hardening. However, as shown in Table 5.5, supersaturated solid solutions of Mg–Al alloys transform directly to the equilibrium precipitate β on artificial ageing, without the appearance of GP zones or an intermediate precipitate. The β phase mainly forms as coarse laths on the basal planes of the α Mg matrix and the response to age hardening is relatively poor. Accordingly, alloys based on this system are generally used in the as-cast or as-cast and annealed conditions.

The addition of zinc to Mg–Al alloys causes some strengthening although the amount of zinc is limited because of an increase in susceptibility to hot cracking during solidification. Actual levels of zinc are inversely related to the aluminium contents and two examples of alloys are AZ63 (Mg–6Al–3Zn–0.3Mn) and AZ91 (Mg–9Al–0.7Zn–0.2Mn). AZ91 is the most widely used of all magnesium alloys.

As mentioned earlier, the corrosion resistance of this and related alloys is adversely affected by the presence of cathodic impurities such as iron and nickel and, for some purposes, strict limits have now been placed on these elements. Higher-purity versions such as AZ91E (Fe 0.015% max., Ni 0.001% max. and Cu 0.015% max.) have corrosion rates in salt fog tests which are as much as 100 times lower than for AZ91C so that they become comparable with those for some aluminium casting alloys.

Requirements for specific property improvements have stimulated the development of alternative diecasting alloys. For applications where greater ductility and fracture toughness are required, a series of high-purity alloys with reduced aluminium contents are available. Examples are AM60, AM50 and AM20 (Table 5.3) The improved properties arise because of a reduction in the amount of $Mg_{17}Al_{12}$ around grain boundaries. Such alloys are used for automotive parts including wheels, seat frames and steering wheels. Figure 5.14 shows a seat frame for a Mercedes Benz roadster which has been produced from five AM20 and AM50 alloy diecastings and has a weight of 8.3 kg. It is claimed that a comparable steel seat would weigh an estimated 35 kg and require between 20 and 30 stampings and weldments. An added advantage of the magnesium alloy seat is its extra stiffness which has allowed the safety belt and belt mechanism to be attached and incorporated in the back rest.

Cast Mg–Al and Mg–Al–Zn alloys show some susceptibility to microporosity but otherwise have good casting qualities and resistance to corrosion is generally satisfactory. They are suitable for use at temperatures up to 110–120 °C above which creep rates become unacceptable (Figs. 5.15 and 5.16). This behaviour is attributed to the fact that magnesium alloys undergo creep mainly by grain boundary sliding and the phase $Mg_{17}Al_{12}$, which has a melting point of approximately

Fig. 5.14 Diecast magnesium alloy seat frame.

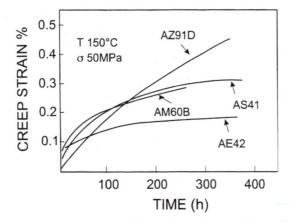

Fig. 5.15 Creep strain versus time for commercial magnesium die casting alloys based on the Mg–Al system. Tests at 150 °C and a stress of 50 MPa (from Han, Q. *et al, Phil. Mag.*, **84**, No. 36, 3843, 2004).

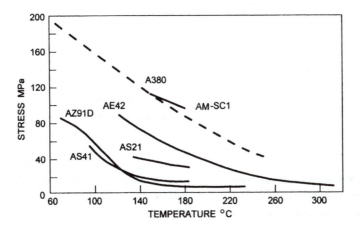

Fig. 5.16 Stress for 0.1% creep strain in 100 h for commercial magnesium die casting alloys based on the Mg–Al system, the Mg–Zn–RE sand cast alloy designated AMC-SC1, and the aluminium die casting alloy 380 (from Waltrip, J.S., *Proc. 47th Ann. World Magnesium Conf.*, 1990, Inter. Magnesium Assoc., p. 124, 1990; Bettles, C.J. *et al, Magnesium Technology 2003*, Ed. H. I. Kaplan, TMS, Warrendale, Pa, USA, p. 223, 2003).

460 °C and is comparatively soft at lower temperatures, does not serve to pin boundaries. Accordingly, commercial requirements have led to the investigation of other alloys based on the Mg–Al system.

Creep properties of Mg–Al alloys may be improved by lowering the aluminium content and introducing silicon which has the effect of reducing the amount of $Mg_{17}Al_{12}$. Providing solidification rates are fast, such as in pressure die casting, the silicon combines with some of the magnesium to precipitate fine and relatively hard particles of Mg_2Si in grain boundaries, rather than forming this phase in its coarse, Chinese script morphology that is detrimental to mechanical properties. Small additions of calcium (e.g. 0.1%) also help suppress this undesirable form of Mg_2Si. Two examples are the commercial alloys AS41 (Mg–4.5Al–1Si–0.3Mn) and AS21(Mg 2.2Al–1Si–0.3Mn). As shown in Figs. 5.16, alloy AS21 with the lower content of aluminium performs better than AS41, but it is more difficult to cast because of reduced fluidity. These alloys were exploited on a large scale in the rear engine of various models of the famous Volkswagon Beetle motor car in which the replacement of the cast iron crank case and transmission housing with magnesium alloys saved some 50 kg in weight. Such a saving was critical for the road stability of such a vehicle.

The creep properties of Mg–Al–Si alloys still fall well below those of competing die cast aluminium alloys such as A380 (Fig. 5.16) and attention has been directed at other alloying additions, notably the rare earth (RE) and alkaline earth elements that form relatively stable intermetallic dispersoids in their as-cast microstructures. While the individual solubility of some of these elements in molten magnesium is quite high, their mutual solubility is usually much less if several elements are present. The challenge in developing more creep resistant die casting alloys is to retain enough of the elements in solution to give some solid solution hardening, while maximizing both the volume fraction and spatial distribution of dispersoids. At the same time, the alloys must show acceptable castability and corrosion resistance.

As shown in Table 5.1, magnesium forms solid solutions with a number of RE elements and magnesium-rich sections of the respective phase diagrams all contain simple eutectics. Because of this, the binary alloys have good casting characteristics since the presence of the relatively low melting point eutectics as networks in grain boundaries tends to suppress microporosity. The presence of the grain boundary phases reduces grain boundary sliding under creep conditions and ageing treatments, if permitted (e.g. for sand castings), promote general strengthening of the matrix. RE elements are comparatively expensive and the most economical way of making these additions is to use naturally occurring misch metals based on cerium (e.g. 50%Ce–20%La–15%Nd–5%Pr) or neodymium (e.g. 80%Nd–16%Pr–2%Gd–2% others) although the use of these compounds still imposes a significant cost penalty.

Individual RE elements have differing effects on the response of magnesium alloys to heat treatment with the mechanical properties generally increasing in the order of lanthanum, misch metal, cerium and didymium (72% neodymium). This effect is demonstrated for creep extension curves (Fig. 5.17) and probably corresponds to an increasing solid solubility of the elements in magnesium. It should also be noted that cerium is in fact marginally better than misch metal over the useful addition range of these elements, i.e. below 3%, but the expense of separating individual RE elements from misch metal is normally not justified.

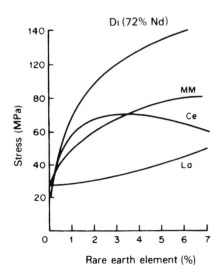

Fig. 5.17 Effect of various rare earth metals on the stress for 0.5% extension in 100 h at 205 °C. Fully heat-treated Mg–RE alloys (from Leontis, T. E. *TAIMME*, **35**, 968, 1949).

Generally Mg–Al–RE alloys are only suitable for die casting since fast cooling is needed form a relatively fine dispersion of the compound $Al_{11}RE_3$, rather than coarser particles of Al_2RE. $Al_{11}RE_3$ forms as lamellae within the interdendritic regions of the matrix. In addition, nucleation of the stable phase $Mg_{12}Ce$ in the grain boundaries has been observed to occur during creep which is also considered to reduce grain boundary sliding. One composition, AE42 (Mg–4Al–2RE–0.3Mn) has a good combination of mechanical properties including creep strength that is superior to the Mg–Al–Si alloys (Figs. 5.15 and 5.16). However, the creep properties of AE42 deteriorate rapidly above 175 °C, apparently because $Al_{11}RE_3$ decomposes to Al_2RE which releases aluminium to form the undesirable and much softer phase $Mg_{17}Al_{12}$.

Alkaline earth additions of calcium and strontium to Mg–Al alloys have been studied, with and without the presence of RE elements, as another way to improve creep resistance. For example, it has been claimed that, with the die cast alloy AM50 (Mg5Al–0.3Mn), the minimum creep rate at 150 °C may be reduced by three orders of magnitude by the addition of 1.7%Ca. A more detailed comparison of the effects of (a) aluminium and (b) alkaline earth and RE elements on the properties and cost of magnesium die casting alloys is shown schematically in Fig. 5.18. Both calcium and strontium have low solubilities in magnesium and form stable particles of Al_2Ca or Al_4Sr, mainly in grain boundaries, that reduce grain boundary sliding. When RE elements are also present, the $Al_{11}RE_3$ phase forms within the grains that reduces dislocation slip. Calcium is cheaper than strontium although the latter element is claimed to have the advantage of promoting higher tensile properties at elevated temperatures. Both elements have some

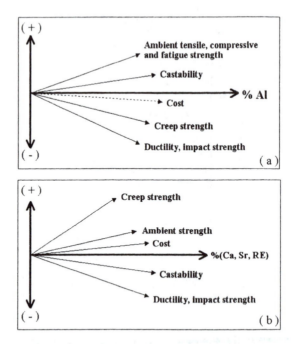

Fig. 5.18 General effects of (a) aluminium and (b) alkaline and rare earth elements on the properties and cost of die cast magnesium alloys (from Aghion, E. *et al, Materials Science Forum*, **419–422**, 407, 2003.).

adverse effects on castability and may also increase the susceptibility of castings to hot tearing.

A number of new alloys are being produced, some of which are now finding industrial applications. Examples are AX81 (Mg–8Al–1Ca), AXE522 (Mg–5Al–2RE–2Ca), and AJ62 (Mg–6Al–2.3r). (The letters "X" and "J" have been used to represent calcium and strontium.) Some, such as MRI 153 and 230D that have been developed by collaboration between the companies Dead Sea Magnesium and Volkswagon AG are known only by their proprietary designations. Properties of some of these alloys are summarized in Table 5.6 in which comparisons can be made with the commonly used magnesium alloys AZ 91, AS21, AS41 and AE42, and with the aluminium casting alloy 380. The new magnesium alloys have room temperature tensile properties similar to, or exceeding those of other alloys based on the Mg–Al system, but show improved creep strengths at temperatures in the range 150 to 175 °C, and are generally more resistant to corrosion. Small additions of zinc are known to improve metal flow during casting which has led to the development of another series of creep-resistant alloys of which one designated ZACE05613 (Mg–0.5Zn–6Al–1Ca–3RE) is an example. An important innovation has been the selection of the strontium-containing alloy AJ62, that was

Table 5.6 Summaries of properties of new and experimental die cast magnesium alloys, the magnesium alloys AZ91D, AS41, and AE42 and the aluminium casting alloy 380 (from Pekguleryuz, M. O. and Kaya, A. A., *Proceedings of 6th International Conference Magnesium Alloys and Their Applications*, Ed. K. U. Kainer, Wiley-VCH Verlag GmbH, Germany, 74, 2004; Luo. A.A., *Materials Science Forum*, **419-422**, 59, 2003; and Aghion, E, et al, *ibid.* p. 409)

Alloy	Composition	Tensile Properties 20 °C			Tensile Properties 150 °C			Tensile Properties 175 °C			%Tensile Creep at 50 MPa, 200h		Salt Spray Corrosion Rate
		YS MPa	TS MPa	Elong. %	YS MPa	TS MPa	Elong. %	YS MPa	TS MPa	Elong. %	150 °C	175 °C	Mg/cm²/day
A380	Al–8.5Si–3.5Cu	168	290	3	155	255	6	151	238	8.5	0.08	–	0.34
AZ91D	Mg–9Al–1Zn	150	222	3	105	160	18	89	138	21	2.7	(i)	0.10
AS21	Mg–2Al–1Si	121	210	5.5	87	120	27	78	110	23	0.19	1.27	–
AS41	Mg–4Al–1Si	138	206	5	94	154	20	85	127	22	0.05	2.48	0.16
AE42	Mg–4Al–2RE	137	225	10.5	100	160	22	86	135	23	0.06	0.33	0.21
AX51	Mg–5Al–1Ca	128	192	7	102	161	7	–	–	–	–	–	–
AX52	Mg–5Al–2Ca	161	228	13	–	–	–	133	171	23	–	–	–
AJ52	Mg–5Al–2Sr	134	212	6	108	164	14	100	141	18	0.04	0.05	0.09
AJ62	Mg–6Al–2.2Sr	143	240	7	116	166	27	103	143	19	0.05	0.05	0.11
MRI153M	–	170	250	6	135	190	17	112	139	3.5	0.18	–	0.09
MRI230D	–	180	235	5	150	205	16	–	–	–	–	–	0.10

(i) Failed after 80 h

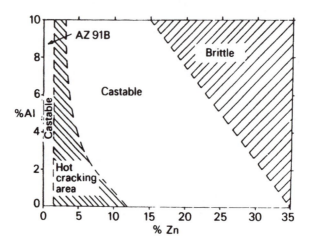

Fig. 5.19 Potential composition ranges of Mg–Al–Zn alloys for diecasting (from Foerster, G. S., *Proceedings of 33rd Annual Meeting International Magnesium Association*, Montreal, 1976).

developed in Canada, for use in the lightweight composite engine for one model of a motor car being produced by BMW in Germany. As shown earlier in Fig. 4.7, this magnesium alloy is used for the engine block which is cast around an aluminium-silicon alloy A390 core containing the cylinders.

It has been shown the Mg–Al alloys with high zinc contents may have attractive die casting characteristics combined with tensile and corrosion properties that exceed traditional alloys such as AZ91. Composition ranges for these potential alloys are shown in Fig. 5.19 and alloys with zinc contents as high as 12%, e.g. Al–12Zn–4Al, have been investigated. Studies of the alloy Mg–8Zn–4Al have shown that the as-cast microstructure contains a high volume fraction of primary intermetallic compounds, most of which are a quasi-crystalline phase with the approximate composition $Mg_9Zn_4Al_3$. Such phases lack the translational long range order of the crystalline state and may display unique properties; in particular a very high hardness. The amount of this phase decreases on homogenising at 325 °C, and quenching from this temperature followed by ageing at 150–200 °C promotes age hardening through precipitation of a finely dispersed, diamond-shaped phase that also appears to have a quasi-crystalline structure (Section 7.4). The addition of 0.5% calcium is reported to result in the replacement of the primary quasi-crystalline phase with a crystalline cubic phase in the as-cast microstrucure, and to stimulate precipitation of an additional rod-like phase when the alloy is aged. Although the presence of calcium does not increase maximum hardness, the rate of overageing at 200 °C is much reduced and creep resistance is improved. For example, the creep strain after 100 h at 150 °C and a stress of 60 MPa is 0.05% which compares with 0.21% for the ternary Mg–Zn–Al alloy.

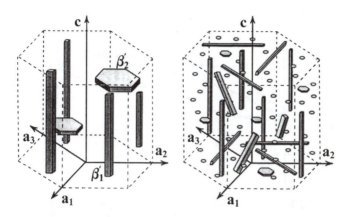

Fig. 5.20 Schematic representations of the precipitate sizes, morphologies and habit planes with respect to the α Mg matrix for (a) Mg–4Zn and (b) Mg–4Zn–Sr aged for 2.7 h at 177 °C (courtesy of C. J. Bettles and M.A. Gibson).

5.4.2 Magnesium–zinc alloys

As mentioned in Section 5.1, zinc causes more solid solution strengthening than an equal atomic per cent of aluminium but its solubility is much less (Table 5.1). Mg–Zn alloys also respond to age hardening and, as shown in Table 5.5, the ageing process is complex and may involve four stages. The GP zones solvus for the alloy Mg–5.5Zn lies between 70 and 80 °C, and preageing below this solvus before ageing at a higher temperature (e.g. 150 °C) refines the size and dispersion of rods of the coherent phase $MgZn_2$ that form from the GP zones. Maximum hardening is associated with the presence of this coherent phase. However, binary Mg–Zn alloys are not amenable to grain refining by superheating or inoculation, and are prone to microporosity. As a consequence, they are not used for commercial castings.

Minor additions may modify precipitation in Mg–Zn alloys. Examples are the elements calcium and strontium which accelerate the rate of ageing but delay overageing, refine the sizes, and increase the number densities of precipitates that form. The resulting microstructure is shown schematically in Fig. 5.20.

5.4.3 Magnesium–zinc–copper alloys

Recent work has shown that the addition of copper to binary Mg–Zn alloys causes a marked improvement in ductility and induces a relatively large response to age hardening. The copper-containing alloys exhibit tensile properties similar to alloy AZ91 (e.g. 0.2% P.S. 130–160 MPa, T.S. 215–260 MPa and ductility 3–8%) but have the advantage that these properties are more reproducible. Elevated temperature stability is also improved.

One typical sand casting alloy is designated ZC63 or ZCM630 (Mg–6Zn–3Cu–0.5Mn). It has been found that progressive addition of copper to Mg–Zn alloys raises the eutectic temperature (Fig. 5.21) which is important

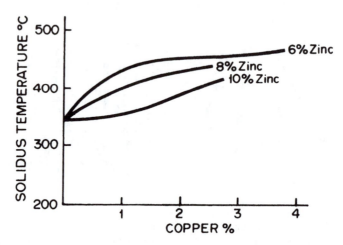

Fig. 5.21 Effect of copper in raising the eutectic temperature of Mg–Zn alloys (from Unsworth, W. and King, J. F., *Magnesium Technology*, p. 25, Institute of Metals, London, 1987).

because it permits the use of higher solution treatment temperatures, thereby ensuring maximum solution of zinc and copper. The structure of the eutectic is also changed from being completely divorced in binary Mg–Zn alloys, with the Mg–Zn compound distributed around grain boundaries and between dendrite arms, to truly lamellar in the ternary copper-containing alloys (Fig. 5.22).

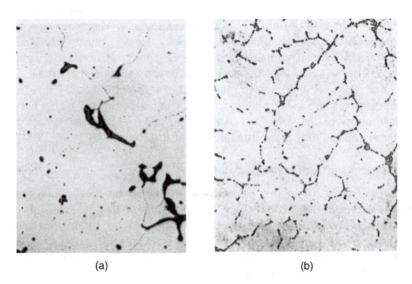

Fig. 5.22 Effect of the addition of copper on the morphology of the eutectic in the alloy Mg–6Zn (× 70): (a) binary alloy solution treated 8 h at 330 °C; (b) Mg–6Zn–1.5Cu solution treated 8 h at 430 °C (courtesy G.W. Lorimer).

On solution treatment, partial dissolution of the eutectic occurs leaving rounded rods or platelets within the α matrix. It is this structure that is believed to improve the ductility of alloy. A typical heat treatment cycle involves solution treatment for 8 h at 440 °C, hot water quench, followed by ageing for 16–24 h at 180–200 °C. Two main precipitates have been identified, β'_1 (rods) and β'_2 (plates), which appear to be similar in structure to the phases observed in binary Mg–Zn alloys. However, the density of precipitates is much greater and more uniform when copper is present (Fig. 5.23). Peak hardening occurs around 200 °C when both these precipitates are present. Manganese is added because it has been found to stabilize the ageing response and to reduce the rate of over-ageing.

The addition of copper to Mg–Al–Zn alloys has a detrimental effect on corrosion resistance but this appears not to be the case in Mg–Zn–Cu alloys. This may be attributed to the incorporation of the copper in the eutectic phase as $Mg(Cu,Zn)_2$. Fatigue strength in the unnotched condition (e.g. endurance limit 10^7 cycles of ± 90 MPa) is better than that of Mg–Al–Zn alloys, while the notched values are comparable.

A number of Mg–Zn–Cu castings have been produced under practical foundry conditions using sand, gravity die, and precision casting techniques. These have confirmed the good casting characteristics of the alloy, notably its freedom from microshrinkage, which enables pressure tight castings to be produced without impregnation. Castings can be welded using the standard tungsten-inert gas technique. Taken overall, these alloys would seem to offer significant improvements over the traditional Mg–Al–Zn alloys for critical high strength castings. It is considered that they may find particular application in the automotive industry.

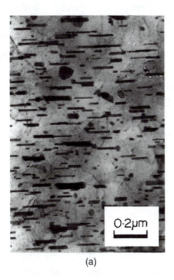

(a)   (b)

Fig. 5.23 Comparison of precipitate densities in (a) Mg–6Zn and (b) Mg–6Zn–1.5Cu alloys aged 8 h at 200 °C (courtesy G. W. Lorimer).

5.4.4 Magnesium–rare earth–zinc alloys

Some time ago, Magnesium Elektron in England developed an alloy designated MEZ (Mg–2.5RE–0.5Zn–0.35Mn) that has potential automotive applications such as gear boxes and oil pans. This alloy was designed for use in high pressure die castings and therefore does not require grain refining. Zinc was added to improve castability. The tensile properties are lower than for the alloy AE42 (Section 5.4.1) but MEZ shows superior creep resistance, especially at temperatures above 150 °C. This latter behaviour is attributed to the presence of a more stable compound in grain boundaries that has the general formula $Mg_{12}RE$ and is probably $Mg_{12}(La_{0.43}Ce_{0.57})$ in which partial substitution of zinc for magnesium may occur.

Although special attention has been given to the development of magnesium alloys for high pressure die casting, this method of manufacture is not necessarily the best for casting an engine block. Design variations can be accommodated more readily when using permanent mould or sand casting. Moreover, because such castings usually contain less porosity, they are amenable to heat treatment to improve their mechanical properties because there in little danger of swelling and blistering. Several programs are in progress to design and evaluate permanent mould and sand cast magnesium alloys for possible use in light weight engines.

One such development has involved groups in Australia, Germany, Austria and England that have focused on producing a cost-effective, sand cast alloy that based on the Mg–RE system for use for the cylinder blocks and crank cases. This alloy has been designated AM–SC1, has the nominal composition Mg–2.7RE–0.5Zn, which is similar to MEZ. However a special combination of RE elements has been used to promote creep resistance. It has a microstructure after a T6 temper that is generally similar to MEZ and has an intermetallic phase dispersed around the grain boundaries and triple points to minimize grain boundary sliding at the operating temperature of around 150 °C.

Since the new alloy must compete with the cast aluminium alloys A319 (Al–6Si–3.5Cu) and A380 (Al–8.5Si–3.5Cu), it was considered that the target values for the critical creep and fatigue properties should at least equal the values for these two alloys. These target values and the actual properties achieved for AM-SC1 aged to a T6 temper were as follows:

- Stress to give 0.1% creep strain after 100 h : target 110 MPa at 150 °C and 90 MPa at 177 °C, actual values 115 and 100 MPa respectively (Fig. 5.16).
- Fatigue limit after 5×10^7 cycles at 25 °C, R = −1 : target 50 MPa, actual value 75 MPa.

While engine blocks do not require a high level of tensile yield strength, it is vital that the compressive yield strength and creep resistance of an alloy are adequate to minimize relaxation, especially in bolted assemblies. The compressive yield strength of AM–SC1, which is 133 MPa at 20 °C, remains unchanged up to 177 °C. Bolt load retention properties at 150 °C were found to be similar to aluminium alloy A319 and superior to the die cast magnesium alloy AE42.

The alloy has been successfully road tested in the engine block of a small three cylinder diesel engine installed in a Volkswagon Lupo automobile. The block, that has separate hypereutectic Al–Si alloy cylinder liners and reinforced main bearing housings, weighs only 14 kg. In 2004, the AM–SC1 alloy was selected by the United States Automobile Materials Partnership for the sand cast cylinder block in their Magnesium Intensive Engine Research Project. For this purpose a V6 engine block has been specially designed which incorporates cast iron cylinder liners.

5.5 ZIRCONIUM-CONTAINING CASTING ALLOYS

The maximum solubility of zirconium in molten magnesium is 0.6% and as binary Mg–Zr alloys are not sufficiently strong for commercial application, the addition of other alloying elements has been necessary. The selection of these elements has been governed by three main factors:

1. compatability with zirconium
2. founding characteristics
3. properties desired of the alloy.

With respect to (1), a pre-requisite is the absence of elements such as aluminium and manganese that effectively suppress the liquid solubility of zirconium in magnesium. Elements that do not interfere with the grain refining effect of zirconium include zinc, RE elements, thorium, silver and yttrium. So far as (3) is concerned, improved tensile properties including higher ratios of proof strength to tensile strength and enhanced creep resistance, have been the two principal objectives in a alloy development that has been driven by the aerospace industries. Developments in various countries have been rather similar and Table 5.4 and Figs 5.24 and 5.25 compare the elevated temperature proof stresses of some of these alloys.

5.5.1 Magnesium–zinc–zirconium alloys

The ability to grain refine Mg–Zn alloys with zirconium led to the introduction of alloys such as ZK51 (Mg–4.5Zn–0.7Zr) and the higher strength ZK61 (Mg–6Zn–0.7Zr) that are normally used in the T5 and T6 tempers respectively. However the fact that these alloys are susceptible to microporosity, and are not weldable, has severely restricted their practical application.

5.5.2 Magnesium–rare earth–zinc–zirconium alloys

These alloys also show good casting characteristics because the presence of the RE elements promotes formation of relatively low melting point eutectics that improve fluidity and tend to prevent microporosity. In the as-cast condition,

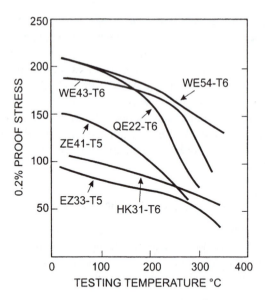

Fig. 5.24 Effect of temperature on the 0.2% proof stress of sand cast magnesium alloys grain refined by zirconium and age hardened (courtesy Magnesium Elektron Ltd.).

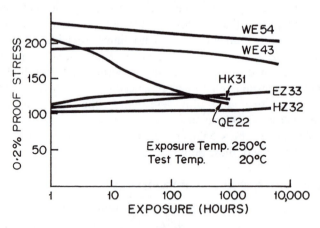

Fig. 5.25 Effect of exposure at 250 °C on the 0.2% proof stress at room temperature for several creep-resistant magnesium alloys containing zirconium.

the alloys generally have cored α grains surrounded by grain boundary networks. Ageing causes precipitation to occur within the grains and the generally good creep resistance they display is attributed both to the strengthening effect of this precipitate and the presence of the grain boundary phases which reduce grain boundary sliding (Fig. 5.26).

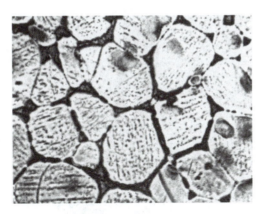

Fig. 5.26 Mg–RE–Zn–Zr alloy EZ33 as-cast and aged 8 h at 350 °C (T5 temper) × 500.

The properties of Mg–RE alloys are enhanced by adding zirconium to refine grain size and further increases in strength occur if zinc is present as well. One commonly used alloy is ZE41 (Mg–4.2Zn–1.3Ce–0.6Zr) that develops moderate strength when given a T5 ageing treatment which is maintained up to 150 °C. One application has been helicopter transmission housings (Fig. 5.27). Higher tensile properties combined with good creep strength at temperatures up to 250 °C have been achieved in the alloy EZ33 (Mg–3RE–2.5Zn–0.6Zr).

The precipitation processes in Mg–RE systems are not completely understood. Particular attention has been paid to Mg–Nd alloys in which four stages have been detected (Table 5.5). Most hardening is associated with formation of the coherent β''-phase, which has the ordered DO_{19} structure and is probably Mg_3Nd. Loss of coherency of this phase occurs close to 250 °C and is associated with a marked increase in creep rate. The β'-phase is thought to have a hexagonal structure and is nucleated on dislocation lines when the alloys are aged in the temperature range 200–300 °C. It is possible that the MgRE (Ce) alloys have a similar ageing sequence which leads to the eventual formation of an equilibrium precipitate that is either $Mg_{12}Ce$ or $Mg_{27}Ce_2$.

The role of zinc with respect to strengthening is uncertain. It is likely that independent formation of Mg–Zn precipitates occurs although part of the zinc is associated with RE elements in constituents that form in grain boundaries. The effect of zirconium is thought to be confined to its role in grain refinement.

It might be thought that by starting with a high zinc content, e.g. 6%, and adding sufficient rare earth metals, e.g. 3%, an alloy could be produced which could combine the high tensile properties of ZK51 with the castability and freedom from microporosity of EZ33. Unfortunately such an alloy composition proves to be unusable since the low solidus temperature precludes solution heat treatment and the elongation, at 3% RE, is almost nil. Realizing that the low ductility was connected with a brittle grain boundary phase, it occurred to P. A.

Fig. 5.27 Cutaway section of the magnesium alloy gearcase in the Sud-Aviation Super Frelon helicopter. Casting is made from the Mg–Zn–RE–Zr alloy ZE41 and weighs 133 kg as-cast and 111 kg finish machined (courtesy Magnesium Industry Council).

Fisher to decompose the latter by diffusing hydrogen into the alloy at a high temperature, thereby precipitating the RE metals as hydrides and enabling the zinc to be taken into solution (Fig. 5.28a, b). After final precipitation treatment in which a needle-like phase forms in the grains (Fig. 5.28c), castings in this alloy show high tensile properties with freedom from microporosity, together with high elongation values and outstanding fatigue resistance, which is an altogether remarkable combination of properties. The effect of the hydrogen treatment on tensile properties is shown for the Mg–6Zn–2RE composition in Table 5.7.

The final alloy developed in this way (ZE63) has the composition Mg–5.8Zn–2.5RE–0.7Zr. It has found limited but important usage in the aircraft industry. The rate of penetration of hydrogen is about 6 mm in 24 h at a temperature of 480 °C and a pressure of 1 atm. Penetration can be accelerated by increasing the gas pressure, but the slowness of the hydriding step has hitherto restricted use of the alloy to castings with fairly thin sections.

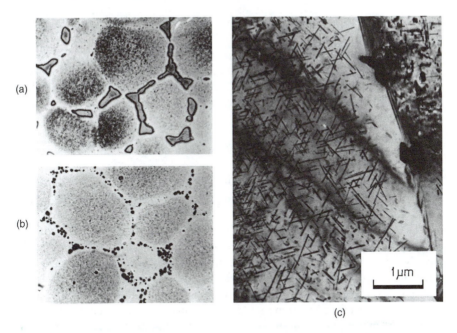

Fig. 5.28 Effect of hydriding on the microstructure of the alloy ZE63: (a) as-cast microstructure; (b) alloy heat treated in hydrogen atmosphere at 480 °C and given T6 temper (courtesy Magnesium Elektron Ltd) (× 300); (c) thin-foil electron micrograph showing massive RE hydrides in the grain boundary and needles within the grains which are probably ZrH_2 (courtesy K. J. Gradwell) (× 13 000).

Table 5.7 Tensile properties of cast test bars of the alloys ZK61 and ZE62 after solution treatment in an atmosphere of (a) SO_2 and (b) wet H_2 (from Fisher, P. A. et al., *Foundry*, **95**(8), 68, 1967)

Alloy		Tensile properties		
	Heat treatment: Solution treatment 24 h 500 °C; Aged 64 h 125 °C	0.2% proof stress (MPa)	Tensile strength (MPa)	Elongation (% in 5 cm)
ZK61 (Mg–6Zn–0.7Zr)	SO_2	186	271	3.5
	H_2	183	252	2.5
ZE62 (ZK61 + 2%RE)	SO_2	94	173	3.8
	H_2	181	306	12.0

5.5.3 Alloys based on the magnesium–thorium system

It has been known for some time that additions of thorium also confer increased creep resistance to magnesium alloys and cast and wrought compositions were developed for service at temperatures as high as 350 °C (Figs. 5.24 and 5.25). As with the RE elements, thorium improves casting properties and, in the as-cast condition, the ternary alloy HK31 (Mg–3.2Th–0.7Zr) has a microstructure similar to the Mg–RE–Zr alloys. Thorium-containing alloys are normally given a T6 ageing treatment and investigations of the precipitation processes has also revealed similarities with those for the Mg–RE–Zr alloys. Although GP zones are likely to occur at low temperatures, none have been identified and the precipitation sequence has been described as

$$\text{SSSS} \rightarrow \beta'' \rightarrow \beta' \rightarrow \beta \ (\text{Mg}_{23}\text{Th}_6)$$

The phase β'' again has an ordered DO_{19} structure and occurs as thin discs which are coherent with the $\{10\bar{1}0\}$ and probably $\{11\bar{2}0\}$ planes of the matrix. There is dispute as to whether the formula of this compound is $MgTh_3$ or Mg_3Th, although the latter would provide the low energy interfaces with the matrix proposed in Fig. 5.4. This phase may transform directly to the equilibrium β-phase although two semi-coherent polymorphs β'_1 and β'_2 have been detected which form on dislocation lines in cold-worked alloys. The presence of zirconium directly favours the formation of one or both of these phases as they precipitate on dislocations generated around zirconium-containing compounds. All these phases, as well as the equilibrium precipitate, appear to be resistant to coarsening at temperatures up to 350 °C.

Again, in parallel with alloys based on the Mg–RE system, thorium-containing alloys have been developed to which zinc has been added, e.g. HZ32A (Mg–3Th–2.2Zn–0.7Zr). The addition of zinc further increases creep strength (Fig. 5.25) and a Th:Zn ratio of 1.4:1 appears to be optimal in this regard. Zinc promotes formation of an acicular phase in grain boundaries and the good creep properties of alloy HZ32 are attributed, at least in part, to the presence of this phase.

Thorium-containing alloys were used in early missiles and spacecraft but they are now generally considered obsolete for environmental reasons. In England, alloys having as little as 2% of thorium are classified as being radioactive and therefore require special handling procedures which increases the complexity and cost of manufacturing products.

5.5.4 Alloys based on the magnesium–silver system

The importance of this class of alloys stems from the discovery by Payne and Bailey that the relatively low tensile properties of age-hardened Mg–RE–Zr alloys could be much increased by the addition of silver. Room-temperature tensile properties were obtained which were similar to those of the high-strength Mg–Zn–Zr alloys, such as ZK51, with the experimental alloys having superior casting and welding characteristics. Substitution of normal cerium-rich misch metal with didymium misch metal (average composition 80%Nd, 16%Pr, 2%Gd, 2% others) gave a further increase in strength which was attributed to the presence of neodymium. Several commercial compositions were subsequently developed having tensile properties which exceeded those of any other magnesium alloys at temperatures up to 250 °C, and which were comparable to the high-strength aluminium casting alloys.

The most widely used alloy has been QE22 (Mg–2.5Ag–2RE(Nd)–0.7Zr) for which the optimal heat treatment is: solution treatment for 4–8 h at 525 °C, cold-water quench, age 8–16 h at 200 °C. If these alloys contain less than 2% silver the precipitation process appears to be similar to that occurring in Mg–RE alloys and involves the formation of Mg–Nd precipitates. However, for higher silver contents, this element apparently modifies precipitation and increases the volume fraction of particles that are formed (Fig. 5.29). Two independent precipitation processes have been reported, both of which lead ultimately to the formation of an equilibrium phase of probable composition $Mg_{12}Nd_2Ag$ (Table 5.5). The presence of a precipitate with the DO_{19} structure has not been confirmed although the phase designated γ has characteristics which suggest that it may be such a phase. Maximum age-hardening and creep resistance are associated with the presence of γ- and β-precipitates in the microstructure.

Alloy QE22 has been used for a number of aerospace applications, e.g. aircraft landing wheels, gearbox housings and helicopter rotor fittings. Its superior tensile properties over most magnesium alloys are maintained to 250 °C (Fig. 5.24) although it is only considered to be resistant to creep at temperatures up to 200 °C. The alloys are relatively expensive and attempts have been made to replace at least some of the silver with copper. This work has met with some

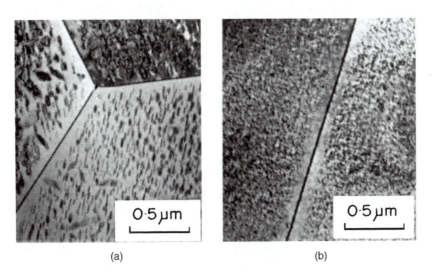

(a)   (b)

Fig. 5.29 Thin-foil electron micrographs showing the effect of silver on precipitate size in Mg–RE(Nd)–Zr alloys aged to peak hardness at 200 °C: (a) EK21 (Mg–2.5RE(Nd)–0.7Zr); (b) QE22 (Mg–2.5Ag–2RE(Nd)–0.72Zr) (courtesy K. J. Gradwell) ($\times$ 32 000).

success despite the relatively low solubility of copper in magnesium (maximum of 0.55% at the eutectic temperature) although no practical alloys have so far been produced.

5.5.5 Magnesium alloys containing yttrium

Further improvements in the creep resistance of magnesium alloys have been achieved by taking advantage of the high solid solubility of yttrium (maximum 12.5 wt.%). Yttrium is particularly effective in solid solution strengthening and Mg-Y alloys also respond to age hardening at relatively high ageing temperatures.

A series of My–Y–Nd–Zr alloys have been developed which combine high strength at ambient temperatures with good creep resistance at temperatures up to 300 °C. At the same time, the heat-treated alloys have a resistance to corrosion which is superior to that of other high-temperature magnesium alloys, and comparable with many aluminium-based casting alloys (Table 5.8). From a practical viewpoint, pure yttrium is expensive; it is also difficult to alloy with magnesium because of its high melting point (1500°) and its strong affinity for oxygen. Subsequently, it was found that a cheaper yttrium-rich misch metal containing around 75% of this element, together with heavy RE metals such as gadolinium and erbium, could be substituted for pure yttrium. Melting practices were also changed because the standard fluxes based on alkali and alkaline earth halides resulted in the loss of yttrium by reaction with $MgCl_2$. It is therefore necessary to process the alloys in an inert atmosphere of argon and 0.5 vol.% SF_6.

Precipitation in Mg–Y–Nd alloys is again complex (Table 5.5). Extremely fine β'' plates having the DO_{19} structure are formed on ageing below 200 °C. However, the T6 treatment normally involves ageing at 250 °C, which is above the solvus for β'' and leads to precipitation of three other metastable phases. Initial decomposition involves precipitation of fine plates of an as yet unidentified phase on the $\{11\bar{2}0\}_{mg}$ planes, and globular particles of the body-centred orthorhombic phase β' ($Mg_{12}NdY$) that also form on the $\{1\bar{2}10\}_{mg}$ planes.

Table 5.8 Comparative corrosion resistance of WE54–T6 and other magnesium and aluminium alloy castings after immersion in sea water for 28 days (Unsworth, W. and King, J. F. *Magnesium Technology*, p. 25, Institute of Metals, London, 1987)

Alloy type/designation		Weight loss (mg cm^{-2} day^{-1})
Magnesium	WE54–T6	0.08–0.2
casting	ZE41–T5, EZ33–T5	2–4
alloys	AZ91–T6	6–10
Aluminium	A356, A357	0.04–0.08
casting		
alloys		

With continued ageing, the first precipitate is gradually replaced by relatively coarse plates of a face-centred cubic phase that forms on the prismatic $\{1\bar{1}00\}_\alpha$ planes in contact with the β' particles. This phase, which has been designated β_1, is claimed to be the major hardening precipitate in Mg–Y–Nd alloys aged to peak strength at 250 °C. It has the composition Mg_3 (Nd,Y) and is similar to other Mg_3X phases (X = Nd, La, Ce, Pr, and Sm) that may form when alloying magnesium with the RE elements. The observation that β_1 plates always form heterogeneously at the sites of β' particles suggests that it is energetically difficult for β_1 to nucleate, and it has been found that cold work prior to ageing (T8 temper) promotes the formation of β_1 and the expense of β'. This increases the response to age hardening. Continued ageing at 250 °C leads to gradual precipitation of the equilibrium β phase on the $\{1\bar{1}00\}_\alpha$ planes, which may form by *in situ* transformation from β_1,. It is reported to have the composition $Mg_{14}Nd_2Y$. The fact that some of these phases contain such a high amount of magnesium helps promote high volume fractions of precipitates.

Maximum strengthening combined with an adequate level of ductility was found to occur in an alloy containing approximately 6% yttrium and 2% neodymium and the first commercially available alloy was WE54 (Mg–5.25Y–3.5RE(1.5–2Nd)–0.45Zr). In the T6 condition, typical tensile properties at room temperature are 0.2% P.S. 200 MPa, T.S. 275 MPa, elongation 4% and it showed elevated temperature properties superior to existing magnesium alloys (Figs 5.24 and 5.25). It was revealed, however, that prolonged exposure to temperatures around 150 °C led to a gradual reduction in ductility to levels that were unacceptable and this change was found to arise from the slow, secondary precipitation of the β'' phase throughout the grains. Subsequently it was shown that adequate ductility can be retained with only a slight reduction in overall strength if the yttrium content is reduced and the neodymium content is increased. On the basis of this work, an alternative composition WE43 (Mg–4Y–2.25Nd–1Heavy R.E.−0.4 min.Zr) was developed. Because of its high strength at elevated temperatures, WE43 is being used for some cast components in racing car engines and for aeronautical applications such as helicopter transmission casings.

Recent work has revealed that microalloying additions of zinc (e.g. 0.05%) may increase the creep strength of Mg–Y alloys because zinc appears to suppress non-basal slip that occurs at high temperatures. This suggests that it may be possible to decrease the cost of these alloys by reducing the yttrium content without sacrificing their excellent creep properties. Another interesting observation is that the crystal structures, atomic radii and electronegativities of yttrium and gadolinium are identical. Gadolinium is present in small quantities in yttrium-containing misch metal, and a study has been made of some alloys with larger additions of this RE element. Binary Mg–Gd alloys show a rapid response to age hardening at high ageing temperatures which is attributed to nucleation of a higher volume fraction of relatively stable precipitates that appear to be isomorphous with those formed in Mg–Y system. A proprietary

Mg–Nd–Gd–Zn–Zr alloy, Elektron 21, has been developed which has mechanical and corrosion properties similar to WE43 and is claimed to have better castability because it is less prone to oxidation during melting. It shows excellent properties up to 300 °C. The heavy RE element terbium also has a high solid solubility in magnesium and binary alloy Mg–20Tb has shown a 0.2% proof stress in the range 220–250 MPa at 300 °C.

The comparatively rare and costly light element, scandium (S.G 3.0) is another element having a particularly high solubility in magnesium (maximum of ∼24.5 wt.% or 15 at.%) and it also increases the melting point of the α-Mg solid solution. Furthermore, because of its relatively high melting point (∼1540 °C), scandium is assumed to have a low diffusivity in magnesium. Creep properties much superior to the alloy WE43 at high temperatures (e.g. 350 °C) have been reported for experimental alloys such as Mg-Sc-Mn.

5.6 WROUGHT MAGNESIUM ALLOYS

5.6.1 Introduction

Interest in wrought magnesium alloys peaked in the 1930s and 1940s when significant amounts were used in military aircraft. For example, magnesium alloy sheet was once used for some 50 per cent of the fuselage of two large American bombers amounting 5500 kg in weight. Seven hundred kg of forgings were also used. Since then there has been very little interest in wrought alloys which have consumed less than 1% of the annual output of magnesium metal. The largest application of flat products has been for AZ31 alloy photoengraving plate because of the high reactivity of magnesium to acid etching, whereas sacrificial anodes for cathodic protection of steel structures has been the main extruded product. Only recently has there been a large scale, global interest in developing new wrought products and this has been stimulated mainly by the potential applications in the automotive industry.

Because of its hexagonal crystal structure, magnesium possesses fewer slip systems than fcc aluminium which restricts its ability to deform, particularly at low temperatures. At room temperature, deformation occurs mainly by slip on the basal planes in the close-packed ⟨1120⟩ directions and by twinning on the pyramidal {1012} planes (Fig. 5.1). With stresses parallel to the basal planes, twinning of this type is only possible in compression whereas, with stresses perpendicular to the basal planes, it is only possible in tension. Above about 250 °C, additional pyramidal {1011} slip planes become operative so that deformation becomes much easier and twinning is less important. Production of wrought magnesium alloy products is, therefore, normally carried out by hot-working.

Currently wrought materials still account for only about 1% of magnesium consumption and are produced mainly by rolling, extrusion and press forging at temperatures in the range 300–500 °C. As mentioned in Section 5.2.3, rolling is usually required to be carried out in a number of stages, and extrusion speeds are five to ten times slower than is possible with aluminium alloys. Some general remarks can be made concerning the way properties vary in different directions in the final wrought products.

1. Since the elastic modulus does not show much variation in different directions of the hexagonal magnesium crystal, preferred orientation has relatively little effect upon the modulus of wrought products.
2. Extrusion at relatively low temperatures tends to orient the basal planes and also the $\langle 10\bar{1}0 \rangle$ directions approximately parallel to the direction of extrusion. Rolling tends to orient the basal planes parallel to the surface of sheet with the $\langle 10\bar{1}0 \rangle$ directions in the rolling direction.
3. Because twinning readily occurs when compressive stresses are parallel to the basal plane, wrought magnesium alloys tend to show lower values of longitudinal proof stress in compression than in tension. The ratio may lie between 0.5 and 0.7 and, since the design of lightweight structures involves buckling properties which, in turn, are strongly dependent on compressive strength, the ratio is an important characteristic of wrought magnesium alloys. The value varies with different alloys and is increased by promoting fine grain size because the contribution of grain boundaries to overall strength becomes proportionally greater.
4. Strengthening of wrought products by cold-reeling in which alternate tension and compression occurs can cause extensive twinning through compression, with a marked reduction in tensile properties.

As with cast alloys, the wrought alloys may be divided into two groups according to whether or not they contain zirconium. However, it is proposed here to consider the alloys with regard to the form of the wrought product. Compositions and mechanical properties are summarized in Table 5.4. Discussion of the ageing behaviour is included only where the compositions differ significantly from the cast alloys.

Wrought thorium-containing alloys such as HM21 (Mg–2Th–0.8Mn) have been used for manufacturing missile and spacecraft components which require creep resistance at temperatures up to 350 °C. However, as mentioned earlier, these alloys are now considered to be obsolete for environmental reasons.

5.6.2 Sheet and plate alloys

The early sheet alloys were AZ31 (Mg–3Al–1Zn–0.3Mn), which is still the most widely used magnesium alloy for applications at room or slightly elevated temperatures, and the now little used alloy M1A (Mg–1.5Mn). AZ31 is strengthened by strain hardening and is weldable, although weldments should be stress-relieved to minimize susceptibility to stress-corrosion cracking. Higher room temperature properties can be obtained with the British alloy ZK31 (Mg–3Zn–0.7Zr) but weldability is limited. Two lower strength alloys

ZM21 (Mg–2Zn–1Mn) and ZE10 (Mg–1.2Zn–0.2RE) are weldable and do not require stress-relieving. ZE10 has the highest toughness of any magnesium sheet alloy.

In general, sheets are still produced by reversing, multi-stage hot rolling of cast slabs that are first homogenised for several hours. Intermediate heating and annealing is required due to the low thermal capacity of magnesium. At one stage there was only one company world-wide that was producing sheet on a commercial scale. Now a German company is producing thick plate and sheet with a minimum thickness of 1 mm to a width of 1850 mm.

The Mg–Li system has attracted attention as a basis for very light-weight sheet and plate. Lithium with a relative density of 0.53 is the lightest of all metals and the Mg–Li phase diagram (Fig. 5.30) shows this element to have extensive solid solubility in magnesium. Moreover, only about 11% lithium is needed to form a new β-phase, which has a body-centred cubic (bcc) structure, thereby offering the prospect of extensive cold-formability. Finally, the slope of the α + β/β phase boundary suggested that selected compositions may show age-hardening. Early work on binary alloys revealed that traces of sodium caused grain boundary embrittlement but this problem was overcome with the availability of high-purity lithium. A second difficulty was that the binary alloys became unstable and over-aged at slightly elevated temperatures (50–70 °C) resulting in excessive creep under relatively low loads. Greater stability has since been achieved by adding other elements and one composition LA141 (Mg–14Li–1Al), which was developed in the USA and is weldable, has been used for armour plate

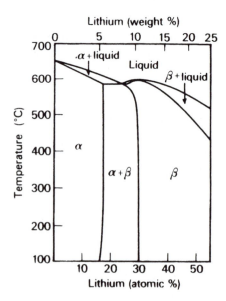

Fig. 5.30 Section of binary Mg–Li equilibrium diagram (from Freeth, W. A. and Raynor, G. V., *J. Inst. Metals*, **82**, 575, 1953–54).

Table 5.9 Properties of the Russian Alloys MA18 and MA21. (from Elkin F.M. and Davydov, V.G., *Proceedings of 6th International Conference on Magnesium Alloys and Their Applications*, Wiley-VCH, p. 94, 2004)

Alloy	Density g/cm^3	0.2% Proof Stress MPa	Tensile Strength MPa	Elongation %	Elastic Modulus GPa
MA18	1.48	90–180	140–210	10–30	45–46
MA21	1.60	130–220	200–260	6–20	45–46

and for aerospace components. Elevated temperature stability can be improved by the addition of 0.5% Si. A number of wrought alloys were developed in Russia including two designated MA18 and MA21 that were registered in 1983. MA18 (Mg–10.8Li–2.25Zn–0.75Al–0.25Ce) has a bcc lattice, whereas MA21 (Mg–8.75Li–4.8Al–4.5Cd–1.5Zn–0.08Ce) has a mixed hcp/bcc structure. These alloys are also weldable and some properties are summarised in Table 5.9.

Minimum strength properties in Table 5.9 concern large forgings, die forgings and extruded bars, whereas maximum properties are for warm rolled thin sheets and other thin shapes and tubes. Billet weighing up to 3 tonnes have been cast in special gas-shielded melting facilities.

Only limited cold-forming can be carried out with sheet alloys and typical minimum bend radii vary from 5 to 10T, where T = thickness of sheet, for annealed material, and from 10 to 20T for the hard-rolled condition. Thus, for even simple operations, hot-forming within the temperature range 230–350 °C is preferred. Under these conditions sheet can be formed by pressing, deep drawing, spinning and other methods using relatively low-powered machinery. A number of deep drawn automotive sheet panels in alloys such as AZ31 have been produced for prototype testing. Indicative weight savings are 50% compared with the same panels made from steel, and 15% when compared with aluminium. This alloy has also been shown to exhibit superplasticity if processing is carefully controlled. If the overall cost of producing magnesium alloy sheet can be reduced, this feature may offer unique opportunities for forming complex automobile body panels.

5.6.3 Extrusion alloys

A wide range of magnesium alloys can be warm or hot extruded at temperatures in the range 300 to 450 °C to produce solid and hollow profiles. Mg–Al–Zn extrusion alloys are used with aluminium content between 1 and 8%, the strongest alloy, AZ81 (Mg–8Al–1Zn–0.7Mn) showing some response to age hardening if heat treated after fabrication, AZ61 is often selected as a general purpose extruded alloy. One alloy was specifically developed as a canning material for use in the British gas-cooled Magnox nuclear reactors. This has the composition Mg–0.8Al–0.005Be and the fuel element cans are impact extruded with integral cooling fins, as shown in Fig. 5.31, or machined from a finned

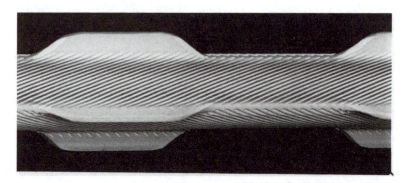

Fig. 5.31 Part of extruded magnesium alloy fuel can from a British Magnox nuclear reactor.

extrusion. The spiral shape is obtained by hot-twisting after extrusion. The selection of magnesium alloy was made because the metal has a relatively low capture cross-section for thermal neutrons (0.059 barns), is resistant to creep and corrosion by the carbon dioxide coolant at the operating temperatures (180–420 °C) and, contrary to aluminium, does not react with the uranium fuel. The addition of aluminium provides some solid solution strengthening whereas the trace amount of beryllium improves oxidation resistance.

The alloy ZK61 (Mg–6Zn–0.7Zr), which is normally aged after extrusion, develops the highest room temperature yield strength of the more commonly used wrought magnesium alloys. It also offers the advantage that tensile and compressive yield strengths are closely matched. The lower zinc alloys ZK21 and ZM21 (Mg–2Zn–1Mn) are widely used where higher extrusion rates (e.g. 40 m min^{-1}) are desired. The highest strength recorded for a wrought magnesium alloy is believed to be for the composition Mg–6Zn–1.2Mn (ZM61) in the form of heat-treated extruded bar. The alloy shows a high response to age-hardening, and solution treatment at 420 °C, water quenching and duplex ageing below and above the GP zones solvus (24 h at 90 °C and 16 h at 180 °C) results in the following tensile properties: 0.1% P.S. 340 MPa; T.S. 385 MPa; with an elongation of 8%. Similar properties have been achieved with the alloy that is based on the Mg–Zn–Cu system which has the advantage of faster extrusion speeds than ZM61. This alloy is Mg–6.5Zn–1.25Cu–0.75Mn (ZCM711) which is given a T6 heat treatment (solution treat 8 h at 435 °C, hot-water quench, age 24 h at 200 °C). On a strength/weight basis, these alloys have properties comparable with some of the strongest wrought aluminium alloys.

A typical magnesium alloy must be extruded five to ten times slower than a typical aluminium alloy. In the belief that the availability of cost effective extruded sections is essential if magnesium alloys are to be used more widely for structural applications, a consortium from the European Community is investigating the use of hydrostatic extrusion. In this process, a billet is forced through a die opening by means of a pressurised fluid as is shown in Fig. 5.32.

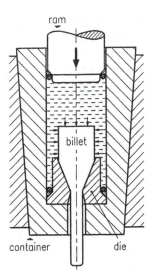

Fig. 5.32 Arrangement for hydrostatic extrusion (after Sillekens, W.H. and Bohlen, J., *Proceedings 6th International Conference on Magnesium Alloys and Their Applications*, Wiley-VCH, p. 1046, 2004).

Unlike conventional direct or indirect extrusion, there is no metallic contact between the ram and the billet so that the only billet-tooling contact is with the die cone Plastic deformation takes place under high hydrostatic pressure and, as lubrication is almost ideal, no significant friction is involved. This contrasts with conventional extrusion is which frictional forces and shearing within the billet add considerably to the mechanical work that is required. Thus the thermal effects associated with the dissipation of this mechanical work are less during hydrostatic extrusion which reduces the danger of incipient melting of the billet material, and allows processing to be done at higher speeds.

Hydrostatic extrusion was first developed in the 1960s as a means for increasing the ability to cold work relatively brittle materials. It is now also used for applications involving warm and hot working including the production of copper tubing, copper-clad aluminium wire, and for compacting ceramic superconductors. Hot hydrostatic extrusion trials involving the magnesium alloys M1, AZ31 and ZM21 have shown that speeds five to ten times faster than those used for conventional extrusion are possible. Grain sizes are finer and more uniform and tensile properties are reported to be comparable with those for alloys prepared by conventional extrusion.

5.6.4 Forging alloys

Forgings represent a comparatively small part of the inventory of wrought magnesium products and can only fabricated from alloys with fine grained microstructures. They tend to be made from the higher strength alloys AZ80 and ZK60 if they are to be used at ambient temperatures, or WE 43 (and formerly HM21) for elevated temperatures applications. Alloys that show rapid grain growth at the forging temperature are subsequently forged in stages at successively lower temperatures Forgings are often specified when a component has an intricate shape and is required to have a strength higher than can be achieved with castings. Press forging is more common than hammer forging and it is often the practice to pre-extrude the forging blanks to refine the microstructure.

5.7 ELECTROCHEMICAL ASPECTS

5.7.1 Corrosion and protection

Magnesium has a normal electrode potential at 25 °C of -2.30 V, with respect to the hydrogen electrode potential taken as zero, which places it high in the electrochemical series. However, its solution potential is lower, e.g. -1.7 V in dilute chloride solution with respect to a normal calomel electrode, due to polarization of the surface with a film of Mg $(OH)_2$. The oxide film on magnesium offers considerable surface protection in rural and most industrial environments and the corrosion rate of magnesium lies between aluminium and mild steel (Table 5.10). Essentially it is the high susceptibility to impurities and lack of passive film stability in solutions below pH 10.5 that accounts for most corrosion problems in magnesium alloys. Tarnishing occurs readily and some general surface roughening may take place after long periods. However, unlike some aluminium alloys, magnesium and its alloys are virtually immune from

Table 5.10 Results of 2.5-year exposure tests (from *Metals Handbook*, Volume 2, 9th edn, American Society of Metals, Cleveland, Ohio, 1979)

Material	Corrosion rate (mm y^{-1})	Tensile strength after $2\frac{1}{2}$ y (% loss)
Marine atmosphere		
Aluminium alloy 2024	0.002	2.5
Magnesium alloy AZ31	0.018	7.4
Mild steel	0.150	75.4
Industrial atmosphere		
Aluminium alloy 2024	0.002	1.5
Magnesium alloy AZ31	0.028	11.2
Mild steel	0.025	11.9
Rural atmosphere		
Aluminium alloy 2024	0.000	0.4
Magnesium alloy AZ31	0.013	5.9
Mild steel	0.015	7.5

intercrystalline attack because the grain centres are usually anodic with respect to grain boundary regions.

Magnesium is readily attacked by all mineral acids except chromic and hydrofluoric acids, the latter actually producing a protective film of MgF_2 which prevents attack by most other acids. In contrast, magnesium is very resistant to corrosion by alkalis if the pH exceeds 10.5, which corresponds to that of a saturated $Mg(OH)_2$ solution. Chloride ions promote rapid attack of magnesium in aqueous solutions, as do sulphate and nitrate ions, whereas soluble fluorides are chemically inert. With organic solutions, methyl alcohol and glycol attack magnesium whereas ethyl alcohol, oils and degreasing agents are inert.

The corrosion behaviour of alloys varies with composition. Where alloying elements form grain boundary phases, as is generally the case in casting alloys, corrosion rates are likely to be greater than those occurring with pure magnesium. As mentioned earlier, the first magnesium alloys suffered rapid attack in moist conditions due mainly to the presence of more noble metal impurities, notably iron, nickel and copper (Fig. 5.3). Each of these elements, or compounds they form, act as minute cathodes in the presence of a corroding medium, creating microgalvanic cells with the relatively anodic magnesium matrix. Nickel and copper are not usually a problem in current alloys due to the very low levels of these elements present in primary magnesium. Iron tends to be more troublesome as there is always a risk of pick-up from crucibles which are made from mild steel. However, the potential detrimental effect of iron in zirconium-free alloys is reduced by adding manganese (as $MnCl_2$) to the melt. This element combines with the iron and settles to the bottom of the melt or forms intermetallic compounds which, depending on the Fe: Mn ratio, reduces the cathodic effect of the iron. This ratio should not exceed 0.032. Zirconium has a similar effect in those alloys to which it is added. As mentioned previously, Mg–Al–Zn and Mg–Al alloys are particularly susceptible to the presence of impurities and the widely used alloy AZ91C has largely been superseded by higher-purity versions known as AZ91D for pressure diecasting and AZ91E for gravity diecasting which have stricter limits for the nickel, iron and copper contents. Further improvements are possible by applying a T6 ageing treatment. Alloys AZ91D and AZ93E exhibit corrosion rates which are similar to those of comparable cast aluminium alloys.

Some magnesium alloys may be susceptible to stress-corrosion cracking (SCC) which is especially severe in chromate-chloride solutions. Special attention has been paid to alloys based on the Mg–Al system. Cracking is usually transgranular and involves discontinuous cleavage on microstructural features that have been identified as twin boundary interfaces and various preferred crystallographic planes. There is general agreement that hydrogen embrittlement is the dominant mechanism. Zirconium-containing alloys are less susceptible and SCC only occurs at stresses approaching the yield stress of the alloy concerned. Wrought products are more likely to undergo SCC than castings and it is desirable to stress-relieve components that may be exposed to potential corrodents.

It is common practice to protect the surface of magnesium and its alloys and such protection is essential where contact with other metals may lead to galvanic corrosion. Methods available for magnesium are given below.

1. Fluoride anodizing—this involves alternating current anodizing at up to 120 V in a bath of 25% ammonium bifluoride which removes surface impurities and produces a thin, pearly white film of MgF_2. This film is normally stripped in boiling chromic acid before further treatment as it gives poor adhesion to organic treatments.

2. Chemical treatments involving pickling and conversion of the oxide coating—components are dipped in chromate solutions which clean and passivate the surface to some extent through formation of a film of $Mg(OH)_2$ and a chromium compound. Such films have only slight protective value, but form a good base for subsequent organic coatings.

3. Electrolytic anodizing, including proprietary treatments that deposit a hard ceramic-like coating which offers some abrasion resistance in addition to corrosion protection, e.g. Dow 17, HEA, and MGZ treatments—such films are very porous and provide little protection in the unsealed state but they may be sealed by immersion in a solution of hot dilute sodium dichromate and ammonium bifluoride, followed by draining and drying. A better method is to impregnate with a high-temperature curing epoxy resin (see (4)). Resin-sealed anodic films offer very high resistance to both corrosion and abrasion, and, in some instances, can even be honed to provide a bearing surface. Impregnation is also used to achieve pressure tightness in castings that are susceptible to microporosity.

4. Sealing with epoxy resins—in this case, the component is heated to 200–220 °C to remove moisture, cooled to approximately 60 °C, and dipped in the resin solution. After removal from this solution, draining and air-drying to evaporate solvents, the component is baked at 200–220 °C to polymerize the resin. Heat treatment may be repeated once or twice to build up the desired coating thickness which is commonly 0.025 mm.

5. Standard paint finishes—the surface of the component should be prepared as in (1) to (4), after which it is preferable to apply a chormate-inhibited primer followed by good quality top coat.

6. Vitreous enamelling—such treatments may be applied to alloys which do not possess too low a solidus temperature. Surface preparation involves dipping in a chromate solution before applying the frit.

7. Electroplating—several stages of surface cleaning and the application of pre-treatments, such as a zinc conversion coating, are required before depositing chromium, nickel or some other metal.

Magnesium alloy components for aerospace applications require maximum protection; schemes involving chemical cleaning by fluoride anodizing, pre-treatment by chromating or anodizing, sealing with epoxy resin, followed by chromate primer and top coat are sometimes mandatory.

5.7.2 Cathodic protection

Due to its very active electrode potential, magnesium and its alloys can be used to protect many other structural materials from corrosion when connected to them in a closed electrical circuit. Magnesium acts as an anode and is consumed sacrificially, thereby offering protection to metals such as steel. Magnesium metal and, more commonly, the alloys AZ63 and M1A (Mg–1.5Mn) which offer higher relative voltages are used for this purpose. Examples of areas where cathodic protection is used are ships' hulls, pipelines and steel piles. It should be noted, however, that magnesium and its alloys are not used to protect oil rigs because of the potential incendive sparking risk. The anodes are usually produced by extruding the magnesium alloys.

5.8 FABRICATION OF COMPONENTS

5.8.1 Machining

Magnesium and its alloys are the most machinable of all structural materials. This applies with respect to depth of cut, speed of machining, tool wear and relative amounts of power required for the equipment being used (Table 5.11). Magnesium is normally machined dry but, where very high cutting speeds are involved and there is a possibility of igniting fine turnings, it may be necessary to employ a coolant. For this purpose, mineral oils must be used because water-based coolants may react chemically with the swarf. Good tool life is experienced providing cutting edges are kept sharp and generous rake clearance angles (usually 7° minimum) are used. Sharp tools also reduce the possibility of fires due to frictional heat.

Magnesium alloys can be chemically machined or milled by pickling in 5% H_2SO_4 or in dilute solutions of HNO_3 or HCl. Some alloys also lend themselves to contour etching and AZ31 is widely used for the production of printing plates.

Table 5.11 Comparative machinability of metals (from *Machining*, Magnesium Elektron Limited Handbook)

Metal	Relative power required[*]	Rough turning speeds (m s^{-1})	Drilling speeds (5–10 mm drill) (m s^{-1})
Magnesium	1	up to 20	2.5–8.5
Aluminium	1.8	1.25–12.5	1–6.5
Cast iron	3.5	0.5–1.5	0.2–0.65
Mild steel	6.3	0.65–3.3	0.25–0.5
Stainless steel	10.0	0.3–1.5	0.1–0.35

[*]1 = lowest

5.8.2 Joining

Early magnesium alloys were gas welded with an oxyacetylene torch and required careful fluxing to minimize oxidation. Apart from the normal difficulties associated with such a process, extensive corrosion of welds was common when the flux was incompletely removed by the cleaning methods applied. Since then, virtually all magnesium welding has been done using inert gas shielded tungsten arc (TIG) or consumable electrode (MIG) processes (Section 3.6.1). Increasing interest in using magnesium alloys in automobiles is requiring more attention to be given to alternative welding methods. Spot welding is possible but so far has been little used. Some success has also been achieved with laser welding and with friction stir welding (Section 3.5.1) and both these methods have been used experimentally to join dissimilar magnesium alloys, or magnesium alloys to aluminium alloys.

Comparisons can be made between the welding characteristics of magnesium and aluminium alloys. As with aluminium, the solubility of hydrogen in magnesium deceases significantly as it solidifies which can lead to porosity in weld beads. The fact that the viscosity and surface tension of molten magnesium are both lower than that of molten aluminium can cause increased sputter during welding and reduce the surface quality of welded regions. Furthermore, the relatively high vapour of liquid magnesium can lead to higher evaporative losses during welding, particularly in alloys containing zinc. On the other hand, the combination of lower heat capacity and heat of fusion of magnesium means that less energy is consumed during welding and offers the potential to achieve higher welding speeds.

Both cast and wrought magnesium alloy products can be welded and the weldability of different alloys has already been compared in Tables 5.3 and 5.4. In general, filler rods are of the same composition as the parent alloy are desirable although the use of a more highly alloyed rod with lower melting point and wider freezing range is sometimes beneficial to minimize cracking. Castings are often preheated to 250–300 °C to reduce weld cracking during solidification, and stress-relieving may be desirable after welding is completed.

The design of mechanical joints in magnesium and its alloys is qualified by the vital consideration of galvanic corrosion. This problem precludes direct contact with most other metals and special coatings or insulating materials must be used as separating media. Care must also be taken in the design of joints to avoid crevices, grooves and such like, where water and other corrosive materials can collect.

5.9 TRENDS IN APPLICATIONS OF MAGNESIUM ALLOYS

As mentioned in Chapter 1, annual production of magnesium in the Western World was relatively constant at close to 250 000 tonnes during much of the 1990s. Of this total, more than half was used as an addition to various aluminium alloys and only some 40 000 tonnes was actually consumed to produce structural magnesium alloys, mainly as die castings. In recent years, consumption of magnesium has been increasing and an estimated 336 000 tonnes was produced worldwide in 2003, some two thirds of which came from China. Annual shipments of die castings were estimated to have risen to 127 800 tonnes in 2002 and some of the products made from magnesium alloys are shown in Fig. 5.33

A comparative analysis of Western World markets for magnesium alloys in 1966 and for the period 1981–1992 by F. Hehmann revealed that the number of individual applications actually fell from an estimated 198 to 161. The latter figure could be reduced further to 125 if automotive applications were classified into representative groups. Major changes included substantial reductions in the aeronautical and missile markets (totals of 96 applications in 1966 and only 23 in 1981–1992) whereas there were marked increases in the surface transport area. This latter trend has continued during the past decade as market demands for magnesium have become linked more and more to developments in the automotive industry.

In the aeronautical industry, airframe applications in new designs have virtually disappeared and significant uses of magnesium alloys has been confined to castings for engine and transmission housings, notably for helicopters. Historically, the so-called Volkswagon Beetle motor car has represented the largest single market for magnesium alloys which were used for engine crank case and transmission housing castings weighing a total of 17 kg. As was mentioned earlier, this resulted in a weight saving of some 50 kg when compared with using traditional cast iron which was critical for improving the stability of

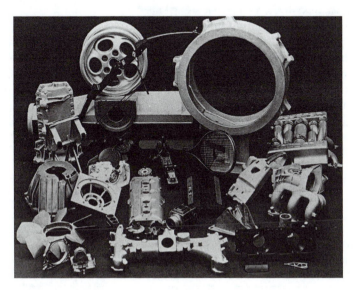

Fig. 5.33 Range of products made from magnesium alloys (courtesy Hydro Magnesium).

this rear-engined vehicle. A total of 21 559 464 vehicles were produced between 1934, when production commenced in Germany, and when it finally ceased in 2003, in Mexico. Large increases in the price of magnesium in the mid-1970s led to its replacement, at least in part, by aluminium alloy castings. Before then a total of more than 400 000 tonnes of magnesium had been consumed.

Current interest in possible applications of magnesium alloys in automobiles has been stimulated by continuing demands to lower weight, thereby reducing fuel consumption and pollution. A target being pursued in some countries is to develop a vehicle that consumes only 3 litres of fuel to travel 100 km. In this regard, the mass-dependent component of fuel consumption is the key factor since it contributes approximately 60% to the total. With the engine block for example, weight savings of 35% and 75%, respectively, are possible if it can be cast from a magnesium alloy rather than from an aluminium alloy or cast iron.

Although there has been an average annual increase in the use of magnesium alloys in automobiles of some 15% during the past decade, this only represents a change from around 1 kg to 4 kg per vehicle. This compares with current consumption of 120 kg each for aluminium and plastics in a vehicle weighing 1 500 kg that is produced in North America. As shown in Table 5.12, cost is a major factor limiting the wider use of magnesium. This table compares the estimated performance cost indices in the year 2003 for magnesium alloys, the aluminium alloys 6061 and 380, and the widely used plastic PC/ABS (a blend of polycarbonate and acrylonitrile-butadiene-styrene) with indices for steel all set as 1. In this regard, a European survey has suggested that the wider use of magnesium in automobiles requires the following cost goals to be achieved : castings < cost of aluminium alloy castings + 30%, and sheet and extruded components to be half their current costs.

Examples of some current global applications of magnesium alloy automotive components are die cast steering wheels and steering column components, instrument panels, seat frames (Fig. 5.14), small motor housings, door handles, pedals, various brackets, engine valve cover and oil pan,. Much larger consumption will be involved with the wider use of magnesium alloys for powertrain components such as the die cast gearbox casing shown in Fig. 5.34. As mentioned in Section 5.4.1, developments are in progress with sand cast engine

Table 5.12 Estimated performance cost indices (2003) for various materials compared with steel (from Luo A. A., *JOM*, **56**, (2), 42, 2002)

Material	Steel (galvanized)	Al (A380)	Al (6061)	Plastic (PC/ABS)	Mg (2002)	Mg (2003)
Cost ratio per unit weight	1	3.36	3.59	6.14	7.5	5.91
Cost ratio per unit volume	1	1.17	1.33	0.83	1.67	1.33
PCI for equal stiffness	1	1.67	1.93	3.74	2.79	2.22
PCI for equal strength	1	1.13	1.13	1.61	1.87	1.49

PCI is Performance Cost Index

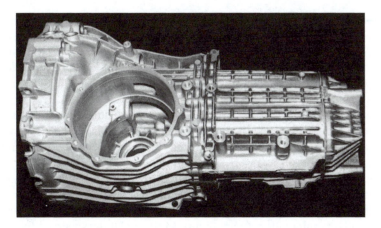

Fig. 5.34 Die cast AZ91 magnesium alloy gearbox casing (courtesy Volkswagen AG).

blocks (Figs. 4.7). Table 5.13 summarizes potential automotive applications for magnesium alloys together with the technical challenges they present.

On the basis of relative densities, it has been generally accepted that magnesium can become an effective substitute for aluminium if the price ratio falls

Table 5.13 Potential magnesium alloy automotive applications and their technical challenges (from Luo A. A., *JOM*, **56**, (2), 42, 2002)

System	Product	Technical Challenges
Interior	airbag housing	improved casting process
	window regulator housing	rapid prototyping
	glove box	design for magnesium
Body	door frame/inner	thin-wall design
	A&B pillar	thin-wall casting
	hatchback frame	joining and welding
	spare tyre jack	
Chassis	wheel	high-strength alloy development
	control arm	new casting processes
	rack and pinion housing	(squeeze and semi-solid metal casting)
	brackets for rail frames	low-cost coatings
	spare tyre rim	
Powertrain	automatic transmission case	creep-resistant alloy development
	engine block	design for magnesium
	crankcase	fastening strategy
	oil pan	engine coolant compatibility
	starter housing	
	oil/water pump housing	
	intake manifold	
	engine mount	

below 1.5 : 1. During the last three decades, this ratio has varied between 1.5 and 2.5. What may favour magnesium in the future is a greater appreciation of the cost savings that are possible through savings in energy in casting (estimated to be up to 30%) and machining (up to 45%) when calculated on a volumetric basis. Improved methods of recycling magnesium alloy scrap is another area in which there is a potential for cost savings in the use of this metal.

Other areas in which the use of magnesium is expanding can be categorized as appliances and sporting goods. As two examples, there has been a trend to use magnesium alloy die castings for producing thin-walled computer housings and mobile telephone casings (Fig. 5.8) where lightness, ability to be thin-wall cast, and the provision of electromagnetic shielding are special advantages. Magnesium alloys have also been used to cast the frames of light weight bicycles.

One unexpected and exciting development has been the discovery, in 2001, that the magnesium alloy (compound) MgB_2 exhibits superconductivity at approximately 40 K (-233 °C). This critical temperature (T_c) is nearly twice that at which more traditional metallic superconductors, such as Nb_3Sn, can operate and offers the prospect of cooling with liquid hydrogen or neon rather than using more expensive liquid helium. Moreover, when doped with carbon or other impurities, MgB_2 has a current-carrying capacity in the presence of magnetic fields which is at least equal or better than that of these other metallic compounds. MgB_2 wires can be formed by reacting magnesium vapour with boron fibres at temperatures of 1000 °C, or by synthesizing powder mixtures of these two elements in thin tubes. Potential applications are thought to include superconducting magnets, power lines and sensitive magnetic field detectors.

FURTHER READING

Emley, E.G., *Principles of Magnesium Technology*, Pergamon, London, 1966

Raynor, G.V., *The Physical Metallurgy of Magnesium and Its Alloys*, Pergamon, London, 1959

Roberts, C.S., *Magnesium and Its Alloys*, Wiley, New York, 1960

Avedesian, M.M., and Baker, H., *ASM Specialty Handbook: Magnesium and Magnesium Alloys*, ASM International, Materials Park, Ohio, USA, 1999

Nayeb-Hashemi A.A. and Clark, J.B., *Phase Diagrams of Binary Magnesium Alloys*, ASM International, Materials Park, Ohio, USA, 1988

Kainer, K.U. Ed., *Magnesium – Alloys and Technologies*, Wiley-VCH, Weinheim, Germany, 2003

Mordike, B.L. and Kainer, K.U. (Eds.), *Proc. 4th Inter. Conf. on Magnesium Alloys and Their Applications*, Werkstoffe Informationsgesellschaft, Frankfurt, Germany, 1998

Luo, A.A, (Ed.) *Magnesium Technology 2004*, TMS, Warrendale, Pa, USA.

Kainer, K.U., Ed. *Proc. 6th Inter. Conf. on Magnesium Alloys and Their Applications*, Wiley-VCH, Weinheim, Germany, 2004

Neite, G., Kubota, K., Higashi, K. and Hehmann, F., Magnesium-Based Alloys, in Cahn, R.W., Haasen, P. and Kramer, E.J. Eds. *Materials Science and Technology – A Comprehensive*

Treatment **Vol. 8**, Ed. Matucha, K.H., *Structure and Properties of Non-Ferrous Alloys*, 113, 1996

Nie, J.-F. *et al*, Viewpoint on phase transformations and deformation in magnesium alloys, *Acta. Mater.*, **48**, 981, 2003

Rokhlin, L.L., *Magnesium Alloys Containing Rare Earth Metals: Structure and Properties*, Taylor and Francis, London, 2003

Pekguleryuz, M.O. and Kaya, A.A., Creep resistant magnesium alloys for powertrain applications, *ibid.* p. 74

Abbott. T.B., Easton, M.A. and Caceres, C.H., Designing with Magnesium: Alloys, Properties, and Casting Processes, *Handbook of Mechanical Alloy Design* Eds, Totten, G.E., Xie, L. and Funatani, K., Marcel Dekker, Inc., New York, 2004

Han, Q., Kad, B.K. and Viswananathan, S., Design perspectives for creep-resistant magnesium die-casting alloys, *Phil. Mag.*, **84**, No. 36, 3843, 2004

Westengen, H., Magnesium: alloying and Magnesium alloys: properties and applications, *Encyclopedia of Materials Science and Technology*, Elsevier Science Ltd., p. 4745, 2001

Makar, G.L. and Kruger, J., Corrosion of magnesium, *Int. Mater, Rev.*, **38**, (3), 138, 1993

Brown, R.F., Magnesium wrought and fabricated products: yesterday, today, and tomorrow, *Magnesium Technology 2002*, Kaplin, H.I. (Ed.), TMS, Warrendale, Pa, USA, 155, 2002

Schumann, S. and Friedrich, H., Current and future use of magnesium in the automotive industry, *Mater. Sci. Forum*, **419–422**, 51, 2003

6

TITANIUM ALLOYS

6.1 INTRODUCTION

Stimulus for the development of titanium alloys during the past 40 years came initially from the aerospace industries when there was a critical need for new materials with higher strength: weight ratios at elevated temperatures. As mentioned in Chapter 1, the high melting point of titanium (1678 °C) was taken as a strong indication that the alloys would show good creep strengths over a wide temperature range. Although subsequent investigations revealed that this temperature range was narrower than expected, titanium alloys now occupy a critical position in the materials inventory of the aerospace industries (see Fig. 1.6) and around 50% of titanium is used in this way. More recently the importance of these alloys as corrosion-resistant materials has been appreciated by the chemical industry as well as by the medical profession which uses titanium alloy prostheses for implanting in the human body. It is proposed to consider the alloys with respect to these applications and to concentrate on wrought products as titanium alloy castings amount to less than 2% of titanium metal.

Titanium has a number of features that distinguish it from the other light metals and which make its physical metallurgy both complex and interesting.

1. At 882.5 °C, titanium undergoes an allotropic transformation from a low-temperature, hexagonal close-packed structure (α) to a body-centred cubic (β) phase that remains stable up to the melting point. This transformation offers the prospect of having alloys with α, β or mixed α/β microstructures and, by analogy with steels, the possibility of using heat treatment to extend further the range of phases that may be formed.
2. Titanium is a transition metal with an incomplete shell in its electronic structure which enables it to form solid solutions with most substitional elements having a size factor within $\pm 20\%$.

3. Titanium and its alloys react with several interstitial elements including the gases oxygen, nitrogen and hydrogen, and such reactions may occur at temperatures well below the respective melting points.
4. In its reactions with other elements, titanium may form solid solutions and compounds with metallic, covalent or ionic bonding.

Alloying of titanium is dominated by the ability of elements to stabilize either of the α- or the β-phases. This behaviour, in turn, is related to the number of bonding electrons, i.e. the group number, of the element concerned. Alloying elements with electron/atom ratios of less than 4 stabilize the α-phase, elements with a ratio of 4 are neutral, and elements with ratios greater than 4 are β-stabilizing. Compared with β, the α phase is characterized by the following properties:

– higher resistance to plastic deformation
– lower ductility
– significant anisotropy of physical and mechanical properties
– diffusion rates that are lower by at least two orders of magnitude
– higher creep resistance.

6.1.1 Classification of titanium alloys

Titanium alloy phase diagrams are often complex and many are unavailable. However, the titanium-rich sections of pseudo-binary systems enables them to be classified into three simple types, as shown in Fig. 6.1. Elements that dissolve preferentially in the α-phase expand this field thereby raising the α/β transus (Fig. 6.1a) and, of the comparatively few elements that behave in this way, aluminium and oxygen are the most important. Zirconium, tin and silicon are regarded as neutral in their effect on either phase. Elements which depress the α/β

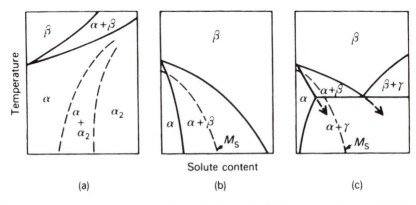

Fig. 6.1 Basic types of phase diagrams for titanium alloys. The dotted phase boundaries in (a) refer specifically to the Ti–Al system. The dotted lines in (b) and (c) show the martensite start (M_s) temperatures. Alloying elements favouring the different types of phase diagrams are (a) Al, O, N, C, Ga; (b) Mo, W, V, Ta; (c) Cu, Mn, Cr, Fe, Ni, Co, H.

transus and stabilize the β-phase may be classified in two groups: those which form binary systems of the β-isomorphous type (Fig. 6.1b) and those which favour formation of a β-eutectoid (Fig. 6.1c). It should be noted, however, that the eutectoid reactions in a number of alloys are very sluggish so that, in practice, the alloys tend to behave as if this reaction did not occur. Examples are the binary systems Ti–Fe and Ti–Mn and these alloys behave as if they conformed to the β-isomorphous phase diagram; hence the arrows shown in Fig. 6.1c.

The main elements that promote the three types of binary phase diagrams are also given in Fig. 6.1. It will be noted that the interstitial elements also exert stabilizing effects: oxygen, nitrogen and carbon favouring the α-phase and hydrogen promoting the β-phase. Of the substitutional elements that stabilize the β-phase, molybdenum and tungsten have the greatest effects although the latter element is little used because of its high density and problems of segregation during alloy preparation. Vanadium is another common β-stabilizer although it is less effective than molybdenum in the higher temperature ranges.

It is customary to classify titanium alloys into three main groups designated α, α + β and β which will each be considered in Sections 6.2–6.4. The compositions and a selection of properties of representative commercial alloys in each group are listed in Table 6.1. In addition, the creep characteristics of a number of these alloys are shown in Fig. 6.2 because this property has dominated much alloy development. Both Table 6.1 and Fig. 6.2 should be consulted in conjunction with the foregoing discussions in which special consideration is given to commercial compositions and the roles of particular alloying elements.

6.1.2 Basic principles of heat treatment

The properties of titanium alloys are determined primarily by the morphology, volume fraction and individual properties of the two phases α and β. Although the first alloys that will be discussed (α-alloys) show little response to heat treatment, it is desirable to examine the general principles that are involved even though they relate mainly to the α/β and β groups. This is possible by considering the effects of alloy content on the β to α transformation in a typical binary β-isomorphous system, as shown in Fig. 6.3. Also included is a schematic diagram which depicts trends in tensile strength with respect to alloy content resulting from different heat-treatment procedures.

It will be seen that the strength of annealed alloys increases gradually and linearly as alloy content, or percentage of β-phase, increases. It should be noted that the β-phase in these alloys does not transform during cooling to room temperature. However, for alloys quenched from the β-phase field, a more complex relationship exists between strength and composition which is dependent upon the transformation of β to the martensitic form of the α-phase, designated α′. For low concentrations of solute, some strengthening occurs as a result of this transformation, but the effect is much less than that traditionally found for martensitic reactions in ferrous materials. Moreover, little change occurs when

Table 6.1 Compositions, relative densities and typical room temperature tensile properties of selected wrought titanium alloys

Common designations	Al	Sn	Zr	Mo	V	Si	Other	Relative density	Condition	0.2% proof stress (MPa)	Tensile strength (MPa)	Elongation (%)
α-alloys												
CPTi99.5%, IMI115, Ti–35A							0	4.51	Annealed 675 °C	170	240	25
CPTi99.0%, IMI155, Ti–75A							0	4.51	Annealed 675 °C	480	550	15
IMI260							0.2Pd	4.51	Annealed 675 °C	315	425	25
IMI317	5	2.5						4.46	Annealed 900 °C	800	860	15
IMI230							2.5Cu	4.56	ST(α)†;duplex aged 400 and 475 °C	630	790	24
Near-α alloys												
8-1-1	8			1	1			4.37	Annealed ‡ 780 °C	980	1060	15
IMI679	2.25	11	5	1		0.25		4.82	ST(α + β) aged ‡ 500 °C	990	1100	15
IMI685	6		5	0.5		0.25		4.49	ST(β) aged 550 °C	900	1020	12
6-2-4-2S	6	2	4	2		0.2		4.54	ST(α + β) annealed 590 °C	960	1030	15
Ti-11	6	2	1.5	1		0.1	0.35Bi	4.45	ST(β) aged 700 °C	850	940	15
IMI829	5.5	3.5	3	0.3		0.3	1Nb	4.61	ST(β) aged 625 °C	860	960	15
Ti1100	6	2.75	4	0.4		0.45		4.50	ST(β), aged 600 °C	895	1000	10
IMI834	6	4	3.5	0.5		0.35	0.7Nb, 0.06C	4.59	ST(α + β), aged 625–700 °C	905	1035	10

α/β alloys

Alloy							Density	Treatment			
IMI318,6–4	6				4		4.46	Annealed 700 °C	925	990	14
IMI550	4	2		4		0.5	4.60	ST(α + β) aged 500 °C	1100	1170	10
IMI680	2.25	11		4		0.2	4.86	ST(α + β) aged 500 °C	1000	1100	14
6–6–2	6	2			6	0.7 (Fe, Cu)	4.54	ST(α + β) aged 550 °C	1190	1310	15
										1275	10
6–2–4–6	6	2	4	6			4.68	ST(α + β) annealed 590 °C	1170	1270	10
IMI551	4	4		4		0.5	4.62	ST(α + β) aged 500 °C	1200	1310	13
Ti8Mn						8Mn	4.72	Annealed 700 °C	860	945	15

β alloys

Alloy							Density	Treatment			
13–11–3	3				13	11Cr	4.87	ST(β) aged 480 °C	1200	1280	8
Beta III		4.5	6	11.5			5.07	ST(β) duplex aged 480 and 600 °C	1315	1390	10
8–8–2–3	3			8	8	2Fe	4.85	ST(β) aged 580 °C	1240	1310	8
Transage 129	2	2	11	11			4.81	ST(β) aged 540 °C	1280	1400	6
Beta C	3		4	4	8	6Cr	4.82	ST(β) aged 580 °C	1130	1225	10
10–2–3	3				10	2Fe	4.65	ST(β) aged 580 °C	1250	1320	8
Timetal 21S	3		15			0.2 · 2.7Nb, 0.2Si	5.34	ST(β) aged 560 °C	1170	1240	8

†ST(α), ST(α + β), ST(β) correspond to solution treatment in the α, α + β, and β-phase fields respectively

‡Annealing treatments normally involve shorter times than ageing treatments

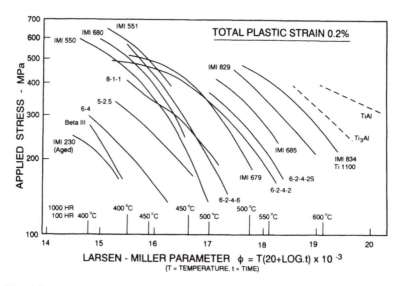

Fig. 6.2 Creep curves of some commercial titanium alloys (courtesy IMI Titanium).

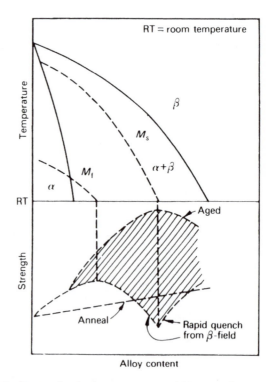

Fig. 6.3 Schematic diagram for the heat treatment of β-isomorphous titanium alloys (from Morton, P. H., *Rosenhain Centenary Conference on the Contribution of Metallurgy to Engineering Practice*. The Royal Society, London, 1976).

martensitic α' is tempered or aged. The maximum strength obtainable from this β to α' transformation occurs at a composition for which the martensite finish (M_f) temperature corresponds to room temperature.

Increasing the solute content above this level results in a progressive increase in the amount of metastable β that is retained on quenching from the β or $\alpha + \beta$ phase fields, and there is a gradual decrease in strength of quenched alloys to a minimum value at a composition at which the martensite start (M_s) temperature occurs at room temperature, i.e. 100% metastable β. On the other hand, these compositions provide the maximum response to strengthening if the quenched alloys are aged to decompose the retained β (see shaded area in Fig. 6.3).

The formation and decomposition of martensitic α', and the ageing of alloys containing metastable β, both involve a number of complex reactions some of which have little significance in the heat treatment of commercial alloys. These reactions are discussed when considering appropriate groups of titanium alloys.

6.2 α-ALLOYS

6.2.1 General

The main substitutional alloying elements which dissolve in the α-phase are the stabilizing elements aluminium and oxygen and the neutral elements tin and zirconium. All cause solid solution hardening and increase tensile strength by between 35 and 70 MPa for each one per cent of the added element. Oxygen and nitrogen, which are normally present as impurities, contribute to interstitial hardening and controlled amounts of oxygen are used to provide a specific range of strength levels in several grades of what is known as commercial-purity (CP) titanium. Small amounts of other elements can be present and α-alloys are often divided into three subgroups depending upon whether they are entirely single phase α, contain up to 2% of β-stabilizing elements (near-α alloys) or respond to a conventional age-hardening reaction (Ti–Cu alloys with $< 2.5\%$Cu).

There is a practical limit to the amount of α-stabilizing elements that may be added to titanium since the alloys tend to embrittle because of an ordering reaction that occurs if the 'aluminium equivalent' exceeds about 9%. This quantity, or ordering parameter, may be calculated empirically from the composition by summing the weight percentage as follows:

$$\text{Al} + \tfrac{1}{3}\text{Sn} + \tfrac{1}{6}\text{Zr} + 10(\text{O} + \text{C} + 2\text{N})$$

The ordering reaction has been widely studied, particularly in binary Ti–Al alloys, but it remains incompletely understood. In this system, what is known is that elevated temperature ageing of alloys with an aluminium content above 5–6% can lead to the formation of a finely dispersed, ordered phase (α_2) which is coherent with the lattice of the α-phase over a wide temperature range. The phase α_2 has the general formula Ti_3X and has the DO_{19} (hexagonal) crystal structure that was noted for precipitates formed in several magnesium alloys (Table 5.5). Continuing ageing gradually causes α_2 to coarsen. In other α-alloys such as Ti–Sn, and in more complex compositions, the misfit between α and α_2 is larger so that nucleation of α_2 becomes more difficult and it tends to form heterogeneously. This non-uniform dispersion has a less deleterious effect on ductility. There is general agreement concerning the location of the $\alpha/(\alpha + \alpha_2)$ phase boundary up to 800 °C, but uncertainty exists as to whether α_2 forms by peritectic reaction involving β and an intermediate compound TiAl, or by phase separation of α at higher temperatures (see Fig. 6.1a).

The poor ductility of α-alloys containing the α_2-phase has been a disappointment as the strengthening of the creep-resistant nickel alloys (superalloys) is based on microstructures containing a rather similar coherent phase $Ni_3(Ti,Al)$ or γ'. The only element reported to improve the ductility of α_2 in titanium alloys is gallium and, although some experimental gallium-containing alloys have been produced, the high cost of this element and problem with melting suggest that its commercial use is unlikely.

In the presence of hydrogen, titanium hydrides may form as long thin plates in pure and alloyed α-titanium. These hydrides may be important in fracture phenomena (see Section 6.7.2). They have the stoichiometric composition TiH_2, although it has been reported that hydrogen: metal ratios may vary from 1.5 to 1.99. The basic (δ) hydride is face-centred cubic (CaF_2 structure) while a face-centred tetragonal (ε) hydride forms below 37 °C, which has a c/a ratio less than one. A third (γ) hydride has been observed to form in the α-phase at low hydrogen concentrations which is metastable and also has a face-centred tetragonal structure with c/a equal to 1.09. Additionally, a strain-induced, body-centred cubic hydride has been reported to form on the $\{10\bar{1}0\}$ slip planes during deformation of α solid solutions. A large volume expansion of as much as 18% accompanies hydride formation and this may cause the generation of dislocations so that it can be accommodated in the surrounding matrix.

6.2.2 Fully-α alloys

The only commonly used alloys in this group are the several grades of CP titanium, which are in effect Ti–O alloys, and the ternary composition Ti–5Al–2.5Sn. As the alloys are single phase, tensile strengths are relatively low although their high thermal stability leads to reasonable creep strengths in the upper temperature range (Fig. 6.2). They display good ductility down to very low temperatures and are readily weldable. As it is usually necessary to hot-work the alloys at temperatures below α/β transus in order to prevent excessive grain growth, formability is limited because of their hexagonal crystal structure and the fact that they exhibit a high rate of strain hardening. For this reason, Ti–5Al–2.5Sn has tended to be replaced by the age-hardenable Ti–Cu alloy which can be more easily fabricated after solution treatment, but prior to ageing when it is relatively soft.

The mechanical properties of α-alloys are comparatively insensitive to micro-structure although it is possible to obtain α in three different forms (Fig. 6.4).

1. Equi-axed grains which are formed when the alloys are worked and annealed in the α-phase field (Fig. 6.4a). Grain sizes tend to be relatively small because grain growth is inhibited due to the comparatively low temperatures that are involved and to the presence of impurities which pin grain boundaries. Yield strength at room temperature can be predicted from the Hall–Petch relationship, e.g. the equation for one grade of CP titanium (Ti–50A) is:

$$\sigma_{YS} = 231 + 10.54d^{-\frac{1}{2}} \text{ (MPa)}$$

2. Quenching from the β-phase field produces the hexagonal martensitic phase α′ in which the original β-grains remain clearly delineated. α′ forms by a massive transformation, i.e. the martensite contains a high density of dislocations but few or no twins, and is composed of colonies of plates or laths separated by low angle boundaries. The transformation is characterized by a habit plane near $\{334\}_\beta$. There is negligible hardening associated with the production of α′ martensite because the grain size is large and there is no supersaturation of the substitutional solute atoms.
3. Slow cooling from β-phase field causes α to form as Widmanstätten plates (Fig. 6.4c). In high-purity alloys this structure is referred to as serrated α, whereas, if β-stabilizing elements or impurities such as hydrogen are present, the α-plates may be delineated producing a 'basket weave' effect (Fig. 6.4d).

The α-alloys that are cooled from the β-phase field exhibit lower values of tensile strength, room temperature fatigue strength and ductility, than those having an equi-axed grain structure. For low-cycle fatigue strength there is an empirical relation:

fatigue strength at 5×10^4 cycles $\propto 0.1\%$ proof stress $\times \log R_A$

where R_A = reduction in area in tensile test. This is a useful guide in rating alloys. On the other hand, cooling from the β-field leads to improved values of fracture toughness and higher creep resistance. These trends in mechanical properties which arise from the shape and size of the grains, and from the structure of the grain boundaries, are important as they are characteristic of many other titanium alloys.

CP titanium is the second most used titanium alloy which finds application in the aerospace, architectural, chemical, and process engineering industries. A famous example is its use for the external cladding of the Guggenheim Museum is Bilbao, Spain. Four grades are commonly available which have increasing levels of oxygen, and have room temperature tensile strengths ranging from 240 to 740 MPa. Grade 1 (0.18O–0.2Fe) has the lowest strength and displays excellent cold formability so that it can be deep drawn. Grade 2 has

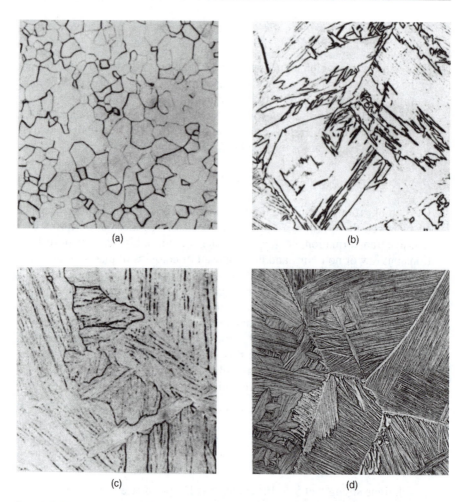

Fig. 6.4 Microstructure of CP titanium: (a) annealed 1 h at 700 °C showing equi-axed grains of α ($\times$ 100); (b) quenched from β-phase field showing martensitic α' ($\times$ 150); (c) air-cooled from the phase field showing Widmanstatten plates of α ($\times$ 100) (courtesy W. K. Boyd); (d) near-α alloy IMI 685 air-cooled from the phase field showing a basket weave configuration of Widmanstatten plates of the α-phase delineated by small amounts of the β-phase (courtesy IMI Titanium) ($\times$ 75).

a tensile strength between 390 and 540 MPa and is the most widely used CP titanium alloy. Examples of applications are skin panels and fire walls in aircraft, and tubing in heat exchangers. Grade 3 (0.40O–0.5Fe) is confined mainly to pressure vessels. In addition, a composition containing 0.2%

palladium (IMI 260) has been developed which has a particularly high resistance to corrosion.

Use of the alloy Ti–5Al–2.5Sn has declined in recent years as alloys with better forming properties and higher creep resistance have become available. One continuing application, however, has been cryogenic storage tanks for which the relatively high strength of titanium alloys at low temperatures is attractive. For this purpose a special grade which is low in interstitial elements (designated ELI) has been developed in the United States to increase the toughness of the alloy and it has been used to store liquid hydrogen (−253 °C). Titanium alloy pressure vessels have become standard for fuel storage in a number of space vehicles as their specific strengths are approximately double those of aluminium alloys and stainless steel at such temperatures.

6.2.3 Near-α alloys

This class of forging alloys was developed to meet demands for higher operating temperatures in the compressor section of aircraft gas turbine engines as part of the continuing quest for improved performance and efficiency. They possess higher room-temperature tensile strength than the fully-α alloys and show the greatest creep resistance of all titanium alloys at temperatures above approximately 400 °C. Early near-α titanium alloys are the American composition Ti-8-1-1 (see Table 6.1) and the British alloy IMI 685 which was used for the forged gas turbine compressor disc or wheel shown in Fig. 6.5.

Fig. 6.5 Forged compressor disc or wheel made from the near-α alloy IMI 685 (courtesy Rolls Royce Ltd).

Bomberger has produced an empirical expression to denote those compositions giving maximum creep strength:

$$36 - 2.6(\% \text{ Al}) - 1.1(\% \text{ Sn}) - 0.7(\% \text{ Zr}) - 27(\% \text{ Si})$$
$$- 3(\% \text{ Mo equivalent}) \leq 10$$

where % Mo equivalent = $1.0(\% \text{ Mo}) + 0.67(\% \text{ V}) + 0.44(\% \text{ W}) + 0.28$ (% Nb) + 0.22(% Ta) + 2.9(% Fe) + 1.6(% Cr) − 1.0(% Al). In practice, the near-α alloys contain up to 2% β-stabilizing elements which both introduce small amounts of β-phase into the microstructure and improve forgeability. However, these additions are normally too small to provide significant strengthening through the decomposition of retained β (Fig. 6.3) and the improvement in mechanical properties arises mainly from the formation of martensitic α' and from the manipulation of α/α' microstructures.

Most near-α alloys are forged and heat treated in the α + β phase field so that primary α-grains are always present in the microstructure. More recently, improved creep performance has been achieved in special compositions by carrying out these operations at higher temperatures which places the alloys in the β-phase field and results in a change in the microstructure. It is interesting to note that the two alloys which currently show the highest creep resistance (maximum operating temperature 590–600 °C) are IMI834 and Ti 1100 and they are forged in the α + β and β phase fields respectively. It is now appropriate to consider the metallurgical synthesis of alloy compositions in each of these two categories.

Alloys heat-treated in α + β phase field One of the first alloys specifically designed to meet creep requirements was the composition Ti–11Sn–2.25Al– 5Zr–1Mo–0.2Si (IMI 679). Development occurred in three well-defined stages. The first was to determine the maximum amounts of α-stabilizing elements that could be added without a severe loss of ductility and it was recognized that, although tin caused less solid solution hardening than aluminium at room temperature, it became a more effective strengthener as the temperature was raised. Tin also had the advantage that higher amounts could be tolerated without causing formation of the embrittling α$_2$-phase, although it was appreciated that such large additions would increase density and prevent the alloy from being welded. In actual practice, the total content of tin and aluminium was limited by a tendency for the alloys to become susceptible to hydrogen embrittlement and a ternary composition Ti–11Sn–2.25Al was selected. Zirconium was added to provide further solid solution strengthening of the α-phase.

The second stage of development was to introduce a β-stabilizing element that would both promote some response to heat treatment and render the alloy more forgeable without adversely affecting creep properties. For this purpose 1% molybdenum was selected. Finally, sufficient silicon was added to further increase strength and creep resistance mainly by dissolving in α in which there is evidence that it segregates to, and reduces the mobility of, dislocations.

A parallel development in the United States led to the alloy Ti–8Al–1Mo–1V (Ti 8–1–1) which had a lower density, was weldable and had better forging characteristics because of the higher content of β-stabilizing elements. However, this alloy has an ordering parameter in excess of 9 which has led to problems of instability and loss of ductility due to the tendency to form the α_2-phase after long time exposure at elevated temperatures. Another American alloy has been developed which is a compromise between the above two alloys. This is known as Ti–6242 and has the composition Ti–6Al–2Sn–4Zr–2Mo. Later, an addition of 0.1% silicon was made and this alloy is designated Ti–6462S. This amount of silicon has an optimal effect in reducing creep deformation (Fig. 6.6) and is presumed to correspond to the limit of supersaturation in this alloy. Nevertheless the minimum in the curve is unexpected. The effect of silicon in improving creep performance can be seen by comparing the curves for Ti–6462 and Ti–6462S in Fig. 6.2 and this latter alloy (Ti–6462S) is widely used in the United States where high creep resistance is required.

All these alloys are forged at temperatures that place them well within the α + β phase field. The recommended heat treatment is to solution treat at a temperature at which the alloy consists of approximately equal proportions of α- and β-phases, e.g. 900 °C for IMI 679 and 1010 °C for Ti 8–1–1. For maximum creep strength, the alloys are then air-cooled to form a microstructure of equi-axed grains of primary α and Widmanstätten α which forms by nucleation and growth from β (Fig. 6.7a). Faster cooling will cause the high-temperature β-phase to transform to martensitic α′, at least in thin sections, which causes some increase in tensile strength although creep resistance is reduced at the

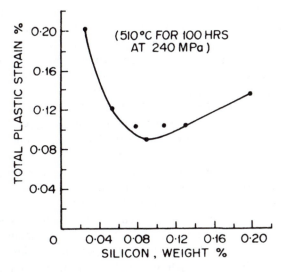

Fig. 6.6 The effect of minor additions of silicon on creep strain in the alloy Ti–6462 (from Seagle, S. R. *et al., Metals Engineering Quarterly,* p. 48, February 1975).

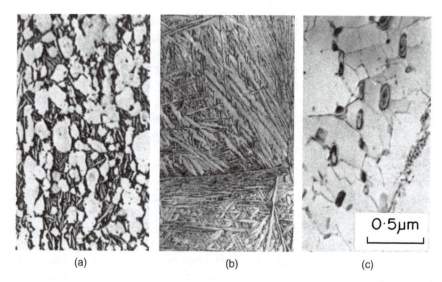

(a)   (b)   (c)

Fig. 6.7 (a) Alloy IMI 679 air-cooled from the $\alpha + \beta$-phase field. The white phase is primary α and the other is Widmanstätten α ($\times$ 500); (b) alloy IMI 685 oil-quenched from the β-phase field showing laths of the martensitic α'-phase delineated by small amounts of the β-phase ($\times$ 75); (c) IMI 685 quenched from the β-phase field and aged at 850 °C showing particles of the phase $(TiZr)_5Si_3$ (courtesy IMI Titanium) ($\times$ 30 000).

upper end of the temperature range (>450 °C). The alloys are then normally given a stabilizing heat treatment within the range 500–590 °C.

β-heat-treated alloys Forging of titanium alloys in the β-phase field offers the advantage of easier deformation because of the higher working temperatures and the fact that the alloys have a body-centred cubic structure. However, this practice, and the subsequent heat treatment of the alloys in this phase field, is normally avoided because excessive grain growth may occur which adversely affects ductility at room temperature.

The near-α alloy IMI 685 (Ti–6Al–5Zr–0.5Mo–0.25Si) is an example of a composition developed to explore the opportunities of both β-forging and β-heat-treatment. The α/β transus is 1020 °C and quenching from 1050 °C produces laths of martensitic α' which are delineated by thin films of β that are retained (Fig. 6.7b). Subsequent ageing at 500–550 °C reduces quenching stresses and causes some strengthening. Martensitic α' transforms to α and the microstructure comprises laths of α bounded by a fine dispersion of particles. Electron diffraction studies have indicated that these particles may be either body-centred cubic ($a = 0.33$ nm) which is the normal β-titanium structure, or face-centred cubic ($a = 0.44$ nm). The β-particles are considered to form by spheroidization of the inter-lath films but little is known of the other phase except that it contains titanium, molybdenum and silicon. If ageing is carried out at higher temperatures, e.g. 850 °C, softening occurs and it is possible to observe the precipitate $(Ti,Zr)_5Si_3$ which forms on dislocation networks in the boundaries between the α-laths (Fig. 6.7c).

Creep resistance is high in the range 450–520 °C and is at a maximum if intermediate quenching rates in the range 1–10 °C s^{-1} are used (Fig. 6.8). In this condition the basket weave morphology, e.g. Fig. 6.4d, is present in the microstructure. In thick sections, or with slow quenching rates, the microstructure can contain coarse, aligned laths of α which reduce room temperature ductility and increase the rate of crack propagation in low-cycle fatigue (Section 6.7.2).

As compared with earlier alloys, the essential features of the composition IMI 685 are as follows.

1. Tin is replaced by a lower amount of aluminium to reduce density whilst maintaining the ordering parameter within safe limits.
2. The content of the β-stabilizing element molybdenum is halved which reduces the amount of the β-phase, the presence of which leads to lower creep resistance.
3. Zirconium is added to provide solid solution strengthening of α.
4. The level of silicon is increased slightly to allow for its greater solubility at the higher β-heat-treatment temperature.

More recently the near α-alloy IMI 829 was developed with enhanced creep strength (Fig. 6.2) and an upper operating temperature of 580 °C. This alloy is also heat treated in the β-phase field, or just below the β-transus,

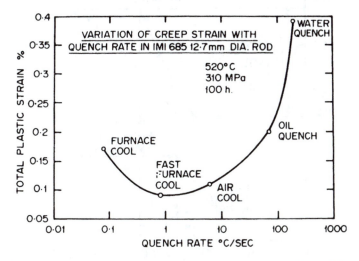

Fig. 6.8 Effect of cooling rate from β-phase field on creep strain of alloy IMI 685 (from Blenkinsop, P. A. et al., in *Titanium and Titanium Alloys*, Proceedings of 3rd International Conference on Titanium, Williams, J. C. and Belov, A. F. (Eds), Plenum Press, p 2003, 1982).

so that a small amount of α (e.g. 5 vol%) is retained to pin grain boundaries and minimize grain growth of the β-matrix. It has the composition Ti–5.5Al–3.5Sn–3Zr–0.3Mo–1Nb–0.3Si. Niobium is considered to confer oxidation resistance superior to that of the earlier near-α alloys, which is important as oxidation is also a limiting factor when considering long-term exposure at such temperatures.

A further marginal increase in operating temperature has been achieved with a later alloy, IMI 834, which has the slightly changed composition Ti–5.5Al–4Sn–4Zr–0.3Mo–1Nb–0.35Si–0.6C. The alloy chemistry has been tailored to allow a greater degree of flexibility in heat treatment and to optimize both creep and fatigue strength. With respect to heat treatment, it is important that such alloys can be held at temperatures very close to the β-transus temperature to retain some α-phase without requiring impractical levels of temperature control. The microstructure is similar to that shown in Fig. 6.7a except that the amount of the primary α-phase is reduced to 5–10 vol. %. As shown in Fig. 6.9, the slope of the so-called β-transus approach curve for IMI 834 is less than for IMI 829 which has the effect of widening the allowable temperature range to obtain the microstructure containing a small amount of the α-phase. This change is attributed to the presence of the minor amount of carbon. The effect of the different proportions of the α- and β-phases on creep and fatigue properties of IMI 834 is shown schematically in Fig. 6.10. It should be noted that this figure also serves as a useful reminder that alloy design is of necessity

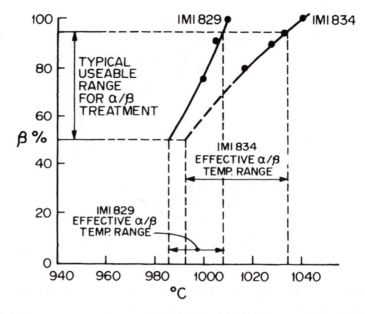

Fig. 6.9 β-transus approach curves for IMI 829 and IMI 834 (courtesy P. A. Blenkinsop).

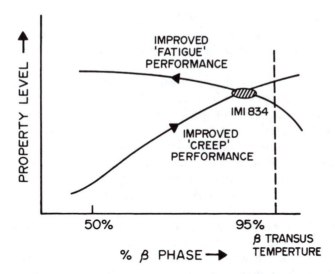

Fig. 6.10 Schematic representation of the effect of α/β-phase proportions on the creep and fatigue properties of IMI 834 (courtesy P. A. Blenkinsop).

a compromise; maximization of one particular property, such as creep resistance, can only be achieved at the expense of some other property.

The alloy Ti 1100 (Ti–6Al–2.75Sn–4Zr–0.4Mo–0.45Si–0.07O) is normally forged in the β region and then directly aged. The microstructure is similar to that shown in Fig. 6.7b. Prolonged exposure of Ti 1100 at relatively high temperatures (e.g. 600 °C) causes two precipitates to form; one is the coherent, ordered phase $\alpha_2(Al_3Ti)$ that nucleates homogeneously within the martensitic α' laths and the other is the silicide $(Ti,Zr)_5Si_3$ that forms at the interfaces. Some increase in proof stress and loss of ductility is observed which is more marked in tests carried out at ambient than at elevated temperatures. This effect is attributed mainly to the presence of α_2. These two precipitates have also been observed in alloy IMI 834 exposed under similar conditions.

It seems probable that, with the development of the alloys IMI 834 and Ti 1100, the limit has been reached in optimizing the composition of the near-α alloys. To achieve further increases in operating temperatures of titanium–based materials, recourse to new approaches to alloy development will be needed, and these are discussed in Section 7.9.

6.2.4 Ti–Cu age-hardening alloy

Although Ti–Cu has no commercial significance as a β-eutectoid system, it was recognized that the titanium-rich end of the phase diagram offered potential for developing an alloy that may respond to age-hardening. This follows because the solubility of copper in α-titanium reduces from 2.1% at the eutectoid temperature of 798 °C to 0.7% at 600 °C and to a very low value at room temperature. Moreover, it seemed possible that such an alloy could be cold-formed after solution treatment when in a relatively soft condition and then strengthened by ageing.

Ti–Cu alloys were investigated in Britain where hot-forming facilities were limited and the composition Ti–2.5Cu (IMI 230) was developed as a heat-treatable sheet material. It is of special interest as it is one of very few titanium alloys that is strengthened by a classical age-hardening reaction. Solution treatment is carried out at 805 °C and is followed by air-cooling (sheet) or oil-quenching to room temperature. A double ageing treatment at 400 and 475 °C promotes precipitation of a fine dispersion of the metastable from of the phase Ti$_2$Cu which is coherent with the β-matrix (Fig. 6.11) and forms on the {1011} planes.

A moderate increase of 150–170 MPa in tensile strength may be achieved by ageing. Strength properties are further enhanced if the alloys are cold-formed prior to ageing and, in this condition, compare favourably with the α-alloy Ti–5Al–2.5Sn which requires hot-forming. Ti–2.5Cu is weldable and strength may be recovered providing the duplex ageing treatment is applied after welding. Figure 6.12 shows as application which is a casing for a gas turbine engine that is constructed by welding together forged rings and vanes formed from IMI 230 sheet.

It will be noted that the commercial composition is a binary alloy and it might be anticipated that additions of other elements, singly or together, may induce a further response to age-hardening. However, extensive investigation has failed to reveal any addition having this desired effect.

Fig. 6.11 Coherent plates of Ti$_2$Cu zones in an aged Ti–2.5Cu (IMI 230) alloy (courtesy IMI Titanium).

Fig. 6.12 Welded IMI 230 alloy casing from a Rolls Royce/SNECMA Olympus engine (courtesy Rolls Royce Ltd).

6.3 α/β **ALLOYS**

The limitations in strength that can be developed in the fully-α alloys because of the ordering reaction occurring at higher solute contents, together with difficulties with hot-forming, led to the early investigation of compositions containing both the α- and β-phases. These α/β alloys now have the greatest commercial importance with one composition, Ti–6Al–4V (IMI 318), making up more than half the sales of titanium alloys both in Europe and the United States. They offer the prospect of relatively high tensile strengths and improved formability, although some sacrifice in creep strength occurs above 400 °C as well as reduced weldability. Their principal use is for forged components, e.g. in the fan blades of jet engines (Fig. 6.13). Closed die forgings weighing 1600 kg and measuring 3.8 m long by 1.7 m wide are produced in the same alloy for use as bulkheads in the new Lockheed/Boeing F22 military aircraft.

Most α/β alloys contain elements to stabilize and strengthen the α-phase, together with 4–6% of β-stabilizing elements which allow substantial amounts of this phase to be retained on quenching from the β or α + β phase fields. The common β-stabilizing elements confer solid solution strengthening of the β-phase although, as shown in Table 6.2, these effects are relatively small. This table also gives the minimum solute concentration needed to give complete retention of metastable β on quenching of binary alloys to room temperature. Strength properties of α/β alloys may be enhanced by subsequent tempering or ageing treatments and room-temperature tensile strengths exceeding 1400 MPa have been achieved. However, few compositions can sustain these levels of strength in thick sections because of hardenability effects on quenching which are aggravated by the low thermal conductivity of titanium (Table 1.1).

Fig. 6.13 Forged IMI 318 (Ti–6Al–4V) blades from the LP rotor stage of the Rolls Royce/SNECMA Olympus 593 jet engine (courtesy Rolls Royce Ltd).

Table 6.2 Solid solution strenthening and β-stabilizing capacity of β-stabilizing alloying elements (from Hammond, C. and Nutting, J., *Metal Science*, **11**, 474, 1977)

	Element							
	V	Cr	Mn	Fe	Co	Ni	Cu	Mo
Solid solution strengthening (MPa wt%$^{-1}$)	19	21	34	46	48	35	14	27
Minimim alloy content to retain β on quenching (%)	14.9	6.3	6.4	3.5	7	9	13	10

It is proposed now to consider structure/property relationships in α/β alloys which are developed by heat treatment. Reference should again be made to Fig. 6.3 although some additional phase transformations will need to be considered.

6.3.1 Annealed alloys

For alloys which have phase diagrams of the β-isomorphous type, uniform properties can be obtained in thick sections by slow cooling from either the β or α + β phase fields, known as the β-annealed and mill-annealed conditions respectively. In the first case, it is usual for the α-phase to form as Widmanstätten laths in a β-matrix, although β may itself transform to martensitic α′. The size of the laths depends on the rate of cooling and the basket weave structure is again obtained when cooling rates are slow (Fig. 6.14a). Annealing in the α + β phase field is usually carried out at about 700 °C and, in addition to providing stress-relief, this treatment results in the formation of an equi-axed structure composed of α-grains and grains of transformed β (Fig. 6.14b). These latter grains transform to Widmanstätten α as is evident

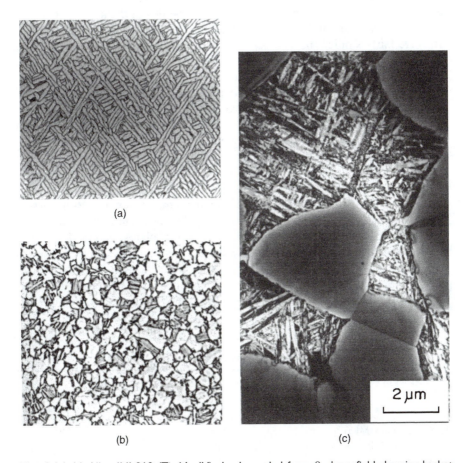

(a)

(b)   (c)

Fig. 6.14 (a) Alloy IMI 318 (Ti–6A–4V) slowly cooled from β-phase field showing basket weave structure of Widmanstätten α-matrix (courtesy Rolls Royce Ltd) (× 320); (b) alloy IMI 318 annealed at 700 °C in α + β phase field. Equi-axed grains of α (white) and transformed β (Widmanstätten α) (courtesy IMI Titanium) (× 500); (c) transmission electron micrograph of (b) showing the structure of the transformed β (Widmanstätten α) (courtesy C. Hammond) (× 7500).

in the transmission electron micrograph shown in Fig. 6.14c. Grain size can be modified by suitable adjustment of working and annealing cycles and the amount of primary α is dictated by the lever rule. Frequently a second, or duplex anneal, is given which causes further partitioning of alloying elements between the α- and β-phases, and the main purpose of this treatment is to enrich the β-phase which increases the stability of the alloys for service at elevated temperatures.

The α/β titanium alloys are most often used in the annealed condition and it should be noted that both microstructure and some mechanical properties may differ depending upon whether or not prior forming was carried out above or below the β-transus. Table 6.3 compares the properties of the alloy Ti–6Al–4V forged in these two conditions. It should be noted that, although tensile properties are fairly similar, the samples forged in the α + β phase field (equi-axed grains) are more ductile, whereas fracture toughness and fatigue strength are both notably higher in β-forged and annealed material (acicular Widmanstätten structure). Work on Ti–6Al–4V rolled plate has indicated that the superior fatigue performance with the β-annealed condition is associated with relatively slower rates of crack propagation (Fig. 6.15a). This effect, in turn, is attributed to the slower progress of cracks through the Widmanstatten microstructure, particularly at stress intensities below a critical value (T in Fig. 6.15a) at which desirable crack branching occurs within packets of the α-laths (Fig. 6.15b). These trends have already been noted when considering α-alloys and appear to apply generally in the α/β alloys. On the other hand, the β-processed, Widmanstätten-α microstructure (Fig. 6.14a) is less resistant to the initiation of fatigue cracks than the mixed-α and transformed-β microstructure (Fig. 6.14b and c). Thus the low-cycle (high-stress) fatigue strength of β-annealed alloys is inferior to that for the mill-annealed (α + β) condition, as is shown in Fig. 6.16. This effect is aggravated at elevated temperatures because enhanced oxidation occurs along phase boundaries which permits premature crack initiation

Table 6.3 Properties of annealed Ti–6Al–4V forgings†

	Forging treatment‡	
	α + β phase field	β phase field
Tensile ultimate (MPa)	978	991
Tensile yield (MPa)	940	912
Tensile elongation (%)	16	12
Reduction in area (%)	45	22
Fracture toughness (MPa m$^{1/2}$)	52	79
10^7 fatigue limit (MPa)§	±494	±744

†Annealed 2 h at 705 °C, air-cooled after forging
‡α/β transus 1005 °C
§Axial loading: smooth specimens, $K_t = 1.0$

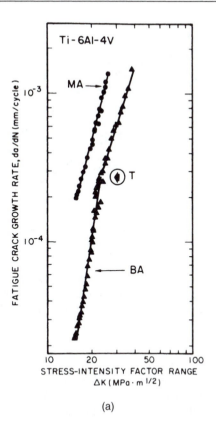

(a)

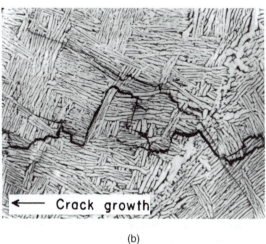

(b)

Fig. 6.15 (a) Fatigue crack growth rates for Ti–6Al–4V rolled plates in the β-annealed (BA) and mill-annealed (MA) conditions. BA = 0.5 h 1038 °C, air-cool to room temperature. Tests conducted at 5 Hz using compact tension specimens. Ratio of minimum to maximum load = 0.1 (from Yoder, G. R. *et al.*, *Met. Trans.*, **8A**, 1973, 1977); (b) branching of fatigue cracks within the Widmanstätten packets of the α-laths (courtesy J. Ruppen and A. J. McEvily) (× 225).

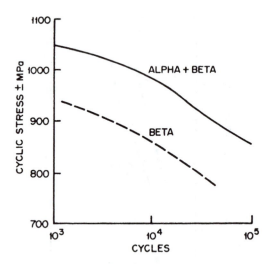

Fig. 6.16 Comparison of low-cycle fatigue lives of Ti–6Al–4V in mill-annealed and β-annealed condition (from Blenkinshop, P. A. et al., in *Titanium Science and Technology*, Proceedings of 5th International Conference on Titanium, Lütjering, G. et al., (Eds) D. G. M., p 2323, 1985).

along surface-connected, acicular-α interfaces. To achieve a good balance of creep and fatigue properties in α/β alloys, a duplex structure consisting of about 30 vol% of equi-axed-α combined with Widmanstätten-α is favoured for rotating components, such as compressor discs, that operate at high temperatures.

In the β-eutectoid alloys, very slow cooling from the β-phase field leads to the formation of a lamellar eutectoid of orthorhombic martensitic α and a compound such as Ti_2Cu, in a manner that is analogous to pearlite formation in steels. However, these structures have so far found no application in commercial titanium alloys because the reactions are sluggish and the phases that form cause embrittlement.

6.3.2 Quenching from β-phase field

The range of properties of α/β alloys can be extended by quenching from the β-phase field and then tempering or ageing at elevated temperatures to decompose the quenched structures. The changes that occur may be complex and it is necessary to study them in some detail.

A distinction can be made between relatively dilute α/β alloys which form hexagonal α' martensite or two orthorhombic martensites α″ and α‴ on quenching, and more concentrated alloys in which the β-phase may be partly or completely retained in a metastable condition. The division between the two types of behaviour can be shown by the M_s (martensite start) line that is

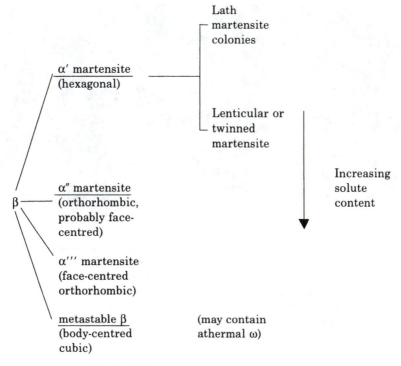

Fig. 6.17 Possible reactions from the β-phase.

included in Fig. 6.1b, c and Fig. 6.3. If the alloys contain a sufficient content of β-stabilizing elements to bring the M_s temperature below room temperature then a fully metastable β-structure can be retained. The possible reactions (with the most important being underlined) and microstructures may be summarized as in Fig. 6.17.

The most common martensite is the hexagonal α′ type. In the more dilute alloys, it forms as colonies of parallel-sided plates or laths (Fig. 6.18a), the boundaries of which consist of walls of dislocations. The internal regions are also heavily dislocated. The structure is similar to that obtained when fully-α alloys are quenched from the β-phase field except that, in the α/β alloys, the laths are separated by thin layers of retained β-phase which is enriched in β-stabilizing solute elements. With increasing solute content and lower M_s temperatures, these colonies decrease in size and many degenerate into individual plates which are randomly oriented. These plates have a lenticular or acicular morphology (Fig. 6.18b) and are internally twinned on $\{10\bar{1}1\}_\alpha$ planes. The orientation relationship of the β-phase and α′ martensite is $(110)_\beta//(0001)_\alpha;[111]_\beta//[11\bar{2}0]_\alpha$, and the habit planes for the untwinned and twinned planes are $\{334\}_\beta$ and $\{344\}_\beta$ respectively.

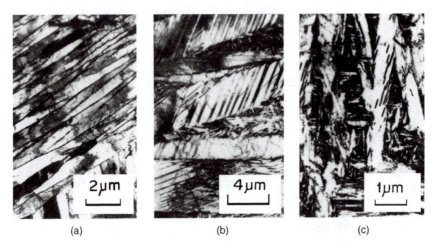

(a)　　　　　　　　　　(b)　　　　　　　　　　(c)

Fig. 6.18 Transmission electron micrographs showing the structure of titanium alloy martensites: (a) hexagonal α' (lath) martensite in a dilute alloy (Ti–1.8Cu) quenched from 900 °C; (b) hexagonal α' (lenticular) martensite containing twins in a concentrated alloy (Ti–12V) quenched from 900 °C; (c) orthorhombic α''-martensite in the alloy Ti–8.5Mo–0.5Si quenched from 950 °C (from Williams, J. C., in *Titanium Science and Technology*, Jaffee, R. I. and Burte, H. M. (Eds), Plenum Press, New York, Vol. 3, 1973).

The second type of titanium martensite (α'') has an orthorhombic structure and a similar lattice correspondence with the β-phase. For this reason it is probably face-centred and its lattice dimensions are $a = 0.298$ nm, $b = 0.494$ nm, $c = 0.464$ nm. It is also internally twinned (Fig. 6.18c), the twins forming on the $\{111\}_{\alpha''}$ planes. It has been proposed that the transition from hexagonal α' to orthorhombic α'' martensite occurs in alloys with increasing solute content as shown above, as well as with decreasing M_s temperature. Moreover, formation of α'' is thought to be strongly composition dependent; e.g. it occurs in Ti–Mo but not in Ti–V alloys, although it may form in the latter alloys if aluminium is added. Although α'' martensite tended to be neglected in earlier studies, its importance is now recognizied. Its presence lowers tensile ductility in alloys although it does have the advantage of being a favourable precursor in producing a very uniform distribution of the α-phase following subsequent heat treatment. α'' martensite can be formed in the following three ways.

1. Decomposition of metastable β during quenching.

$$\beta \rightarrow \alpha'' \, (\beta)$$

2. Decomposition of retained β by intermediate (bainitic) transformation during isothermal ageing.

$$\beta \rightarrow \beta \text{ (lean)} + \beta \text{ (rich)} \rightarrow \alpha'' + \beta \text{ (rich)}$$

3. Stress-induced transformation of retained β.

$$\beta \rightarrow \alpha'' + \text{twinned } \beta$$

Two other martensites (face-centred orthorhombic and face-centred cubic) have been reported in electron microscope studies and the former has been confirmed by X-ray diffraction techniques. It has been termed α''' (or β') and has lattice parameters quite distinct from α'' martensite, i.e. $a = 0.356$ nm, $b = 0.439$ nm and $c = 0.447$ nm. The orientation relationship with the β-phase is $(0\bar{1}1)_\beta//(001)_{\alpha'''}$; $[1\bar{1}1]_\beta//[\bar{1}10]_{\alpha'''}$ and the habit plane is close to the $\{133\}_\beta$ planes. The existence of the second phase has not been confirmed and it may well arise as an artefact due to the use of thin metal foils for examination in the electron microscope.

If the M_s and M_f fall above and below room temperature respectively, then a mixed microstructure containing lenticular α' or α'' (or perhaps α''') martensite may be formed together with retained β. Another feature is that metastable β may contain a fine dispersion of a phase ω, the formation of which cannot be suppressed even at fast quenching rates. The nature of this athermal ω is discussed in Section 6.3.4.

As mentioned earlier, lack of hardenability can cause variations in the strength and microstructure of thick sections that are quenched and, in such cases, solution treating within the $\alpha + \beta$ rather than the β-phase field can be an advantage. Such treatments cause partitioning of solute between α and, more particularly, the reduced volumes of β, thereby increasing the stability of this latter phase so that it is less likely to transform during quenching. This effect increases as the solution treatment temperature is lowered because the volume of β-phase is further reduced, an observation that has led to the practice known as soft quenching, which has been applied to α/β alloys of large section (up to 150 mm). This practice involves solution treating at a temperature high in the $\alpha + \beta$ field to dissolve the alloying elements and slow cooling to 700 °C, at a rate of 50–150 °C an hour, to allow partitioning of solute as the amount of β-phase is reduced. It is then possible to air-cool the alloy to room temperature and still retain the β-phase.

Use is also made of a double solution treatment combined with either water quenching or air cooling to room temperature, between and after these treatments, in order to obtain different microstructures. In this case, the initial solution treatment is carried out just below the β-solvus temperature so that some primary α is present and the range of microstructures that can be obtained is shown schematically in Fig. 6.19. As mentioned earlier, it is possible to gain

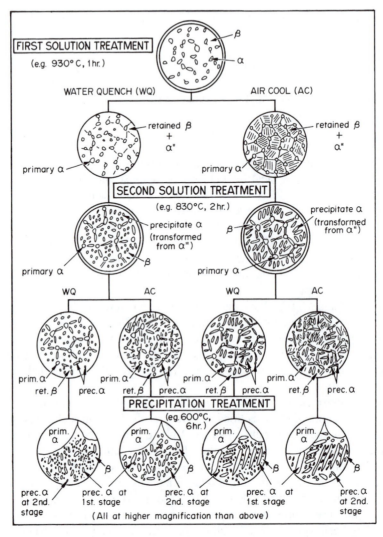

Fig. 6.19 Schematic diagram showing range of microstructures obtained by double solution treatment and either water quenching (WQ) or air cooling (AC) followed by a tempering or precipitation treatment (from Murakami, Y., in *Titanium '80 Science and Technology*, Kimura, H. and Izumi, Q. (Eds), AIME, Warrendale, PA, p. 153, 1980).

further control over mechanical properties. For example, the use of double solution treatment, particularly in conjunction with air-cooling, promotes the formation of coarse, acicular plates of the α-phase which are effective for branching cracks and improving fracture toughness.

Quenched alloys are normally tempered or aged to decompose the retained phases and it is now appropriate to consider the changes that may occur.

6.3.3 Tempering[†] of titanium martensites

Titanium martensites transform on heating at elevated temperatures by several reactions the nature of which depend upon the crystal structure of the martensite and the composition of the alloy concerned. The reactions may be complex and the various types are shown below.

These reactions lead to a wide range of microstructures. In the β-isomorphous alloys, α' decomposes directly to α of equilibrium composition at the tempering temperature and β forms as a fine precipitate that is nucleated heterogeneously at martensite plate boundaries or at internal substructures such as twins. Significant increases in strength may result (Table 6.1). In β-eutectoid alloys, α' may decompose directly into the α-phase and an intermetallic compound, although the formation of this compound may take place in several stages. However, in systems such as Ti–Mn where the normal eutectoid reaction is sluggish, the martensite tempers first by forming α and precipitates of β with the intermetallic compound appearing only slowly at a later stage.

Tempering of α'' martensite may occur by two mechanisms. In alloy compositions in which $M_s(\alpha'')$ occurs at a relatively high temperature, α'' decomposes first by the formation of a fine and uniformly dispersed α-phase in the α''-matrix. Further ageing causes both the coarsening of these particles and the nucleation of a cellular reaction at the prior β-grain boundaries leading to the growth of a lamellar structure of α + β. Growth of these lamellar cells then occurs at the expense of the other regions. In alloys having an $M_s(\alpha'')$ temperature near to room temperature, the α'' reverts to the β-phase, which then decomposes by a mechanism which is characteristic of the particular tempering temperature. Decomposition of the β-phase is discussed in Section 6.3.4.

No information is available concerning the tempering of α''' martensite but its similar crystal structure suggests it may behave like α''.

α' *martensite*

β-isomorphous alloys $\alpha' \rightarrow \alpha + \beta$

β-eutectoid alloys

- Alloys with slow eutectoid reactions e.g. Ti–Mn $\alpha' \rightarrow \alpha + \beta \rightarrow \alpha + \text{compound}$
- Alloys with fast eutectoid reactions e.g. Ti–Cu $\alpha' \rightarrow \alpha + \text{compound}$ (may form in several stages)

[†]Quenched alloys are normally heated for a period of time at an elevated temperature. This treatment is referred to as tempering in the case of martensites, and ageing when retained β is being decomposed, although the actual heat treatments are similar.

α″ martensite

$$
\left\{
\begin{array}{l}
\begin{array}{l}
\text{Alloys with high} \\
M_s(\alpha'') \text{ temperature}
\end{array}
\quad
\begin{array}{l}
\alpha'' \rightarrow \alpha'' + \alpha \rightarrow \alpha'' + \alpha + (\alpha + \beta) \rightarrow \alpha + \beta \\
\qquad\qquad\qquad\qquad\quad \text{(cellular)} \\
\qquad\qquad\qquad\qquad\quad \text{reactions)}
\end{array} \\[2em]
\begin{array}{l}
\text{Alloys with low} \\
M_s(\alpha'') \text{ temperature}
\end{array}
\quad
\begin{array}{l}
\alpha'' \rightarrow \beta \rightarrow \text{products} \\
\text{(see Section 6.3.4)}
\end{array}
\end{array}
\right.
$$

6.3.4 Decomposition of metastable β

Decomposition of the β-phase that is retained on quenching occurs on ageing at elevated temperatures. It is frequently the dominant factor in the heat treatment of α/β and β-alloys, particularly when the aim is to develop high tensile strength (Fig. 6.3). The direct transformation of β to the equilibrium α-phase occurs only at relatively high temperatures probably because of the difficulty of nucleating the close-packed hexagonal α-phase from body-centred cubic β-matrix. Accordingly, intermediate decomposition products are usually formed and the possible reactions are summarized and discussed below.

Medium alloy content
   100–500 °C $\beta \rightarrow \beta + \omega \rightarrow \beta + \alpha$
Concentrated alloys
   200–500 °C $\beta \rightarrow \beta + \beta_1 \rightarrow \beta + \alpha$
   >500 °C $\beta \rightarrow \beta + \alpha$

ω-phase As mentioned earlier, athermal ω may form in the β-phase during quenching of some compositions and it occurs by a displacement reaction. More commonly, however, ω precipitates isothermally as a very fine dispersion of particles when alloys containing metastable β are isothermally aged at temperatures in the range 100–500 °C. The ranges of stability of both types of ω are shown schematically for a β-isomorphous phase diagram in Fig. 6.20, but there is some evidence that athermal ω can also form during heating to the isothermal ageing temperature.

The ω-phase has attracted special attention because its presence can cause severe embrittlement of the alloy concerned. In a more positive vein, ω particles have been beneficial in the specialized field of superconducting titanium alloys as they are effective in flux-pinning, with a consequent large improvement in the critical current densities that may be sustained in the presence of an external magnetic field.

Studies of the isothermal ω-phase have revealed the following characteristics:

1. ω forms rapidly as homogeneously nucleated, coherent precipitates with particle densities that may exceed 80% by volume (Fig. 6.21).

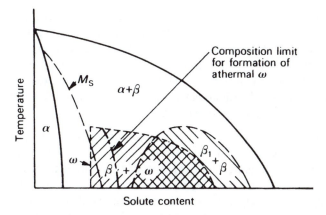

Fig. 6.20 Schematic β-isomorphous alloy phase diagram showing an M_s curve and the ranges of stability of ω, β and $β_1$.

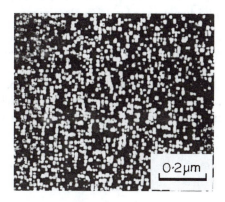

Fig. 6.21 Dense dispersion of cuboids of the ω-phase in a Ti–11.5Mo–4.5Sn–6Zr alloy (from Williams, J. C., in *Titanium Science and Technology*, Jaffee, R. I. and Burte, H. M. (Eds), Plenum Press, New York, Vol. 3, 1973).

2. The ω particles are cuboidal in shape (cube face//{100}$_β$) if there is high misfit and ellipsoidal (long axes //⟨111⟩$_β$) if misfit is low.
3. Partitioning of solute occurs during ageing leading to depletion of ω and enrichment of the β-matrix. The terminal composition of ω in aged binary titanium alloys is related to the group number of the solute in the periodic table because the electron:atom ratios of all ω-phases have been found close to 4.2:1. Thus the possibility exists that ω is an electron compound.
4. Most results suggest that ω has a hexagonal structure with a constant c/a ratio of 0.613 for all systems in which it is found.

5. Dislocations have little or no mobility in ω which accounts for the embrittlement of alloys having high volume fractions of this phase. It is interesting to note however that, even in alloys displaying no macroscopic ductility, the fracture surfaces show exceedingly small dimples which are indicative of some ductility at a microscopic scale. Thus the possibility exists that the potent hardening associated with ω may be used to practical advantage although no progress has been made in this regard.

The formation of isothermal ω may be minimized or avoided by control of the ageing conditions, as well as by varing alloy composition. The significance of both ageing temperature and composition is apparent from Fig. 6.20. The upper temperature limit of stability of ω in most binary alloys is close to 475 °C and the range of stability decreases as the solute content is raised. This latter effect is attributed to a relative increase in the stability of the β-phase and it should be noted that this effect can arise from the presence of both α- and β-stabilizing elements. For example, ω is formed in binary Ti–V alloys, but is absent in the important ternary alloy Ti–6Al–4V. This is one reason why most α/β and β-titanium alloys contain at least 3% aluminium.

β-phase separation Separation of the β-phase into two bcc phases of different compositions is favoured in alloys which contain sufficient β-stabilizer to prevent ω formation during low temperature ageing, and which transform only slowly to the equilibrium phase α under these conditions (Fig. 6.20). This transformation is thought to occur during ageing of a wide range of alloys but has received much less attention than the β → ω reaction because it is not considered to be important in commercial alloys. The phases that form have been designated as β (matrix) and $\beta_1^\dagger$, which occurs as a uniformly dispersed, coherent precipitate. Again there is partitioning of solute between the two phases which leads to enrichment of the β-matrix and depletion of β_1 during treatment.

Formation of equilibrium α-phase Ageing of alloys containing metastable β can, under certain circumstances, result in the direct nucleation of the α-phase. Alternatively this phase may form indirectly from either the ω- or β_1-phases. The route that is followed controls morphology and distribution of α and thus has a marked effect on properties. α that forms directly from β can have two distinct morphologies. It may occur as coarse Widmanstätten plates in a β-matrix in both the relatively dilute binary alloys aged at temperatures above the range for ω formation, and in more complex alloys containing substantial amounts of aluminium, e.g. Fig. 6.14b and c. In such cases, ductility may be adversely affected and deformation prior to ageing is desirable so as to obtain a more uniform distribution of α. Alternatively, a fine dispersion of

---

†Occasionally these phases are designated as β_1 and β_2, respectively.

α in a β-matrix is obtained when alloys containing high concentrations of β-stabilizing elements are aged at temperatures above which phase separation of β occurs.

When α forms in alloys comprising β + ω microstructure, the mechanism of nucleation depends upon both the relative misfit between these two phases as well as the ageing temperature. If the misfit is low, α nucleates with difficulty and it forms by a cellular reaction that occurs heterogeneously at the β-grain boundaries. If the misfit is high then α nucleates at the β/ω interfaces. High ageing temperatures encourage α to form directly from ω.

Continued ageing of alloys that have undergone β-phase separation into β + β₁ leads to the nucleation of the α-phase within the β-particles. Thus the final α-phase distribution is determined by the distribution of β_1, and so is characteristically uniform and closely spaced. Although this reaction may, inadvertently, play a part in the heat treatment of a number of commercial α/β titanium alloys, it has not been studied in detail.

6.3.5 Fully heat-treated α/β alloys

As mentioned earlier, the most commonly used titanium alloy is Ti–6Al–4V, which has many applications as a general-purpose alloy and is usually used in the annealed condition. For applications such as fasteners that require higher strength, the alloy is solution treated high in the α + β phase field, quenched, and aged or annealed at around 700 °C. The resultant structure then comprises equi-axed α-grains in a matrix of fully transformed β (Fig. 6.14b and c). The alloy combines a minimum tensile strength of about 960 MPa with good creep resistance up to 380 °C. It is particularly used in forgings but is also available in plate, sheet, rod or wire forms.

For applications such as aircraft engine mounting brackets and under-carriage components, greater strength is required and a number of other α/β forging alloys have been developed. More use has been made of the heat treatment potential of the alloys and, in general, the content of stabilizing elements has been increased so that the M_f temperature is depressed well below room temperature and significant amounts of β are retained, e.g. the alloys Ti–6Al–6V–2Zr–0.7(Fe, Cu)(Ti–662), Ti–6Al–2Sn–4Zr–6Mo(Ti–6246), Ti–4Al–4Sn–4Mo–0.5Si (IMI 551) and Ti–4.5Al–5Mo–1.5Cr (Corona-5). Full strengthening is again usually obtained by solution treating in the α + β phase field, quenching and ageing to transform the retained β. On the other hand, fracture toughness is higher if the alloys are first cooled from the β-phase field (see Table 6.2). For example, the alloy Corona-5 has a plane strain fracture toughness as high as 155 MPa m$^{1/2}$, for a tensile strength of 950 MPa if it is first annealed in the β-phase field. This value for fracture toughness is double that normally obtained with Ti–6Al–4V and is again attributed to the slower rate of crack propagation through the microstructure containing elongated Widmanstätten laths of the α-phase, e.g. Fig. 6.14(a) and Fig. 6.15(b).

6.4 β-ALLOYS

β-titanium alloys are defined as those containing enough total alloying elements that enable the β phase to be retained in either a metastable or stable condition after cooling to room temperature during heat treatment (Fig. 6.22). This implies that the amount of these elements is sufficient to avoid passing through the martensite start line (M_s). Alloy compositions that place them between the critical minimum level (β_c) of these stabilizing elements (i.e. where the M_s line intersects the room temperature axis) and the similar intersection point of the β-transus line (β_s), are commonly referred to as metastable β-titanium alloys because they will precipitate a second phase (usually α) upon ageing. More highly alloyed compositions to the right of β_s are considered to be stable β-alloys and should not able to be hardened by heat treatment. Such alloys are less commonly used.

Molybdenum is well known as a β-stabilizing element and the general class of β-alloys has been arbitrarily defined as those having a molybdenum equivalent content ≥10. A formula for calculating the molybdenum equivalent value was shown in Section 6.2.3 and, for its application to β-titanium alloys, the aluminium content is subtracted to reflect the opposite effect of this element in stabilizing the α-titanium phase. Other elements that promote β-phase formation are shown in Fig. 6.1. In addition to molybdenum, those most widely used in β-titanium alloys are vanadium, chromium and iron.

β-titanium alloys attracted early attention because of superior forming characteristics anticipated from the body-centred cubic structure. Moreover, they offered the prospect of being cold-formed in a relatively soft condition and then strengthened by age-hardening. Another potential advantage was that the

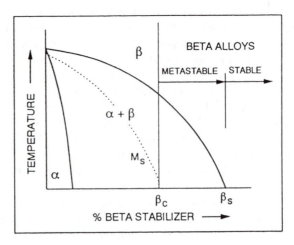

Fig. 6.22 Pseudo-binary β-isomorphous phase diagram showing locations of metastable and stable β-titanium alloys.

presence of a high content of solute elements increased hardenability which may allow the through hardening of thick sections during heat treatment. Such high contents of β-stabilizing elements did, however, cause problems with ingot segregation and tended to increase density, some alloys having relative densities in excess of 5.

The first alloy to be used commercially was the composition Ti–13V–11Cr–3Al which may be solution treated, quenched, cold-formed and aged at 480 °C to give a tensile strength as high as 1300 MPa. Strengthening is due mainly to a combination of solid solution hardening of the β-phase and age-hardening, the latter arising from precipitation of a fine dispersion of the α-phase in the β-matrix. This alloy has limited weldability due mainly to formation of the ω-phase in the heat-affected zones. Moreover, it has proved to be unstable if exposed for long times above 200 °C as such a treatment causes precipitation of the compound $TiCr_2$ which reduces the toughness of the alloy. It should be noted, however, that the above alloy was successfully used for the skin and some structural members of the American SR-71 aircraft designed to fly at speeds around 3200 km h^{-1} at which aerodynamic heating precludes the use of aluminium alloys. Another relatively early β-alloy was the medium-strength British composition Ti–15Mo (IMI 205) which has found application in the chemical industry.

More recently a second generation of metastable β-titanium alloys has been developed for which the formulation of compositions has been controlled mainly by these guidelines:

1. The addition of elements, e.g. aluminium, zirconium and tin, which tend to suppress or restrict the formation of the ω-phase during heat treatment or welding by promoting nucleation of the α-phase.
2. Limiting the amount of elements, e.g. chromium, which tend to stabilize the β-eutectoid transformation and cause embrittlement because of the formation of $TiCr_2$ or other compounds. It should also be noted that excessive eutectoid stabilization leads to a sluggish response to age hardening which is undesirable.
3. Promotion of plastic strain by slip rather than by a mechanism involving a strain-induced transformation to martensite. Alloys which deform primarily by slip have been found to possess better forming characteristics.

Examples of more recent β-titanium alloys are shown in Table 6.4. A high level of strengthening is possible in each alloy which again arises from a combination of solid solution and age hardening. Some cold deformation of alloys such as Beta III can be carried out at room temperature prior to ageing, which may further enhance strength properties due to the nucleation of finely dispersed α-precipitates on dislocations.

A typical treatment cycle for β-titanium alloys is that used for the alloy Beta III. This involves solution treatment at 750–775 °C, water quench or air cool to room temperature and age 24 hour at 475 °C to precipitate the phase α

Table 6.4 Compositions and applications of β-titanium alloys

Alloy	Country of Origin	Applications
Ti–11.5Mo–6Zr–4.5Sn (Beta III)	U.S.A.	Aircraft fasteners, rivets, steel springs for orthodontic devices.
Ti–8V–6Cr–4Mo–4Zr–3Al (Beta C)	U.S.A.	Aircraft fasteners, springs, tubular casings for oil, gas and geothermal wells.
Ti–10V–2Fe–3Al (Ti 10–2–3)	U.S.A.	High strength forgings for airframes, landing gear, bar and plate.
Ti–15V–3Al–3Cr–3Sn (Ti 15–3)	U.S.A.	Airframe sheet, tubing and fasteners.
Ti–15Mo–2.7Nb–3Al–0.2Si (Ti–21S)	U.S.A.	Oxidation and corrosion resistant sheet and foil for use at elevated temperatures. (e.g. 260–425 °C) e.g. jet engine nacelles.
Ti–5Mo–5V–8Cr–3Al (TB2)	P.R. China	High speed rotors and other forgings.
Ti–5Al–5V–5Mo–1Cr–1Fe (VT22)	C.I.S.	High strength forgings, aircraft landing gear.

phase in a β matrix. The volume fraction and morphology of the α precipitates control the strength whereas the β grain size has a major effect on ductility. Tensile strengths exceeding 1400 MPa have been achieved combined with fracture toughness values above 50 MPa m$^{1/2}$. These microstructural features can be changed by deforming before ageing which facilitates heterogeneous nucleation of the α precipitates on dislocations. On the other hand, precipitation combined with recrystallization of the β grains can occur if the alloys are heated at controlled rates to temperatures of 750 to 900 °C after being deformed.

Until recently, β-titanium alloys were little used and they accounted for only one percent of the total titanium market in the United States in 1993. Now it is realised that these alloys are the particularly versatile offering very high strength combined with good toughness and fatigue resistance. Furthermore they show high hardenability so that components with large cross-sections can be heat treated. Sheet materials may also show good cold formability. As an example, the good forging characteristics and high strength of the alloy Ti–10–2–3 has led to its replacement of steel for most of the members of the landing gear of the Boeing 777 aircraft with a weight saving of 270 kg (Fig. 6.35). Another notable feature is the relatively low flow stresses exhibited by the β-alloys during hot-working over a wide range of strain rates when compared with those required for the common α/β-alloy Ti–6Al–4V. This characteristic is shown in Fig. 6.23 and it will be noted that the flow stresses for all three alloys decrease to plateau values as the strain rates fall below approximately 10^{-5} s^{-1}. These values correspond to the onset of superplastic behaviour (Section 6.5.1).

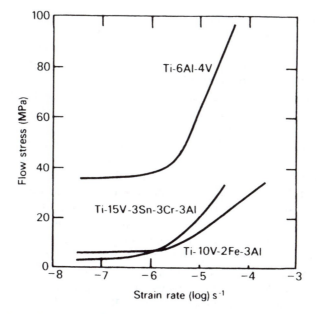

Fig. 6.23 Relationship between flow stress and strain rate for titanium alloys hot-worked at 810 °C (courtesy H. W. Rosenberg).

6.5 FABRICATION

6.5.1 Hot-working

The as-cast structure of consumable arc-melted ingots is sensitive to cracking and the initial working is normally done by hot-press forging. Deformation is carried out at a relatively slow rate in large hydraulic presses and the ingots may be press forged to slabs about 150 mm thick for subsequent rolling to plate or sheet. Alternatively, they may be pressed to round or square billets for processing to bar, rod, tube, extruded sections or wire. Rough forgings for components such as gas turbine compressor discs can be pressed directly from ingots.

Above 550 °C, titanium will absorb oxygen to form both an oxide scale and a brittle, sub-surface layer which can initiate surface cracks. Titanium also absorbs hydrogen and it is necessary to ensure that furnace atmospheres are hydrogen-free. For this reason, electric preheating furnaces are preferable to those fired by oil or gas. Chemical descaling, abrasive cleaning and even machining, frequently combined with careful surface inspection, may be necessary before any further working operations are carried out. Further descaling and cleaning may also be required after later stages of hot-working and heat treatment.

Titanium alloys can be hot-worked to produce most of the shapes that can be obtained with steels and other metals and they are often equated with stainless steels when comparing their hot-working characteristics. Power requirements are usually relatively high, particularly when microstructural considerations dictate that the working operations be conducted below the α/β transus temperature. Another factor is the narrow temperature range over which most hot-working operations must be carried out.

Although some pre-forming of titanium alloys may be undertaken by open-die forging, most forging operations normally involve closed dies in which the shaping of hot metal occurs completely within the walls or cavities of the two dies as they come together. Dies for closed-die forging are normally preheated to 200–250 °C for rapid operations involving hammers or mechanical presses, and to around 425 °C for slower working in hydraulic presses.

Heating of dies is particularly critical when forming thin sections, e.g. sheet, otherwise the poor thermal conductivity of titanium can lead to localized chilling which causes uneven metal flow or even cracking in a workpiece. Some novel hot-forming techniques have been developed. For example, assemblies of quite large dimensions can be produced by slow, isothermal forging (or creep forming) between heated dies, some of which can be made cheaply from cast ceramics. Metallic tooling is much more expensive, particularly as the die faces may need to be made from nickel or cobalt alloys in order that hardness and oxidation resistance are adequate at the relatively high forming temperatures, e.g. 850 °C.

Titanium alloys are particularly susceptible to galling, i.e. wear due to friction, during hot- or cold-working, which causes surface damage. Lubrication is thus essential and care must be taken in selecting materials that do not react with titanium when heated. Suspensions of graphite or molybdenum disulphide are suitable for both types of operation, whereas glass may be required for more severe processes such as hot extrusion.

6.5.2 Superplastic forming

If creep forming is carried out at controlled strain rates of around 10^{-5} to 10^{-6} s^{-1}, and at temperatures close to 0.6 T_M (where T_M is the melting temperature of the alloy in degrees K), then alloys having stable, small grain sizes may exhibit superplasticity. Flow stresses can be very low (e.g. see Fig. 6.23) and components can be produced by simple methods similar to those used for thermoplastics. α/β-alloys such as Ti–6Al–4V were early materials recognized as being super-plastic and typical process parameters are 100–1000 kPa pressure applied at 900–950 °C for 0.25–4 h. Such pressures, which are even lower for β-alloys, can be achieved using an inert gas, e.g. argon, that may be introduced into one part of the mould cavity with titanium sheet serving as a deformable diaphragm that flows into the other part.

One disadvantage with the α/β-alloys is the relatively high temperatures at which superplastic forming has to be carried out in order that the required distribution of the α- and β-phases is obtained. One novel method for reducing

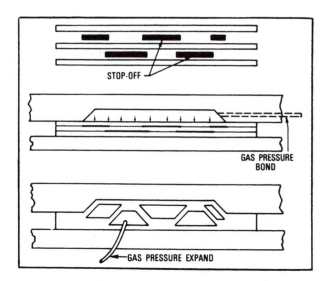

Fig. 6.24 Advanced manufacturing techniques using superplastic forming combined with diffusion bonding to produce a truss section. The stop-off material has been inserted where joining is not wanted (from Tupper, N. G. *et al., J. Metals*, **30**(9), 7, 1978).

this temperature is to exploit the fact that titanium and its alloys readily absorb hydrogen at elevated temperatures. This element has the effect of lowering the β-transus so that the required number of α- and β-phases for super-plastic forming can be achieved at a reduced temperature. In practice, the alloy is first heated in an atmosphere of 4 vol. % hydrogen in argon and later vacuum degassed to remove the hydrogen after forming is completed.

Clean titanium can readily diffusion bond to itself under processing conditions very similar to those used for superplastic forming (Section 6.5.7). Thus forming and joining can be combined in one comparatively simple operation to produce special products such as the truss section shown diagrammatically in Fig. 6.24.

6.5.3 Cold-working

CP titanium and most alloys in the annealed condition have a limited capacity to be cold-worked. For example, minimum bend radii for sheet are commonly 1–3 times the gauge thickness (T) for CP titanium, 2–4 T for β-alloys and 3–6 T for most other alloys. One major problem is excessive springback, which is a consequence of the low moduli and relatively high flow stresses of titanium and its alloys. To improve dimensional accuracy, cold-forming is generally followed by hot-sizing and stress-relieving for periods of 0.25 h at temperatures around 650–700 °C. Such treatments may also help to restore strength properties that are reduced in certain directions of some deformed alloys containing the hexagonal β-phase in the manner described for wrought magnesium alloys in Section 5.6.1. The treatments may also cause changes to fine-scale features of the microstructure, but little is known of these effects.

6.5.4 Texture effects

The major cause of property anisotropy in aluminium alloys, that of aligned, coarse intermetallic compounds (Fig. 2.39), is normally absent in titanium alloys. However, those titanium alloys which contain substantial amounts of the hexagonal α-phase may show marked elastic and plastic anisotropy if fabrication procedures produce a preferred orientation (or texture) in the grain structure. Such anisotropy will also be reflected in the mechanical properties and there is considerable interest in the prospect of controlled texture strengthening of titanium alloys. Such a process introduces a third dimension to alloy development as an adjunct to composition and microstructure.

Three easy slip modes and six twinning modes exist in α-titanium and the former are shown in Fig. 6.25(a). The slip vectors in each case are parallel to the basal planes {0001} so that, if stress is applied in the direction of the c-axis, there will be no critical resolved shear stress acting in this plane. There are, however, other slip systems which can operate so that the requirement of having five independent slip systems for general plasticity is fulfilled. The elastic moduli in the c- and a-directions of single crystals of α-titanium show a

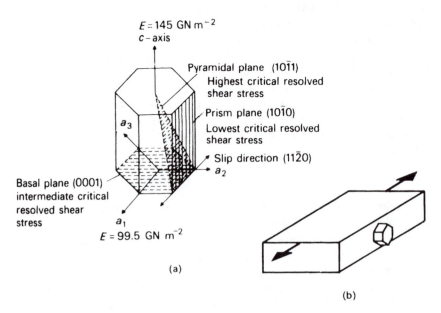

Fig. 6.25 (a) Slip planes in α-titanium; (b) alignment of hexagonal unit cell in α-titanium showing strongly preferred orientation after rolling.

Table 6.5 Mechanical properties of 57 mm thick ×235 mm wide forged and annealed Ti–6Al–4V bar (from Bowen, A. W., in *Titanium Science and Technology*, R. I. Jaffee and H. M. Burte (Eds), Plenum Press, New York, Volume 2, 1973; p. 1271)

Testing directions	0.2% proof stress (MPa)	Tensile strength (MPa)	Elastic modulus (GPa)	Elongation (%)	Approximate fatigue strength at 10^7 cycles ($\pm$MPa)
Longitudinal	834	910	114	17.5	496
Long transverse	934	986	128	17.0	427
Short transverse	893	978	114	12.5	565

large variation from 145 GPa to 99.5 GPa respectively. Although this difference is less in polycrystalline alloys showing preferred orientation, the elastic moduli in the longitudinal, long transverse and short transverse directions (Fig. 2.38) can still differ by as much as 30%. An example of the variation in tensile properties with stressing direction is shown for the alloy Ti–6Al–4V in Table 6.5. In this case, maximum texture strengthening has occurred in the long transverse direction and is associated with the alignment of basal planes normal to the forging plane (Fig. 6.25(b)), which is one of the two favoured textures that may develop in rolled or forged titanium alloys. Fracture toughness and elastic modulus are also maximal in the long transverse direction.

Table 6.5 and Fig. 6.26 show that fatigue properties are lowest in the long transverse direction. This result has been attributed to the fact that Poisson's ratios are also sensitive to crystal orientation, these ratios being higher in the longitudinal and short transverse directions because stressing occurs parallel to the basal planes. Higher ratios imply greater constraint, which means that the levels of strain will be reduced and the fatigue strength enhanced in these two directions. The differences observed in fatigue strengths in the longitudinal and short transverse directions have been attributed to relative changes in grain shapes that also occur during processing.

The other type of texture that may be developed in rolled or forged titanium alloys involves alignment of basal planes parallel to the rolling or forging plane, i.e. the *c*-axis is parallel to the short transverse direction. This texture can be beneficial in sheet metal forming involving biaxial tension as thinning by simple slip becomes difficult. Measurements have shown that *R*-values (Section 2.1) are high and may lie in the range 1–5, whereas they are less than 1 for aluminium alloys. This form of texture strengthening is also useful in applications such as pressure vessels which require high biaxial strength.

6.5.5 Machining

Titanium alloys have unique machining characteristics. Whereas the cutting forces may be only slightly higher than those required for steels of equivalent

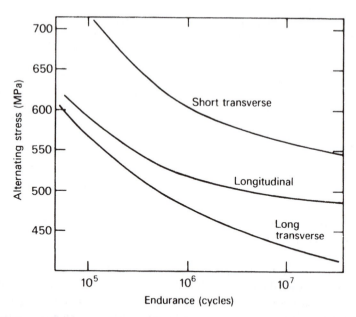

Fig. 6.26 Rotating-cantilever fatigue (*S/N*) curves for three testing directions in 57 mm thick, forged and annealed Ti–6Al–4V bar (from Bowen, A. W., in *Titanium Science and Technology*, Jaffee, R. J. and Burte, H. M. (Eds), Plenum Press, New York, Vol. 2, 1973).

hardness, there are other features that make these alloys relatively difficult to machine. In making comparisons, titanium alloys tend to be ranked together with austenitic stainless steels.

One basic problem arises because of the low thermal conductivity of titanium alloys which is only one-sixth that of steels, and which hinders the conduction of heat away from the region being machined. This feature, together with the fact that the characteristic shape of the chip allows only a small area of contact between chip and tool, means that high cutting temperatures may be generated during machining. For example, tests conducted at similar cutting speeds revealed that the temperature developed at the cutting edge of a tungsten carbide tool was 700 °C for titanium alloys and only 540 °C for a steel. Tool lives will thus be drastically shortened in the former case unless slower cutting speeds are used. Another feature associated with the machining of titanium alloys is the tendency for chips to stick to the cutting edge (galling), particularly once the tool becomes warm. This, in turn, may also reduce tool life as fracture of the cutting edge may occur if the titanium chip is removed when the tool re-enters the workpiece on the next pass. Alternatively, cutting forces may be increased by a factor of several times which, when combined with the relatively low elastic modulus of titanium, can cause serious deflection of the workpiece.

With these considerations in mind it is clear that the machining of titanium must be carried out with high quality, temperature-resistant tools operating at

Table 6.6 Ratios of machining times for various titanium alloys compared with the alloy steel 4340 having a hardness of 300 BHN (from Zlatin, N. and Field, M., in *Titanium Science and Technology*, R. I. Jaffee and H. M. Burte (Eds), Plenum Press, New York, Volume I, 1973; p. 409)

Titanium alloy	Hardness (BHN)	Turning (WC tool)	Face milling (WC tool)	Drilling (High speed steel tool)
CP titanium	175	0.7:1	1.4:1	0.7:1
Near-α: Ti–8Al–1Mo–1V	300	1.4:1	2.5:1	1:1
α/β: Ti–6Al–4V	350	2.5:1	3.3:1	1.7:1
β: Ti–13V–11Cr–3Al	400	5:1	10:1	10:1

comparatively low speeds, and cooled with copious amounts of cutting fluid. Moreover, because of the tendency of titanium to gall or weld to other metals, sliding contact should be avoided, which means that deep cuts with sharp tools are required. Finally, it is necessary to ensure that both tool and workpiece are rigidly supported.

The above features apply to all titanium alloys although the actual machining conditions vary with different categories. Table 6.6 compares the actual ratios of machining times for CP titanium and typical α, α + β and β-alloys with those for the commonly used alloy steel AISI 4340.

The difficulties associated with mechanical machining of titanium and its alloys can be overcome if unwanted metal is removed by chemical dissolution. Thus chemical and, more particularly, electrochemical milling techniques are sometimes used to produce shapes, the latter employing profiled cathodes.

6.5.6 Surface treatments

As mentioned in the previous section, titanium and its alloys show poor tribological properties because they have a strong tendency to gall or weld to themselves, or to other metals, under conditions of sliding contact. Traditional lubricants are often ineffective in overcoming this problem and treatments such as anodizing or electroplating are also of limited value because surfaces treated in these ways are not able to withstand more than light loads.

More success has been achieved with newer, but more costly techniques such as physical vapour deposition (ion plating) and plasma processing, in which particular use has been made of surface coatings of TiN. Physical vapour deposition involves reacting titanium vapour, that is sputtered from a separate source, with nitrogen ions generated by glow discharge in a nitrogen atmosphere. The operating temperature is commonly 500 °C and the deposited coatings, which are typically 3 μm thick, provide a very hard, smooth and low-friction surface capable of withstanding modest loading conditions. Thicker coatings (e.g. 30 μm) may be obtained by plasma nitriding at a higher temperature of

700–850 °C. In this technique, the titanium component is made the cathode in a low-pressure nitrogen glow discharge and the positive nitrogen ions accelerate towards the surface where they react to form TiN.

The factor limiting depth of hardening by plasma nitriding is the rate of diffusion of nitrogen in titanium which is several orders of magnitude less than in iron or steel. If deeper layers are required it is necessary to carry out the process in the molten state, i.e. by melting and alloying the surface of a component by means of a high-energy beam source. One method is laser nitriding in which the laser beam is focused on to the metal surface and the self-quenching effect of the substrate cools the molten surface metal at a rate that may be as high as 10^5 °C s^{-1}. In order to cover a complete surface, the process must be repeated to give a series of overlapping strips. The hard layer can be 0.1–1 mm thick and light grinding is then necessary if a smooth surface is required, Electron beam melting is an alternative technique but, since a vacuum is required, a gas cannot be used for alloying. In this case, appropriate additions have first to be applied to the surface by spraying, painting or plating prior to melting.

Recently a potentially lower cost method for improving the tribological properties of titanium alloys has been developed that is analogous to surface carburising of steel components. The process involves a thermal oxidation treatment that simultaneously produces a thin, hard and adherent surface film of rutile (TiO$_2$) supported by a thicker oxygen-rich, α-titanium sub-surface zone. For components made from the alloy T–6Al–4V, oxygen in concentrations of 100–200 ppm were added to an argon carrier gas in which components were heated to 900 °C for up to 24 hours. The respective depths of the two layers were found to be approximately 2 μm and 20 μm. The surface hardness was found to exceed 1000 H$_v$ which compares with 400 H$_v$ in the interior of the alloy. Experiments with another α–β alloy Timet 550 (Ti–4Al–2Sn–4Mo–0.4Si) have shown that thicker duplex surface layers may be developed by heating for 40–100 hours in air at lower temperatures in the range 600–650 °C. The rutile surface film reduced the coefficients of friction and made both alloys more resistant to wear under dry or lubricated conditions.

6.5.7 Joining

Fusion welding The capacity of titanium alloys to be fusion welded is related to microstructure and composition. General weldability is restricted to α-, near-α- and to α/β-alloys containing less than 20% of the β-phase. Fusion, resistance and flash-butt welding have all been used for titanium. Oxyacetylene and atomic hydrogen fusion welding are unsuitable because of gaseous contamination, but non-consumable (TIG) and consumable (MIG) electrode techniques (Section 3.5.1), as well as electron beam and plasma arc processes are applicable. Open air techniques can be used but extra attention must be given to shielding the area being welded with an inert gas such as argon. In this regard it is necessary to take special protective measures, e.g. supplying argon to the

Fig. 6.27 Polished and etched section of an electron beam butt-weld in 50 mm thick Ti–6Al–4V plate (courtesy Rolls Royce Ltd).

underside of the weld, which are not required with aluminium or magnesium alloys. In the case of complex assemblies it is often preferable to weld in a glove-box type of chamber that can be filled with the inert gas. It is important to note that titanium cannot be fusion welded to the other conventional structural materials such as steel, copper, and nickel alloys. Spontaneous cracking or extreme brittleness are inevitable due to the formation of brittle intermetallic compounds.

Electron beam welding, which is carried out in a vacuum chamber, is especially suitable for titanium alloys, although the capital cost of the equipment is high. Specific powers are greater than those usable by other processes so that deep penetration combined with a narrow heat-affected zone is possible, e.g. Fig. 6.27 shows an electron beam butt-weld in 50 mm thick, rolled titanium alloy plate. Both electron beam and laser beam welding processes produce high integrity, narrow welds which minimizes the risk of distortion.

Diffusion bonding Titanium-based materials are also amenable to joining by diffusion-bonding techniques since surface oxide and minor contaminants are readily dissolved at the temperatures used, which are usually in the range 850–950 °C, i.e. below the transus temperature in α/β-alloys. It is again necessary to exclude air during bonding and the components being joined are held under low pressures of about 1 MPa for periods of 30–60 min. Diffusion readily occurs across the interface and it is usual for localized recrystallization to take place there. This is evident in Fig. 6.28 which is a micrograph of a diffusion-bonded joint between two different titanium alloys. Joint strengths are commonly better than 90% those of the alloys being bonded, providing that the process has no detrimental effect on the microstructure.

Diffusion bonding provides significant cost reductions by eliminating machining operations in the manufacture of shapes having complex geometry. The technique also enables assemblies to be produced in alloys that are not

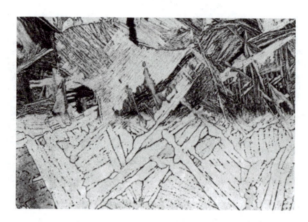

Fig. 6.28 Micrograph of a diffusion-bonded joint between two different titanium alloys. The grain structure shows that localized recrystallization has occurred. The conditions were: temperature 980 °C, pressure 1.5 MPa, time 30 min (from Blanchet, B., *Revue de Métallurgie*, **71**, 99, 1974) (× 100).

normally regarded as being weldable. As mentioned in Section 6.5.2, diffusion bonding can be combined with superplastic forming in certain fine-grained alloys.

Brazing Titanium and its alloys can be brazed at around 1000 °C in a protective atmosphere using materials such as silver, copper, and the composition Ti–15Cu–15Ni. An important use of brazing has been in the production of honeycomb structures (see Section 7.1.2) which offer a unique combination of stiffness and corrosion resistance at elevated temperatures.

6.5.8 Powder metallurgy products

As mentioned in Section 3.6.6, considerable economies in both the use of material and costs of fabrication are possible if components can be produced from powders. These techniques are being applied to titanium alloys now that powders of sufficient purity, i.e. a particularly low content of interstitial elements, are available. Products can be produced close to final size (near net shape) with a consequent large saving in machining and fabricating costs.

Two techniques are usually practised to prepare titanium alloy powders. These are known as the 'elemental' and 'pre-alloyed' approaches. In the former, use is made of so-called sponge fines which are small, irregular particles of titanium that are rejected when sponge is being converted to solid ingots during titanium production (Chapter 1). These particles are blended with alloying elements to achieve the desired alloy chemistry and consolidated at room temperature using pressures of up to 400 MPa to produce shapes with 'green densities' of 85–90%. Methods of preparing pre-alloyed powders are limited because of the high reactivity of molten titanium with gases and refractories.

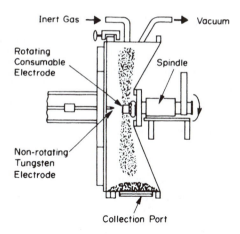

Fig. 6.29 Schematic representation of rotating electrode process for producing spherical powders (from Savage, S. J. and Froes, F. H., *J. Metals*, **36**(4), 20, 1984).

Usually some form of centrifugal atomization in vacuum or an inert atmosphere is employed, one example being the rotating electrode process (REP) in which an electric arc is struck between a rotating consumable electrode and a non-rotating tungsten electrode as shown schematically in Fig. 6.29. The liquid metal droplets that are thrown off by centrifugal force have time to spheroidize and solidify in flight (cooling rates 10^2–10^3 °C s^{-1}). This is an advantage because spherical particles flow readily and pack with a consistent density (~65%) which facilitates dimensional control of components during subsequent consolidation of the powders.

A promising new approach to powder production is the Metal Hydride Reduction (MHR) method that has been developed in Russia. In this process, titanium powder is produced from titanium dioxide by reacting with calcium hydride at temperatures in the range 1100–1200 °C. The chemical reaction is:

$$TiO_2 + 2CaH_2 = Ti + 2CaO + H_2$$

Alloyed powders can be produced directly by using a mixture of oxides of the appropriate elements. Because this method does not require the intermediate production of titanium tetrachloride, the MHR powders contain very low levels of chloride. They do retain significant amounts of hydrogen, however, which may assist in the subsequent sintering operation and can later be removed by annealing.

The two common methods for final powder consolidation, which aim to achieve close to 100% final density, are hot pressing and sintering and hot isostatic pressing (HIP) respectively. Sintering is usually carried out at temperatures in the β-phase field followed by forging or extrusion in the (α + β)-phase field which results in a very fine, duplex microstructure. Hot isostatic pressing, although an expensive process, does offer the advantage that pressing, sintering and sometimes forging can be combined in a single operation. Again the temperature is usually controlled so that the alloy is processed in the ($\alpha + \beta$) field. Another possible method, which offers the prospect of reducing costs, is to injection mould the titanium alloy powders in ways similar to those used by the plastics.

Apart from economies in material and fabrication costs, powder metallurgy techniques can offer other advantages. Uniformity of composition is greater because chemical heterogeneity in cast billets is avoided. In addition, components produced from powders show no crystallographic texture or anisotropy of grain shape, and so are much more uniform in respect of mechanical properties. Significantly higher values of proof stress and tensile strength at room and elevated temperatures are also possible, although ductility and toughness are less than that achieved for conventional wrought alloys. As shown in Fig. 6.30, the fatigue properties are intermediate between conventional wrought alloys and castings.

Generally the powder metallurgy approach is most attractive for large, complex parts where the weight of the incoming mill product compared with the weight of the final component is high when fabricated by conventional means. However, a limit is imposed by the size of the largest autoclave available for hot pressing unless approaches such as subsequent welding are used to join parts so as to produce larger components. Present estimates indicate that cost savings of 20–50% are possible if titanium alloy components are produced by powder metallurgy techniques rather than forging.

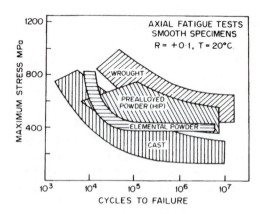

Fig. 6.30 Comparative scatter bands for results of fatigue tests on annealed Ti–6Al–4V products fabricated by different processes (from Kelto, C. A. et al., *Powder Metallurgy of Titanium Alloys*, Froes, F. H. and Smugeresky, J. E. (Eds), Met. Soc. AIME, Warrendale, Pa, p. 1, 1980).

6.5.9 Shape memory alloys

A number of alloys in the martensitic condition may exhibit the phenomenon of shape-memory whereby, if deformed (up to a limit of about 10% strain) below the M_s temperature, they are able to recover their original shape when reheated. The best-known example is the alloy nitinol which has approximately equi-weight amounts of titanium and nickel (Ti–55 to 58Ni). Shape memory has also been observed in a Ti–45Nb alloy which undergoes the transformation $\beta \rightarrow$ martensitic α'' on quenching through approximately 175 °C.

Nitinol has been used for a number of engineering applications including fasteners, actuators and couplings. For example nitinol couplings have been designed to connect aircraft hydraulic lines, plumbing in submarines and fittings to join steel pipes installed under the sea. The couplings are expanded about 4% in the martensitic condition at liquid nitrogen temperatures and placed around tubes to be joined. After warming to room (or sea) temperatures, they contract producing a tight seal thereby providing a convenient alternative to welding or brazing. Nitinol also has application in certain prosthetic devices (Section 6.8.3).

6.6 TITANIUM ALLOY CASTINGS

The high affinity of molten titanium for oxygen, nitrogen and hydrogen requires that melting and pouring be carried out under vacuum. Consumable electrode vacuum arc or electron beam melting furnaces are commonly used. The range of moulding materials is limited because of the reactivity of titanium and rammed graphite is commonly used for this purpose. Lost wax investment casting is used to produce precision parts for the aerospace industry, one example being the intermediate compressor casings for several gas turbine engines. The cost of each part is usually 15–35% lower than for the equivalent wrought component because of savings in materials and the costs of fabrication. The largest castings are prepared for chemical equipment where rammed graphite moulds are used. In the United States, for example, spherical valves up to 2.5 m in diameter and weighing as much as 1 tonne have been produced from titanium alloys. Impellors for pumps is another common application.

Since titanium castings are produced under vacuum, they are inherently clean although conventional defects such as shrinkage porosity will be present. Because of this, it is now standard practice to hot isostatically press (HIP), high-quality castings in order to remove porosity. Static strength values for titanium castings are similar to those for wrought material having the same composition. Low-cycle fatigue strength is usually within the normal range of distribution for wrought material although, above 10^6 cycles, the smooth bar fatigue strength is usually a little lower. Notched fatigue strength (e.g. $K_t = 2.5$) is the same for both cast and wrought alloys.

Two widely used alloys for castings have been the general purpose α/β alloy Ti–6Al–4V and the α-alloy Ti–5Al–2.5Al. Other alloys are now routinely used including the α/β alloy Ti–6242 (Ti–6Al–2Sn–6Zr–6Mo) and the near-β alloy Ti–5553 (Ti–5Al–5V–5Mo–3Cr) which have lower castabilities but develop higher tensile and fatigue strengths. For example, the following properties have been reported for the alloy Ti5553 in the cast, HIPed and heat treated condition:

0.2% PS	TS	Elongation	Fatigue Endurance Limit (10^7 Cycles, R = 0.1)
MPa	MPa	%	Max. Stress MPa
1055	1158	9	780

These properties compare well with those for many wrought titanium alloys and some high strength titanium castings are now used for structural applications in aircraft such as the Boeing 777.

6.7 ENGINEERING PERFORMANCE

6.7.1 Tensile and creep properties

The uniquely high strength:weight ratios of titanium alloys over a wide temperature range were shown in Fig. 1.6. Reference has also been made to the fact that the development of titanium alloys has been dominated by the desire to improve creep behaviour at progressively higher temperatures. This theme has been central to the preceding discussion of titanium alloys, with tensile properties and comparative creep behaviour being summarized in Table 6.1 and Fig. 6.2 respectively. In Section 6.2.2, reference was also made to the high specific strength and toughness of α-titanium alloys at very low temperatures which has led to their use for cryogenic storage vessels in space vehicles. Consequently it is not proposed to consider these properties any further.

Near-α alloys such as Ti 1100, IMI 829, IMI 834 are available for service for long times at 550–600 °C. These temperatures are close to the limit at which conventional titanium alloys can be used in air because oxidation normally becomes significant above 600 °C. This applies with respect to both surface scaling and internal embrittlement, since oxygen is relatively soluble and mobile in titanium above this temperature. There are, however, titanium aluminide intermetallic compounds that show promise of being used for components operating at temperatures as high as 1000 °C. (Section 7.9).

6.7.2 Fatigue behaviour

Titanium and titanium alloys which have isotropic microstructures may display fatigue properties comparable with those obtained with ferrous materials. Curves relating cyclic stress with number of cycles (*S/N* curves) for smooth specimens tend to show true endurance limits and the ratio of this value to tensile strength is commonly 0.50–0.65. Reductions in these ratios for notched specimens are also characteristic of ferrous materials. Thus it can be said generally that factors which increase the tensile (and yield) strength of titanium alloys also raise the fatigue strength.

The fatigue performance of titanium and its alloys is greatly influenced by microstructure and texture. In general, fatigue crack propagation in titanium and its alloys parallels fracture toughness in that conditions which favour highest toughness tend also to give the lowest cyclic growth rates under fatigue loading. Particular attention has been paid to the α/β-alloys, notably Ti–6Al–4V. Effects of microstructure on fatigue crack propagation have already been shown in Fig. 6.15 and Table 6.3, whereas the influence of grain direction (and texture) was demonstrated in Fig. 6.26 and Table 6.5. As mentioned in Section 6.3.1, a coarse, basket-weave type microstructure is resistant to crack propagation, whereas a finer microstructure tends to be beneficial in delaying crack initiation. This applies at both room and elevated temperatures and, for α/β-alloys, the best compromise for overall fatigue strength seems to be a fine microstructure comprising a mixture of small grains of α and transformed β (e.g. Fig. 6.14b and c). It has also been noted that fatigue strength is highest in the short transverse direction of wrought products and this arises primarily as a result of textures developed in the grains (Section 6.5.4).

β-titanium alloys generally display good fatigue performance. Under high cycle, low stress conditions, fatigue strength increases with yield strength up to levels at which fatigue cracks tend to initiate in soft regions in the microstructure, such precipitate-free zones or α phase particles in grain boundaries, where deformation becomes concentrated. In contrast to other mechanical properties, fatigue crack propagation is less sensitive to processing, microstructure or even chemical composition. Crack growth rates are comparable with those for mill annealed Ti–6Al–4V at high stresses, but slightly higher at low stresses.

A phenomenon which may be unique to certain titanium alloys is the effect of dwell periods, at high loads, on rates of growth of fatigue cracks. This effect is shown schematically in Fig. 6.31 and increases in the rates of growth, immediately after the dwell period, of as much as 50 times may occur as compared with results obtained in tests on the same alloy subjected to continuous sinusoidal stress cycles. Dwell effects are greatest in alloys containing substantial amounts of the α-phase which have a preferred texture such that stressing is normal to the basal planes. On the other hand, they appear to be insignificant if stressing occurs parallel to the basal planes of the α-phase, or if the microstructure is homogeneous and fine grained. Once again, special attention has been paid to the α/β-alloy Ti–6Al–4V in which dwell effects have been found to decrease with increasing amounts of the β-phase in the microstructure. In all cases, dwell effects disappear when stressing occurs above 75 °C. They are generally considered to arise because of the presence of relatively high contents of hydrogen (more than 100 ppm) which, during the dwell period, diffuses to regions of localized hydrostatic tension ahead of advancing cracks. Such an accumulation of hydrogen apparently embrittles these regions, and it has even been suggested that brittle plates of TiH_2 may be formed.

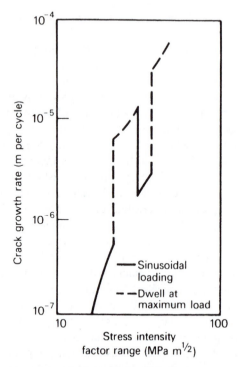

Fig. 6.31 Schematic representation of the effect of dwell periods at maximum load on the rate of crack growth during fatigue tests on certain titanium alloys.

It should also be noted that dwell periods may have the opposite effect of increasing fatigue strength if alloys contain low levels of hydrogen. In this case it has been suggested that strain ageing may occur which is beneficial because it results in pinning of dislocations by atmospheres of hydrogen and/or other interstitial atoms.

As mentioned in Section 6.5.5, titanium alloys are particularly sensitive to contact due to sticking or galling, a characteristic which may lead to fretting fatigue. Fretting itself is a form of wear that occurs when two surfaces, pressed together by an external load, are subjected to transverse cyclic loading so that one contacting face is cyclically displaced relative to the other face. Extremely small displacements may be involved and small fragments of metal and oxidation products may break off leading to surface pitting at which stresses may be concentrated. Fretting damage in titanium alloys can reduce fatigue life by factors of as much as eight times which compares with a reduction factor of around three for similar damage in aluminium alloys.

One example where fretting can be a major concern is at the root fittings of titanium alloy blades in gas turbine engines. Much attention has been given to alleviating the problem and one method involves the application of two thin coatings. One is a Cu–In–Ni alloy that is plasma sprayed on to the surface; after this the second, a phenolic resin which contains graphite and molybdenum disulphide as lubricants, is baked on.

6.7.3 Corrosion

Titanium is a highly reactive metal but, because of the presence of a very stable, self-healing oxide film, it exhibits excellent corrosion resistance in a wide variety of environments. This is a feature shared with aluminium, although the corrosion resistance of titanium is normally much greater. Titanium is also more resistant to attack than stainless steels and copper alloys in a wide range of conditions, which has led to an expansion of its use by the chemical industry.

The surface oxide film of titanium resists attack by oxidizing solutions, particularly those containing chloride ions which are normally difficult to handle. Titanium shows outstanding resistance to atmospheric corrosion in both industrial and marine environments, and its resistance to sea water is virtually unsurpassed by other structural metals. It is also resistant to many acids and salt solutions. However, titanium is attacked in reducing environments in which the oxide film is unstable and cannot be repaired.

One of the most successful applications for titanium alloys is in handling wet chlorine gas, bleaching solutions containing chlorides, hypochlorates and chlorine dioxide. However, titanium suffers catastrophic attack in dry chlorine gas, although as little as 50 parts per million of water will prevent this attack.

For most chemical applications, strength is a secondary requirement and CP titanium is used. In order to cope with non-oxidizing acids such as sulphuric, hydrochloric and phosphoric, or with some reducing conditions, several methods have been developed to improve the corrosion resistance of titanium. An alloy of titanium with 0.2% palladium (IMI 260) was developed in which the role of the noble metal was to induce anodic passivation such that corrosion rates in some solutions have been reduced by a factor of as much as 1000 times. A similar effect may be achieved by external anodic protection in which the potential of titanium is made more positive by connecting to either an electrical power source or to a more noble metal. In this regard titanium has the particular advantage over competing materials such as stainless steel in that it remains passive over a much wider range of potential.

A more recent use of a noble metal as a means of corrosion protection has been the application of the new but rather specialized technique of ion plating to provide a very thin ($\sim$1 μm) coating of platinum. Ion plating is a method of coating in which some of the deposited particles are ionized, e.g. by passing through a plasma which assists in cleaning the surface of the substrate and improving the adhesion of the coating. In addition to raising the general corrosion resistance of titanium, ion plating may also increase wear resistance and fatigue strength.

The major corrosion problem with titanium alloys is crevice corrosion which may take place at joints, seams, welds, etc., where circulation of the corroding medium is restricted. Progressive acidification occurs in the crevice because of hydrolysis of corrosion products and the protective oxide film is destroyed. The problem becomes worse at elevated temperatures. Again the addition of the noble metal palladium to titanium has been found to be beneficial, as has ion plating of surfaces with platinum.

Contrary to crevice corrosion, surface pitting may occur at any location, e.g. inclusions, where weak points exist in the passive oxide film. However, this problem occurs rarely in titanium alloys and may only be significant in halide-containing aqueous solutions which are at elevated temperatures.

6.7.4 Stress-corrosion cracking

The apparent stability and integrity of the oxide film in environments that cause stress-corrosion cracking in more common structural alloys suggested that titanium alloys may be very resistant to this phenomenon. For example, early investigators were unable to crack specimens made from those materials that were stressed and exposed to boiling solutions of 42% $MgCl_2$ or 10% NaOH, both of which induce cracking in many stainless steels. However, a susceptibility to this phenomenon was recognized in 1953 with the cracking of CP titanium in red fuming nitric acid, and in 1955 when the unexpected failure of a titanium alloy undergoing a hot tension test was supposedly traced to chloride salts deposited from finger marks. Since then it has been demonstrated in the laboratory that most titanium alloys will undergo stress-corrosion cracking in one or more environments (Table 6.7) and special attention has been given to the comparative performance of alloys in aqueous halides, organic fluids, e.g. methanol, and hot salts. However, it must be emphasized that actual failures in service have been rare.

It can now be appreciated that the reason why titanium alloys were considered to be immune from stress-corrosion cracking in mild environments, and at low temperatures, is the difficulty in initiating cracks. Many titanium alloys are highly resistant to pitting corrosion whereas, in many other materials, the pits provide the stress concentration necessary to initiate stress-corrosion cracking. Although this characteristic of titanium alloys is very desirable, it is now standard practice to use notched or pre-cracked specimens and to apply fracture mechanics techniques when evaluating and comparing the stress-corrosion resistance of different titanium alloys. This follows because of the use of titanium alloys for a number of critical aerospace applications.

It will be apparent from Table 6.7 that stress-corrosion cracking in most alloys is specific to a particular environment. However, it is possible to identify several trends which are related to composition and microstructure. The compositional features are as follows.

1. α-titanium alloys generally show the greatest susceptibility to stress-corrosion cracking and even CP titanium will crack in some environments if the level of oxygen in the metal is high. Aluminium contents in excess of 5–6% are considered detrimental and this is attributed primarily to the formation of the ordered phase α_2, the presence of which changes the dislocation substructure

Table 6.7 Some environments in which titanium alloys may be susceptible to stress-corrosion cracking (from Boyd, W. K., in *Proceedings of Conference on Fundamental Aspects of Stress-Corrosion Cracking*, R. W., Staehle *et al.* (Eds), Nat. Assoc. Corrosion Eng., 1969; p. 593)

Medium	Temperature (°C)	Examples of susceptible alloys
Cadmium	> 320	Ti–4Al–4Mn
Mercury	Ambient	CP Ti(99%), Ti–6Al–4V
	370	Ti–13V–11Cr–3Al
Silver plate	470	Ti–5Al–2.5Sn, Ti–7Al–4Mo
Chlorine	290	Ti–8Al–1Mo–1V
Hydrochloric acid (10%)	35	Ti–5Al–2.5Sn
	345	Ti–8Al–1Mo–1V
Nitric acid (fuming only)	Ambient	CP Ti, Ti–8Mn, Ti–6Al–4V, Ti–5Al–2.5Sn
Chloride salts	290–425	All commercial alloys
Methanol	Ambient	CP Ti(99%), Ti–5Al–2.5Sn, Ti–8Al–1Mo–1V, Ti–6Al–4V, Ti–4Al–3Mo–1V
Trichloroethylene	370	Ti–5Al–2.5Sn, Ti–8Al–1Mo–1V
Sea water	Ambient	Cp Ti(99%), Ti–8Mn, Ti–5Al–2.5Sn, Ti–8Al–1Mo–1V, Ti–6Al–4V, Ti–11Sn–2.25Al–5Zr–1Mo–0.25Si, Ti–13V–11Cr–3Al

that forms during deformation. The onset of limited co-planar slip is observed which results in coarse slip steps and localized rupture of the oxide film at the surface, thereby exposing unprotected metal to the corrosive medium. It is proposed that such a mechanism may lead to rapid pitting and the subsequent formation of stress-corrosion cracks, as has been observed in tests on some austenitic stainless steels.

2. The addition of elements such as molybdenum and vanadium, which favour formation of the β-isomorphous type of phase diagram, enhances the resistance to stress-corrosion cracking. These elements increase the amount of β-phase and, as this phase appears to be immune from stress-corrosion cracking in a number of α/β alloys, it has been proposed that it serves to arrest cracks that may be propagating in the more susceptible α-phase. It should be noted, however, that elements such as manganese which stabilize β-eutectoid systems increase susceptibility to stress-corrosion cracking.

3. Hydrogen is readily absorbed by titanium and is detrimental because a number of alloys are susceptible to hydrogen embrittlement, particularly in aqueous environments. Although the precise mechanism for embrittlement is uncertain, it seems likely to arise either from the formation of brittle plates of a titanium hydride, or from the directed diffusion of hydrogen to highly stressed regions such as crack tips, thereby assisting crack propagation. In this regard, it may also be noted that hydride formation occurs most readily in alloys containing aluminium.

Although it is more difficult to propose general rules regarding the role of microstructure, the following trends have been observed.

1. Quenching from the β-phase field, which often gives acicular microstructures, generally confers a greater resistance to stress-corrosion cracking than slow cooling from the α/β-field. This result is similar to that described for fatigue cracking in Section 6.3.1 and is again associated with crack branching, as shown in Fig. 6.15(b). However, it should be noted that the reverse seems to hold for alloys exposed to halides at high temperatures, i.e. so-called 'hot salt stress-corrosion cracking'.
2. A combination of plastic deformation and heat treatment is usually beneficial in reducing susceptibilty because it refines the micro-structure and reduces grain size.
3. Ageing leading to precipitation of a second phase, e.g. ω or α, within β-grains can lower resistance to stress-corrosion cracking.
4. As with aluminium alloys, wrought titanium products show greatest susceptibility to stress-corrosion cracking in the short transverse direction.

6.7.5 Corrosion fatigue

The combination of the generally high fatigue strength of titanium alloys and their good resistance to corrosion suggests that they should perform well under conditions of corrosion fatigue and this has been found to be so. Special attention has been given to comparisons between the α/β-alloy Ti–6Al–4V and 12% chromium steel, because of the possible replacement of this material by titanium alloys for the low-pressure sections of large steam turbines. The strength level at which titanium alloys could be put into service is some 25% higher than that for a 12% chromium steel. Taking into account this factor and the difference in relative density, the endurance limit for Ti–6Al–4V tested in air is more than double that for the steel (Fig. 6.32). Moreover, whereas the titanium alloy is unaffected by testing in steam or in a solution of 3.5% NaCl, the endurance limit for the steel fatigue tested in only 1% NaCl solution falls to approximately one-eighth that of the titanium alloy.

6.8 APPLICATIONS OF TITANIUM ALLOYS

6.8.1 Aerospace

Although the development of titanium alloys only commenced in the late 1940s, their uniquely high strength:weight ratios led to their introduction in aircraft gas turbines as early as 1952 when they were used for compressor blades and discs in the famous Pratt and Whitney J57 engine. Immediate weight savings of approximately 200 kg were achieved. At the same time, in Britain, an alloy Ti–2Al–2Mn was used for the equally famous Rolls Royce Avon engine that powered such aircraft as the Comet and Canberra. Since then, aircraft gas

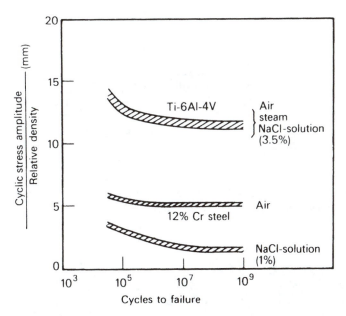

Fig. 6.32 Fatigue properties (compensated for differences in relative density) of Ti–6Al–4V and a 12% chromium steel tested in different environments (from Jaffee, R. I., *Met. Trans*, **10A**, 139, 1979).

turbines have continued to provide the major application for titanium alloys which now make up 25–30% of the weight of most modern engines. In particular, titanium alloys have played a major role in by-pass (fan-jet) engines for which a large front fan is required (Figs. 6.13 & 6.34). Some fans are more than 2 m in diameter and solid blades may weigh as much as 6 kg which means that, when rotating in service, each blade exerts a pull of approximately 75 tonnes on the turbine disc or wheel. This has led to a recent innovation in design whereby the so-called wide-chord, hollow blade has been introduced to reduce weight. This blade comprises a honeycomb core covered by titanium alloy skins which are diffusion bonded together to form an integral structure (Fig. 6.33).

In addition to the fan, titanium alloys are used for most of the blades and discs in the low and intermediate sections of the compressors of modern jet engines as shown, for example, in Fig. 6.34. Selection of the disc material is particularly critical as it is subjected to thermal stresses arising because of temperature differences between the hotter rim and cooler core which are additional to the high loads imposed by the rotating blades. A good disc material should have low values of coefficient of expansion and elastic modulus as well as a high value of low-cycle fatigue strength such that the ratio of modulus × expansion/LCF strength is a minimum.

High-strength titanium alloys are the best materials for this application. In general, the α/β-alloy Ti–6Al–4V is favoured for the fan and cooler parts of

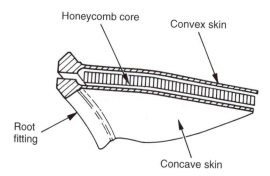

Fig. 6.33 Expanded section showing construction of a wide-chord fan blade for an advanced gas turbine engine (courtesy J. F. Coplin, Rolls Royce Ltd).

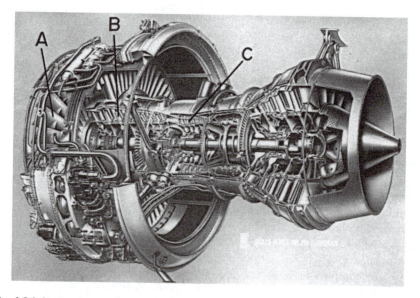

Fig. 6.34 Sectional view of one model of the Rolls Royce RB 211 gas turbine engine. A = fan blades, B = lower pressure compressor, C = intermediate pressure compressor (courtesy Rolls Royce Ltd).

the compressor whereas the near-α alloys are specified where greater strength and creep resistance are required. Other engine applications include the use of the CP titanium sheet for casings and ducting.

The use of titanium alloys for structural members in aircraft developed steadily but more slowly because of their high cost relative to aluminium alloys. For example, the Boeing 707 commercial aircraft, which first came into service in 1958, contained only 80 kg (0.5% of the structural weight). Subsequently the Boeing 727 aircraft (1963) had 290 kg (1%), the Boeing 747 (1969) 3850 kg (2.8%), and the McDonnell-Douglas DC 10 (1971) 5500 kg (10%). In the more recent Boeing 777 aircraft (1994), titanium alloys amount to approximately 9% of the total structural weight. Titanium alloys were first used mainly as sheet for engine nacelles, exhaust shrouds, and fire walls where heating was significant. Now they are used many other purposes including thin straps wrapped around aluminium alloy fuselages to prevent the propagation of possible fatigue cracks, hydraulic tubing, kitchen and toilet flooring where high corrosion resistance is required, and particularly for forgings made mainly from the α-β alloy Ti–6Al–4V. These forgings are used for critical components such as engine mountings, flap and slat tracks in wings and undercarriage components. As mentioned in Section 6.6, most of the main landing gear of the Boeing 777 aircraft is made from high strength, β-titanium alloy Ti 10V–2Fe–3Al forged parts (Fig. 6.35) which resulted in a weight saving of 270 kg when compared with using a high strength steel. In this aircraft, use has also been made of the stronger, and highly cold formable β-titanium alloy Ti–15V–3Al–3Sn–3Cr which relaces CP-titanium for some 49 m of ducting. This allows thinner sheet to be used which enables further weight savings to be obtained.

Much greater use is made of titanium alloys in military aircraft which may vary from 35 to 50% of the weight of modern fighter aircraft which may operate temperatures that exceed the capabilities of high-strength aluminium

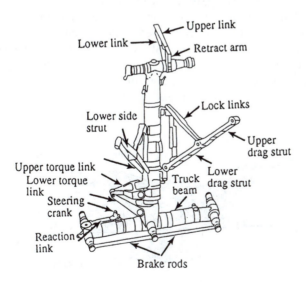

Fig. 6.35 Main landing gear of the Boeing 777 aircraft made from β-titanium alloy Ti–10–2–3 alloy forgings (courtesy Boeing Commercial Aircraft, Seattle, USA).

alloys. Here the most common location for the titanium alloys is in the engine bay as shown for the early McDonnell-Douglas F-15 aircraft in Fig. 6.37. In this example, some 7000 kg of titanium alloys is used which represents 34% of the structural weight compared with 48% for aluminium alloys. Newer alloys such as Ti–6Al–2Zr–2Sn–2M0–2Cr–0.25Si are used in the airframe of the United States F-22 fighter aircraft. The mid-fuselage bulkhead, which makes up part of the wing box of this aircraft and measures 4.9 m by 1.8 m by 0.2 m, is one of the largest titanium alloy forgings ever made. Although this component only weighs 150 kg, it is forged from a cast ingot weighing almost 3000 kg. Complex structures such as the wing box of the swept-wing European Panavia Tornado fighter aircraft are also made largely from welded and machined titanium alloy forged plate and the critical wing-pivot lug is a Ti–6Al–4V forging. Considerable use is being made of high-strength titanium alloy fasteners and the large American military transport aircraft, the Lockheed C5A, uses 1.5 million such units out of a total of 2.2 million. This has resulted in a direct reduction in weight of 1 tonne, with an additional 3.5 tonnes being saved through consequent structural modifications that are possible because by using titanium alloys fasteners. For recent military helicopters, titanium alloys are now also used for the fasteners. Titanium alloys also have critical applications in helicopters. For example, the use of the high-strength, fatigue-resistant β alloy Ti–10–2–3 for the rotor head of the Westland Super Lynx helicopter has enabled this vehicle to operate at gross weights some 45% higher than was intended in the original design. As shown in Fig. 6.36, this rotor head is assembled by bolting together three different forged components.

Some of the high costs associated with titanium alloys have arisen because of the use of fabrication techniques developed for aluminium and its alloys. For example, the sheet and rivet methods were initially used to fabricate the encircled area of the F-15 aircraft shown in Fig. 6.37. Subsequently, cost savings have been achieved with later models with new designs that used superplastic forming combined with diffusion bonding, thereby eliminating more than 700 parts and 10 000 fasteners.

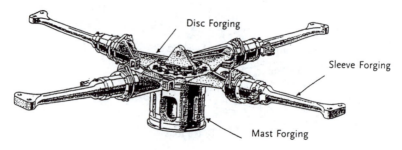

Disc Forging

Sleeve Forging

Mast Forging

Fig. 6.36 Super Lynx helicopter rotor head assembly made from forgings of the β-titanium alloy Ti–10–2–3 (courtesy AgustaWestland).

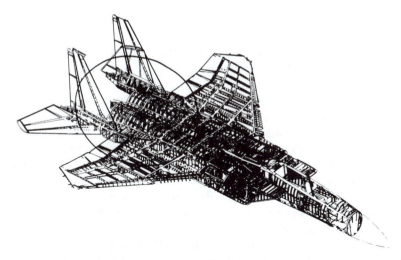

Fig. 6.37 A structural diagram of the McDonnell-Douglas F15 aircraft in which the encircled area is made almost entirely from titanium alloys (from Tupper, N. G. *et al., J. Metals,* **30**(97), 1978).

6.8.2 General applications

For the first time, more than half of the world's production of titanium is now used for non-aerospace purposes. As discussed in Section 6.7.3, it is the outstanding corrosion resistance of titanium and its alloys in many environments that is the prime reason for their selection for chemical engineering and architectural applications. Although titanium alloys are usually more expensive than the materials they replace, their adoption is based on expected cost savings over the planned lifetimes of the particular equipment or building. In areas such as for automotive components, military hardware, and sports equipment, high strength:weight ratios are usually the main attraction.

Chemical engineering The high resistance to general corrosion, pitting and crevice corrosion, and stress-corrosion cracking in the presence of chlorides is the major reason for the selection of titanium and its alloys for most chemical engineering applications. These materials are uniquely suited for piping, valves and pump casings for handling hot brine, bleaching agents, and chlorinated hydrocarbons, and they generally perform better than stainless steels. Tubing for heat exchangers and steam condensers is another common application because titanium alloys have a higher corrosion and erosion resistance than the traditional copper alloys they relace. CP-titanium with thinner wall thicknesses, e.g. 0.5 mm, is normally used which may also result in improved heat transfer despite the relatively poor thermal conductivity of titanium. Titanium can also be used for handling nitric, acetic, and organic acids, as well as acetone and wet bromine. In addition, it is normally stable in alkaline solutions up to a pH 12 at temperatures below 75 °C.

One of the earliest industrial uses of titanium followed the discovery that, whereas this metal rapidly passivates under anodic conditions, current will continue to flow through the surface oxide film if other metals are in contact with it. This effect was exploited by the metal finishing industry with titanium being used for jigs that support aluminium components during anodizing, and for anode baskets to hold parts to be plated with copper or nickel. Sheets of CP titanium are also used in the electrolytic refining of copper where they serve as starter blanks. These are cathodes on to which a thin copper sheet is electro-deposited and then stripped off for transfer to the main production line in order to produce a thick copper electrode. Previously the material traditionally used for starter blanks was copper itself but this had the disadvantage that corrosion occurred at the electrolyte level requiring the copper surface to be coated with oil to act as a parting agent. Titanium is not attacked by this copper sulphate/sulphuric acid solution because of the passivating effect of the cupric ions. Moreover, the oxide film on the titanium surface serves as the parting agent thereby eliminating the need for oil.

Power generation An area of likely future expansion is the use of titanium alloys for blading in the low pressure section of large steam turbines. For many years, these blades have been forged from a 12% chromium stainless steel and their size has become limited by the centrifugal loads imposed on the supporting rotors. This in turn limits the size of the turbines thereby preventing more efficient operation. As shown in Fig. 6.32, Ti–6Al–4V has corrosion-fatigue properties that are superior to this steel and, for comparable stresses on the rotors, titanium alloy blades can be some 40% longer. One German company is now producing blades that are 1.65 m in length.

Erosion by wet steam is a particular problem in this section of the turbine and titanium alloys also perform better than the stainless steel, although they are not as resistant to wear as the stellite shields that are commonly bonded to the leading edges of the steel blades. Apart form increased cost, a disadvantage of the titanium alloy blades is their reduced stiffness arising from a lower elastic modulus although it is possible to alter their shape so that more rigid positioning can be achieved.

Automotive There are many automobile parts that can be made from titanium alloys and which would lead to significant weight reductions. However, because of high costs, their use in mass produced vehicles can only be justified if the technical advantage is exceptional. Examples of components are connecting rods, valves, valve springs, and turbocharger rotors in the engine and, suspension springs and exhaust systems in the body.

Reducing the weight of reciprocating and rotating components in the engine can lower fuel consumption and consequent exhaust emissions. Connecting rods require high tensile and fatigue strength, stiffness and wear resistance. Titanium alloys present problems with the last two of these properties and need both reinforcement with high modulus particulates, and the use of special coatings on wear surfaces. Special attention has been paid to cylinder valves. For one production engine, Toyota in Japan has used the alloys Ti-6Al-4V for inlet valves and IMI 834 for the exhaust valves because they operate in a much hotter regime. These alloys are reinforced by titanium boride particles. The valves are 40% lighter than conventional steel valves and 16% lighter valve springs could also be used. As a result, the maximum engine revolutions increased by 700 rpm and the noise generated in this range was reduced by 30%. In Germany, the lighter titanium aluminide alloy, γ-TiAl, (Section 7.9) is being evaluated for automobile engine valves.

The relatively low modulus of titanium is an advantage for springs and a cold wound, high-strength β-titanium alloy has been used for the two rear coil springs of one model of the Volkswagon Lupo motor car. A lower cost alloy, Ti–4.5Fe–6.8Mo–1.5Al, was developed for this purpose and the spring weighs one half of the equivalent steel spring. Exhaust systems in some production vehicles are consuming the largest amounts of titanium. Mufflers and exhaust pipes, which are mostly made from CP-titanium alloys, are also approximately half the weight of the equivalent mild steel or stainless steel components and have a life expectancy of at least 12 to 15 years. One improved CP titanium alloy has been developed for this purpose which has the composition Ti–0.5Fe–0.6Si–0.15O and is known as TIMETAL Exhaust XT.

Marine Both the high corrosion resistance in the presence of sea water and the high strength:weight ratios of titanium and its alloys, makes them an attractive prospect for use in marine environments such as off-shore oil and gas platforms. Russia has been a leader in this regard and has had several decades of experience using titanium alloys in this way. Stringent requirements that must be met include:

(i) high strength and toughness under static and dynamic loading conditions at temperatures as low as −50 °C,
(ii) resistance to erosion by sea ice,
(iii) fire resistance under conditions of hydrocarbon combustion,
(iv) resistance to the adsorption of hydrogen when in contact with other metals or with cathodic protection systems,
(v) a capacity for repairs to be made without the need for post-weld heat treatment.

Russia has developed several α and near-α alloys especially for marine applications, examples being Ti–5.3Al–2Mo–0.6Zr and Ti–5.5Al–1.5V–1.4Mo. During the last 10 years several thousands of tonnes of titanium alloys, mostly CP-titanium and Ti–6Al–4V, have also been used on oil platforms in the North Sea. Applications include heat exchangers, drilling risers, pipelines,

Fig. 6.38 Tubing and baffles for oil/gas product coolers fabricated from CP-titanium tube and plate (from *Metallurgist and Materials Technologist*, **9**, 543, 1977).

valve castings, and fasteners. In the example shown in Fig. 6.38, sea water is used as the coolant and the tubing must resist attack by sulphide contaminants in the oil and gas.

During the 1970s and 1980s, the Russian navy operated submarines with a hull completely manufactured from titanium. This allowed the thickness and weight of the hull to be reduced which enabled these vessels to travel faster and dive deeper than any other in the world. Deep sea submersible vessels are now made with titanium pressure hulls that are capable of diving to 6000 m.

Architectural As mentioned in Chapter 1, the use of panels of thin (0.35 mm) CP-titanium sheet to clad the Guggenheim Museum in Bilbao, Spain has drawn attention to its architectural attributes. A more recent application has been the use of 200 tonnes of titanium sheet to clad the Beijing Grand National Theatre. Apart from its pleasing silver-grey colour, titanium has the advantages of low values for coefficient of thermal expansion and heat conductivity. The former, which is half that of stainless steel and one third that of aluminium, minimizes thermal stresses whereas the latter provides some opportunity to improve energy efficiency in buildings. Immunity from corrosion guarantees long life and minimal maintenance requirements.

Military hardware Titanium alloy armour is being incorporated into the design of some combat vehicles in order to save weight. In this regard, reductions of 30–40% as compared with using steel while maintaining the same protection from penetration by projectiles.

Sports The constant desire by golfers to hit balls greater distances has exposed a profitable niche market for titanium alloys. Having a larger head on woods and drivers is a considered to be an advantage and this has been achieved without increasing weight by using hollow designs made by lost wax, investment

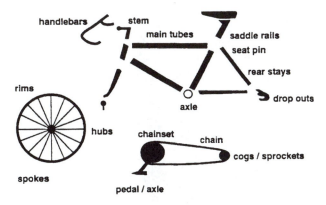

Fig. 6.39 Use of titanium alloys for bicycles (courtesy IMI Titanium Ltd.).

casting. Another application has been for light weight bicycles and components that can be made from titanium alloys are shown in Fig. 6.39. In this regard a cheaper α/β alloy Ti–3Al–2.5V has been specially developed for making the welded frames.

6.8.3 Biomaterials

Titanium and titanium alloys show an outstanding resistance to corrosion by body fluids which is superior to that of stainless steels. They also have a lower elastic modulus which makes them more compatible with respect to the natural elasticity of bone. These factors, together with good stress-corrosion resistance, high mechanical properties and acceptable tissue tolerance has led to their use for prosthetic devices. One feature of particular importance is that titanium is one of the few materials which will not induce formation of a fibrous tissue barrier when placed in contact with healthy bone. This is desirable since it permits bone to grow close to the surface of an implant and to fill grooves or pores that may have been deliberately introduced to enable a device to become more firmly embedded. This is a particular advantage for dental prostheses and Fig. 6.40 shows a titanium implant embedded into a human cheek bone into which an artificial tooth can be screwed. Yet another advantage of titanium is the fact that the fatigue properties of load-bearing devices are not reduced through contact with dilute saline solutions such as body fluids (0.9% NaCl).

Prosthetic devices are usually made from CP titanium or the alloy Ti–6Al–4V. One notable example has been the early Starr–Edwards aortic heart valve in which a Dacron covered titanium cage contains a hollow, electron beam welded titanium ball (Fig. 6.41a). It should be noted that the ball has been designed to have a relative density similar to that of blood and it is preferred to a heavier silicone ball which has proved less satisfactory due to inertial effects. Figure 6.41(b) shows a selection of other prostheses including an artificial hip joint, plates and pins.

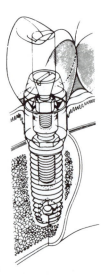

Fig. 6.40 Titanium plug implanted into a human cheek bone into which is screwed an artificial tooth (courtesy Nobel Parma, Sweden).

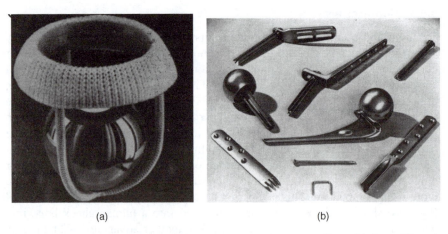

(a)   (b)

Fig. 6.41 Prosthetic devices made from titanium and titanium alloys: (a) Starr–Edwards aortic heart valve; (b) artificial joint, plates and pins.

More recently, concerns about possible long term cytotoxic effects of vanadium has led to the investigation of alternative alloys. One example are the β-titanium alloy Ti–15Mo–5Zr–3Al which has higher tensile and fatigue strength than Ti–6Al–4V and has a reduced elastic modulus of 80 GPa which approaches that of bone. A second alloy is the α/β composition Ti–15Zr–4Nb–4Ta.

The shape memory alloy nitonal, Ti–55Ni (Section 6.5.9), is projected to capture much of the market for orthodontic devices during the next decade because it is also pseudoelastic. This property allows stress to remain constant over a wide range of strain, which means that braces on teeth require less frequent adjustment to compensate for the movement of teeth. Nitonol is also being used for medical splints that can be formed into shapes to meet the unique needs of each individual patient. Another medical application of nitinol is for coiled wire stents that are used to dilate narrowed blood vessels. These stents may be cooled, collapsed, inserted by means of a catheter to the desired location, and then allowed to expand by the heat of the blood stream.

FURTHER READING

Collings, E.W., *Physical Metallurgy of Titanium Alloys*. American Society of Metals, Cleveland, Ohio, USA, 1984

Collings, E.W., Introduction to titanium alloy design, in *Alloying*, Walter, J.L., Jackson, M.R., and Sims, C.T. (Eds.), ASM International, Materials Park, Ohio, USA, P. 257, 1988

Molchanova, E.K., *Phase Diagrams of Titanium Alloys*, Israel Program of Scientific Publications, 1965

Leyens, C. and Peters, M. (Eds.), *Titanium and Titanium Alloys*, Wiley-VCH Verlag GmbH, Weinheim, Germany, 2003

Lütjering, G. and Williams, J.C., *Titanium*, Springer, 2003

Lampman, S., Wrought titanium and titanium alloys, in *Metals Handbook*, 10th Edn., Vol. 2, ASM International, Materials Park, Ohio, USA, 1990

Donachie Jr, M.J., (Ed.), *Titanium and Titanium Alloys : Source Book*, American Society for Metals, Cleveland, Ohio, USA, 1989

McCann, M.L. and Fanning J., Designing with Titanium Alloys, *Handbook of Mechanical Design,* Eds. Totten, G.E., Xie, L., and Funatani, K. Marcel Dekker, Inc., New York, p. 539, 2004

Lütjering, G. (Ed.), *Proc. 10th World Conf. on Titanium Alloys*, Germany, 2003, Wiley-VCH, Weiheim, Germany, 2004

Blenkisop, P.A., Evans, W.J. and Flower, H.M., (Eds.), *Titanium 95 Science and Technology*, Proc. 8th World Conf. on Titanium, The Institute of Materials, London, 1995

Lütjering, G., Property optimization through microstructural control in titanium and aluminium alloys, *Mater. Sci. and Eng.*, **A263**, 117, 1999

Boyer, R., Eylon, D. and Lütjering, G. (Eds.), *Fatigue Behaviour of Titanium Alloys*, TMS, Warrendale, Pa, USA, 1998

Schutz R.W., Stress-corrosion cracking of titanium alloys, in *Stress-Corrosion Cracking – Materials Performance and Evaluation*, Jones, R. H., (Ed.), ASM International, Materials Park, Ohio, USA, 265, 1992

Froes, F.H. and Eylon, D., Powder metallurgy of titanium alloys, *Inter. Mater. Rev.*, **35**, 162, 1990

Boyer, R.R., An overview of the use of titanium in the aerospace industry, *Mater, Sci. and Eng*, **A213**, 103, 1996

Williams, J.C. and Starke Jr, E.A., Progress in structural materials for aerospace systems, *Acta Mater.*, **51**, 5775, 2003

Mantovani, D. Shape memory alloys:properties and biomedical applications, *JOM*, **52**, No. 10, 36, 2000

Niinomi, M., Recent metallic materials for biomedical applications, *Metall. and Mater. Trans A*, **33A**, 477, 2002

7

NOVEL MATERIALS AND PROCESSING METHODS

Service demands for improvements in the properties of engineering materials are unceasing and often exceed the capacity of conventionally processed alloys to respond. This situation has stimulated an interest in new compositions produced by a number of novel methods. Because of ease of handling, aluminium alloys, in particular, have often been chosen to model these new processes and the association of these materials with the advanced aerospace industries tends to place them at the forefront of emerging technologies. Accordingly, it is convenient to consider these new developments with reference to their impact on light alloy metallurgy. Most are experimental or at an early stage of commercial development and all involve cost premiums when compared with conventionally produced alloys which may be substantial.

7.1 COMPOSITES

One method of meeting these new demands, which has considerable historical precedent, is the practice of combining different materials to form composites with properties superior to those of the components either individually or additively. Laminates or sandwich panels and, more recently, metals reinforced with fibres or particulates are common examples.

7.1.1 Laminated composites

Three examples of laminated composites based on aluminium will be mentioned. One is a predominantly sheet product comprising a number of alternating layers of aluminium and plies, or prepregs, of fibres that are bonded together with resin to produce laminates that are notable for their resistance to crack propagation, particularly under fatigue conditions. Two products are

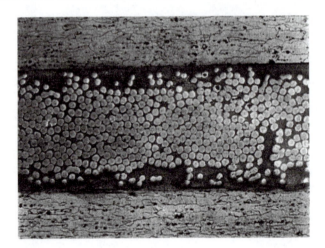

Fig. 7.1 A fibre-metal laminate showing a cross-section of fibres resin-bonded to thin sheets of aluminium (courtesy Delft University of Technology) Approx. × 80.

known as ARALL and GLARE and they use aramid and glass fibres respectively. Both were developed mainly by the Delft University of Technology in conjunction with the Fokker Aircraft Company in the Netherlands. Each is produced using standard bonding procedures and the laminates may be formed, punched, riveted or bolted like a normal metal. A cross-section of a fibre-metal laminate is shown in Fig. 7.1.

One configuration of ARALL had three sheets of aluminium alloy 7075–T6, each 0.3 mm thick, and two 0.2 mm thick internal layers of unidirectional continuous aramid fibres in an epoxy resin prepreg giving a 1.3 mm composite sheet (Fig. 7.2a). In the longitudinal direction, the composite may have a T.S. as

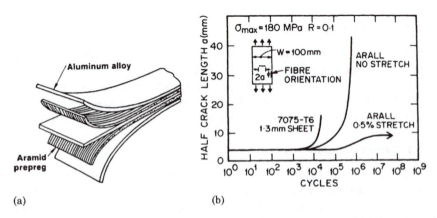

Fig. 7.2 (a) ARALL laminate; (b) fatigue crack propagation behaviour of ARALL compared with 7075–T6 sheet (courtesy R. J. Bucci).

high as 800 MPa and an elongation of 2.5%, which compares with 570 MPa and 11% for monolithic 7075–T6 sheet. The elastic modulus is comparable to 7075–T6 but the density is reduced to 2.35 g cm^{-3}, which is 18% less. The transverse properties are, however, much lower unless some of the fibres are suitably aligned in this direction.

Comparative rates of fatigue crack growth for pre-cracked panels of this ARALL laminate and 7075–T6 sheet of the same thickness (1.3 mm), tested in the longitudinal direction are shown in Fig. 7.2(b). In the unstretched condition (0.2% P.S. 496 MPa), the laminate panel exhibited fatigue lives some 10 times those of the 7075–T6 sheet due to a combination of factors which place restraint on crack opening in individual metal sheets and crack propagation to other sheets. Moreover, the laminate showed greater damage tolerance since it can accommodate a crack which is nearly three times longer before overload failure occurs. Stretching the laminate by 0.5% raises the 0.2% P.S. to 640 MPa and effectively prevents the crack propagating beyond a few millimetres after more than 10^7 test cycles.

ARALL was developed primarily for possible applications in aircraft wings. Further fatigue tests revealed that failure of the aramid fibres tended to occur at low test frequencies that simulated cycles associated with the pressurisation of the fuselage. This fibre failure was attributed to insufficient bonding between fibres and the epoxy adhesive, as well as to damage that occurred under compressive loads. There was also a tendency for the aramid fibres to absorb moisture. Carbon fibres were considered as a possible replacement but there were concerns that attaching carbon to aluminium may lead to galvanic corrosion. The combination of glass fibres and aluminium did not pose such a problem and attention was directed to the sheet product that became known as GLARE (derived from GLAss REinforced laminate).

GLARE is comprised of alternating layers of aluminium foils and continuous, unidirectional or biaxially oriented meshes of high-strength glass fibres impregnated with an epoxy resin adhesive. A laminate can be tailored to suit particular requirements by varying factors such as fibre-resin system, alloy type and thickness, stacking sequence and fibre orientation. The layers are built up in a mould with localised reinforcements being included as required, after which the completed lay-up is bagged and evacuated before curing at 120 °C. Densities range from 2.4 to 2.5 g cm^{-3} and so GLARE is about 10% lighter than conventional aluminium alloys. High static strengths can be achieved, particularly with unidirectional fibres, although elastic moduli are somewhat lower than for monolithic aluminium alloys because of the presence of the glass/epoxy resin layers.

Despite the additional material costs, which may be more than five times that of conventional monolithic aluminium alloy sheet, some 380 m^2 of GLARE is being used for upper fuselage panels in parts the new Airbus A380 aircraft (Fig 7.3). The motivation for this selection is this composite's outstanding resistance to crack growth although there is also a useful weight saving of some 800 kg. Another minor advantage is a capacity to absorb acoustic vibration which is

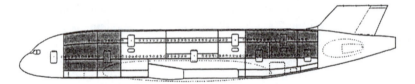

Fig. 7.3 Locations of GLARE laminate sheet in the fuselage of the Airbus A380 passenger aircraft (courtesy Airbus Industrie).

some three times higher then that for conventional aluminium alloys. Further applications are being studied such as the possible use of GLARE for the leading edge of the empennage (vertical tail) to provide protection against bird impact.

An experimental magnesium-based, laminated composite has been developed by sandwiching rolled foils (0.5–0.6 mm thick) of the alloy AZ31 (Mg–3Al–1Zn) between thin sheets of the polymer polyether ether ketone (PEEK), that were produced as a prepreg reinforced with ~60 volume% of continuous 7 μm diameter carbon fibres. Prior to lamination, the metal sheets were etched with chemicals to assist adhesion by roughening the surfaces. Composite panels were then produced by hot pressing. A five-layer panel had a density of 1.7 g/cm^3 and a thickness of 2.7–2.8 mm. In the longitudinal direction the tensile strength was 932 MPa, and the elastic modulus 75 GPa.

A second type of laminate is shown in Fig. 7.4. In this case, layers of high-modulus fibres, such as boron, are strategically placed and bonded to various structural sections to improve their stiffness.

7.1.2 Sandwich panels

Sandwich panels form a third type of laminate and they offer a particular combination of high rigidity and low weight. Such panels comprise thin facings which are secured, usually by adhesive bonding, to a relative thick, low-density core

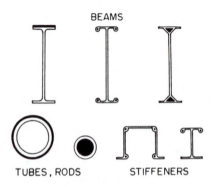

Fig. 7.4 Aluminium alloy sections strategically reinforced with high modulus fibres to increase stiffness (courtesy Avco Systems Division).

material. From the design view-point, a sandwich panel is similar to an *I*-beam with the facings and core corresponding to the flanges and centre web respectively. The facings carry axial compressive and tensile stresses whereas the core sustains shear and prevents buckling of the facings under compressive loading.

Moment of inertia is a direct measure of stiffness or rigidity and it is interesting to compare values for a sandwich panel and an homogeneous isotropic plate of the same material as the facings of the panel. For example, a sandwich panel with two 0.5 mm thick facings of an aluminium alloy and a balsa wood core 6 mm thick would weigh 3.4 kg m^{-2}. A plate of the same alloy of the same size and weight would be 1.25 mm thick. The moment of inertia of the cross-section of the plate I_p about its neutral axis, per unit width, is given by

$$I_p = \frac{t_p^3}{12}$$

where t_p = thickness of the plate. Neglecting the small effect of the core material, the moment of inertia of the sandwich panel, I_s, per unit width, is given by

$$I_s = 2t_f \left(\frac{t_f + t_c}{2} \right)^2$$

where t_f = thickness of the facings, t_c = thickness of the core. For the sandwich panel and plate under consideration, $I_s = 10.5$ mm^4 and $I_p = 0.162$ mm^4. Thus the sandwich panel has the advantage of 65 times the rigidity of a solid plate having the same weight. It can also be shown that, for the aluminium alloy plate to have equal rigidity, it would weigh four times more than the sandwich panel.

Sandwich panels with cores of balsa wood or foamed plastic are now used for applications such as siding for refrigerated trucks and for a variety of aircraft components and containers. In this latter regard, even greater weight savings are possible by using a honeycomb core made from impregnated paper or aluminium alloy foil. The honeycomb is usually made by an expansion method (Fig. 7.5) which begins with stacking of sheets of foil on which adhesive stripes have been printed. The adhesive is cured and the block cut into slices which are then expanded to form the honeycomb panel. Honeycomb can be contoured to desired shapes by high-speed cutters and lightweight sandwich panels are used in aircraft for applications such as fuselage and wing panels, one example being leading-edge wing flaps, a section of which is shown in Fig. 7.6.

Experimental sandwich panels have also been prepared using titanium alloy facings and honeycomb core. In this case it is possible to assemble the components by diffusion bonding (Section 6.5.7).

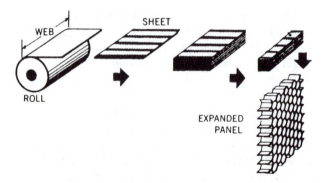

Fig. 7.5 Expansion process for the manufacture of aluminium honeycomb cores for sandwich panels (courtesy Hexel Aerospace).

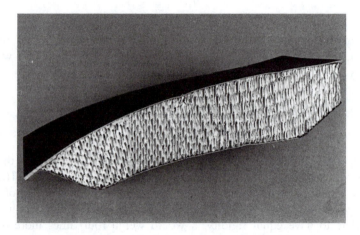

Fig. 7.6 Section of honeycomb sandwich panel from an aircraft wing flap.

7.1.3 Metal matrix composites

Aluminium alloys Fibreglass is the most widely known fibre-reinforced composite material but its use at even moderately elevated temperatures is severely restricted because the polymeric matrix degrades. Another limitation is the relatively low elastic modulus of such a matrix. In this regard, it should be noted that the elastic modulus E_c of a fibre-reinforced composite having continuous, unidirectional fibres is given by the rule of mixtures so that $E_c = E_f V_f + E_M V_M$ where E_f and E_M are the respective elastic moduli of the fibres and matrix in a composite in which the volume fractions of these components are V_f and V_M. Modifications to this relationship are necessary for composites reinforced with discontinuous fibres.

Replacing polymeric matrices with metals would improve both the elevated temperature performance and elastic modulus of fibre-reinforced composites and many attempts have been made to incorporate strong wires and other fibres in aluminium. Examples are hard-drawn stainless steel wires and silica, silicon carbide, boron or carbon fibres. One exotic application has been the use of an aluminium alloy reinforced with continuous carbon fibres for the masts of the Hubble space telescope.

Early attempts to produce metal matrix composites based on aluminium involved interleaving metal foils with silica fibres, fine stainless steel wires or coated boron fibres which were then hot compacted slowly in a press. In the latter case, composites have shown unidirectional T.S. as high as 1200 MPa with elastic moduli as high as steel (220 GPa). They could be used at temperatures up to 320 °C and at one time were considered as candidates for compressor blades in gas turbine engines, as well as for certain structural components in aircraft. Good elevated temperature performance and high stiffness remain key goals in current research and development, but efforts are being directed more at other properties such as wear resistance rather than seeking extreme levels of tensile strength.

Special interest is centred on composites in which short fibres or particulates of a high modulus ceramic are incorporated in a metallic matrix (Fig. 7.7). Such materials can be prepared by compacting powders or by the so-called liquid metallurgy route. In this latter case, a porous ceramic preform may be infiltrated by molten aluminium and suitable alloys, or the fibres or particulates may be stirred into the melt before it solidifies. The integrity of such composites depends critically on the ability of the metal to wet the particulate or fibre surfaces, and after considerable research, significant advances have been made. Usually these composites may be remelted, cast and fabricated by normal processes such as forging and extrusion. Alcan Aluminium Ltd successfully marketed a castable metal matrix composite under the trade name Duralcan and ingots weighing several tonnes have been produced in a plant in Quebec, Canada. Billet prices depend on the size of an order but have been quoted in the

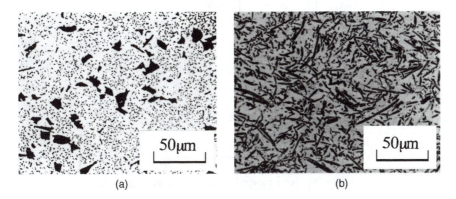

(a)   (b)

Fig. 7.7 Microstructures of metal matrix composites reinforced with (a) particulates and (b) fibres.

range $US6.5–9 per kg. Billets for subsequent fabrication to wrought products can also be produced by other means. One method is based on the Osprey Process (Section 3.6.6) in which ceramic particulates are injected into an atomized stream of molten aluminium in vacuum, or a controlled atmosphere, leading to their co-deposition in solid form on a suitable substrate. Billets weighing several hundred kilograms have been prepared in this way. Other powder metallurgy routes are being pursued in several countries.

Particular attention has been given to composites of aluminium alloys such as 2014 (Al–Cu–Mg–Mn) or 6061 (Al–Mg–Si) with silicon carbide or alumina particles or fibres. These ceramics can be used in various forms: long or short fibres, whiskers or particles. Reactions at the ceramic/alloy interface are limited so that a coating or diffusion barrier is unnecessary, which also reduces costs. The presence of magnesium as an alloying addition in the matrix improves wetting of the reinforcement. Development is well advanced with 6061-SiC composites and extrusions containing 20 vol % SiC as short fibres may have room-temperature tensile strengths as high as 500 MPa combined with an elastic modulus of 120 GPa (cf. 70 GPa for 6061) in the longitudinal direction. However, transverse properties will be much lower unless the fibres are randomly orientated. For this composite, values for fracture toughness may be maintained above 30 MPa m$^{1/2}$ until the volume of fibres reaches approximately 15%, after which it falls rapidly to a value of around 10 MPa m$^{1/2}$ at a fibre volume of 25% because of the greater ease of crack propagation (Fig. 7.8).

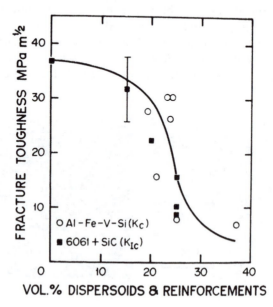

Fig. 7.8 Variations in fracture toughness of aluminium alloys with volume % of dispersoids (Ravichandran, K. S. and Dwaraksdasa, J., *J. Metals*, **39**(5), 28, 1987).

A similar trend has been observed in most other metal–matrix composites. In general, lower values of mechanical properties are obtained when particulates rather than fibres are used but these properties are more isotropic.

It is also necessary to appreciate that the presence of fibres or particulates may modify the ageing behaviour of alloys used for matrices in metal matrix composites. Such effects arise because of the presence of:

1. higher dislocation densities in the matrix, particularly in the vicinity of the reinforcement, that are generated by thermally induced stresses arising from differences between the coefficients of thermal expansion of matrix and reinforcement. These additional dislocations may modify vacancy contents, facilitate pipe diffusion of solutes and provide additional sites for the heterogeneous nucleation of precipitating phases
2. interfaces between matrix and reinforcement which can also serve as sinks for vacancies and facilitate heterogeneous nucleation of precipitates
3. chemical reactions between elements in the matrix and reinforcement. Such reactions can result either in the removal of solutes from the surrounding matrix, or transfer of solutes to the matrix from the reinforcement.

These modifications to the microstructure of the matrix may, in turn, alter:

1. the level of response to age hardening. As may be expected, the presence of the reinforcement fibres or particles can increase the quench sensitivity of alloy matrices during heat treatment (Section 3.1.6). This effect is demonstrated in Fig. 7.9 in which an experimental metal matrix composite, Comral 85, achieved higher hardening compared with the matrix alloy 6061 after water quenching, whereas the reverse occurs if materials are air-cooled before ageing at 175 °C
2. ageing kinetics. In most metal matrix composites, ageing processes at elevated temperatures are accelerated (Fig. 7.9a) so that heat treatment schedules need to be changed from those normally used for unreinforced alloys
3. ageing processes. As well as accelerating the rate of ageing in matrix alloys, the presence of the reinforcement may also modify the actual mechanism of ageing. GP zone formation tends to be suppressed, presumably because of the loss of quenched-in vacancies, whereas the onset of later stages is accelerated.

Most technological barriers to the introduction of aluminium alloy metal matrix composites, or so-called discontinuously reinforced aluminium (DRA), have been overcome and their wider use now depends largely on cost factors. In this regard, Fig. 7.10 compares the earlier total costs (materials plus manufacturing) for a typical automotive component made from steel or DRA. For steel, the material cost is only 14% of the total whereas this figure is estimated to be as high as 63% for DRA. However, the reverse is true for forming, for which the costs for steel are more than four times that of DRA. Typical properties of some commercially available wrought and cast aluminium alloy metal matrix composites are compared with some conventional structural alloys in Table 7.1.

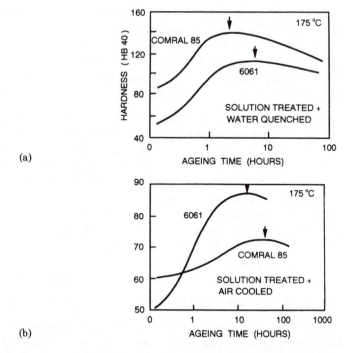

(a)

(b)

Fig. 7.9 Ageing curves at 175 °C showing the greater quench sensitivity of the experimental metal matrix composite Comral 85 (6061 reinforced with 20 vol% fine mullite/alumina spheres) as compared to the matrix alloy 6061 (courtesy Comalco Aluminium Ltd).

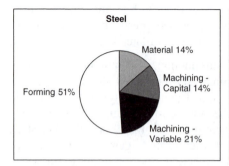

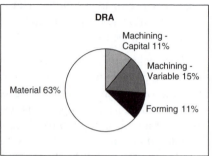

Fig. 7.10 Materials and manufacturing costs for a typical automotive component made from (a) steel or (b) discontinuously reinforced aluminium (DRA) (from Allison, J. E. and Cole, G. S., *J. Metals*, **45**(1), 19, 1993).

Table 7.1 Typical properties of some commercially available metal matrix composites and other structural alloys

Alloy	Tensile Strength MPa	Elastic Modulus GPa	Specific Gravity	Specific Modulus
6061	310	69	2.68	25.7
7075–T6(Al–Zn–Mg–Cu)	570	72	2.8	25.7
8090–T6(Al–Cu–Li–Mg)	485	80	2.55	31.4
Common structural steel	500	210	7.8	26.9
Ti–6%Al–4% V	950	106	4.4	24.1
A356–T6(Al–Si–Mg)	280	76	2.67	28.5
6061 + 20% SiC	500	105	2.78	37.5
7075 + 15% SiC	600	95	2.90	31.7
8090 + 17% SiC	540	105	2.65	39.5
A356 + 20% SiC	357	98	2.77	35.4

Applications for which aluminium alloy metal matrix composites have been evaluated include high volume automotive components such as connecting rods, drive shafts, pump housings, brake calipers and rotors. For connecting rods, the weight saving over steel can be as much as 45% which is critical in reducing undesirable reciprocating forces in the engine. Opportunities for improving wear resistance and elevated temperature properties are being exploited by the Toyota Motor Company which has selectively incorporated short silicon carbide fibres by squeeze casting to reinforce the crown and ring groove of diesel engine pistons. Examples of products are shown in Fig. 7.11.

Fig. 7.11 Extrusions, forgings, sheet and a pressure die casting fabricated from DRA (from Willis, T. C., *Metal. & Mater.*, **4**, 485, 1988).

The high wear resistance of sand and permanent mould cast Duralcan metal matrix composites reinforced with SiC particles has been utilized for large disk brakes used in rail vehicles operating in Europe. Weight reductions compared with using ferrous alloys can amount to around 200 kg. for each axle. The superior thermal conductivity of the aluminium alloy also ensures that thermal stressing of the brake disks never reaches critical levels.

Higher concentrations (20–40 vol.%) of much finer (<1 μm) ceramic particles may be obtained in aluminium and other matrices by a new proprietary process known as XD™ technology which has been developed by the former Martin Marietta Corporation in the United States. In this process, powders of the elemental components of high melting point ceramic or intermetallic phases (X, Y) are heated in the presence of the matrix metal. At some temperature, usually such that the matrix is molten, the component elements X and Y react exothermically forming ultrafine particles of phases such as borides, carbides, nitrides, or mixtures of these (Fig. 7.12). It is claimed that the process is relatively inexpensive and provides the advantage that, after the initial exothermic production step, conventional metallurgical processing (casting, forging etc.) can be used to produce shapes. As an example, Al–TiB$_2$ XD processed alloys have exhibited elastic moduli up to 40% greater than pure aluminium, improved retention of strength at elevated temperatures and useful increases in fatigue and wear resistance. More recent research has been focused on the production of so called 'designer' XD™ microstructures containing hard phases for strength, relatively soft phases for toughness and whiskers for creep resistance.

Magnesium alloys Although most studies of metal matrix composites have been concerned with aluminium alloy matrices, various combinations of magnesium alloys reinforced with ceramic particulates such as SiC, Al$_2$O$_3$ and graphite have been investigated. In this regard, magnesium does offer an advantage over aluminium because it has a greater ability to wet most fibres and particulates. Magnesium does not react with graphite or carbides such as SiC or B$_4$C and Table 7.2 shows the improvements in the properties of extrusions of

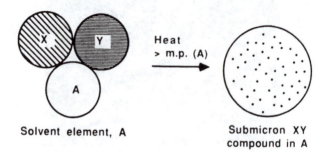

Fig. 7.12 Schematic diagram of process for making XD™ dispersion-hardened composite materials (from Westwood, A. R. C., *Metall. Trans B.*, **19B**, 155, 1988).

Table 7.2 Effect of SiC fibre reinforcement of the extruded magnesium alloy AZ31 (courtesy W. Unsworth and J. F. King)

Alloy	0.2%PS (MPa)	TS (MPa)	Elongation (%)	Elastic modulus (GPa)
AZ31	221	290	15	45
AZ31 + 10% SiC	314	368	1.6	69
AZ31 + 20% SiC	417	447	0.9	100

the medium strength alloy AZ31 (Mg–3Al–1Zn) by incorporating SiC fibres. What is notable is the effective doubling of proof stress and elastic modulus when 20% SiC is added although, as usual, these changes occur at the expense of ductility. As with composites based on aluminium, the presence of the ceramic reinforcement has the added benefit of reducing thermal expansion of the magnesium alloy matrix.

Magnesium does react with oxides such as Al_2O_3 to form the spinel $MgAl_2O_4$ and, if this relatively cheap reinforcement is to be used, it is necessary to process the composite so that liquid metal contact is minimized or avoided. Techniques that have been tried are rapid squeeze casting (Section 4.6.2), spray casting and hot compaction of mixtures of powders.

Results have been obtained from squeeze castings of the alloy AZ91 (Mg–9Al–1Zn) containing a range of different fibres such as glass, carbon and Saffil, which is a proprietary pre-form or woven mat of 14 μm diameter Al_2O_3 fibres. Creep and fatigue test data have been obtained from a 16 vol% Saffil reinforced alloy and compared with standard AZ91 tested under the same conditions. The composite material was found to have a creep life at 180 °C that was an order of magnitude better than AZ91 and the fatigue endurance limit at this temperature was double that recorded for this alloy (Fig. 7.13). The presence of fibres does, however, reduce the fracture toughness to a level of 10 MPa $m^{1/2}$ or less as has been observed for fibre-reinforced aluminium alloys (Fig. 7.8).

The prospect of developing ultra-light composites based on a matrix of Mg–Li alloys has also been investigated. However, it has been found that severe degradation of the reinforcement occurs during processing due to the reaction of lithium with all but SiC whiskers. Moreover, mechanical properties have been found to be unstable at quite low temperatures as a consequence of the abnormally high mobility of lithium atoms and vacancies in the alloy matrices. This has the effect of relaxing the desirable localized stress gradients that normally develop close to the ends of fibres, even at comparatively high strain rates.

Magnesium composites have found some specialized applications in aerospace engineering, examples being trusses, booms and other structural members for space platforms and satellites. Elsewhere these composites are largely at an experimental stage and they are also being evaluated for possible use in various automotive components.

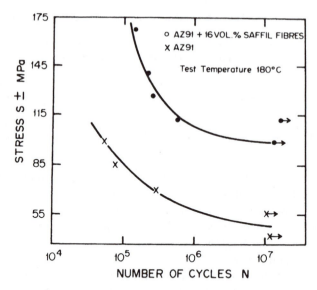

Fig. 7.13 High temperature *S/N* fatigue curves for the cast alloy AZ91 and squeeze cast AZ91 containing 16 vol% of Saffil (Al$_2$O$_3$) fibres (from Chadwick, G. A., *Magnesium Technology*, Institute of Metals, London, 1986).

Titanium alloys Titanium alloy metal matrix composites are being considered for certain sophisticated aerospace and defence applications that must withstand severe thermomechanical environments. One example is leading edge aerofoil structures for proposed hypersonic flight vehicles which may require materials that will maintain adequate stiffness and strength at temperatures of 1100 °C and above.

The composites may be prepared by blending and compacting powders in the conventional way. However, special attention has been given to the incorporation of continuous plies or tapes of relatively coarse (e.g. 140 μm) silicon carbide fibres in matrices such as the β-alloy Ti–15V–3Cr–3Al–3Sn (Ti 15-3). Major problems have been control of interfacial reactions between fibres and matrix and thermal fatigue cracking arising because of their different coefficients of thermal expansion. Vapour deposition of the alloy has been employed as a means of pre-coating the fibres and sections have then been prepared by hot pressing between alloy foils. Such a material system is costly but has the potential to operate at temperatures up to 650 °C.

Because of the inherent problems associated with the use of ceramic fibres, titanium alloy composites have been prepared which are reinforced with metallic rather than ceramic phases. For many years it was known that the addition of boron to titanium and titanium alloys increased strength, stiffness and microstructural stability due to the formation of TiB particles or whiskers. The most economic way of producing such alloys is by blending, compacting and sintering powders. In this case, in situ chemical reactions occur during process-ing which are similar to those described above for the XD™ technology. A commercial application has been the manufacture of two types of titanium alloy composite valves for the engine of a mass produced Toyota motor car that was mentioned in Section 6.8.2. These composites contain 5% by volume of TiB particles and more than 500 000 valves have been manufactured for this purpose which have performed well in service. They are, however, double the cost of steel valves. Composites with higher contents of TiB particles, or with mixtures of TiB and TiC particles, show exceptional wear resistance. Such materials can be prepared by using powder blends made from titanium alloys that contain carbon.

7.2 METALLIC FOAMS

Foamed metals and alloys are a novel class of materials that may have extremely low densities combined with high specific stiffness, reduced thermal conductivity, and a high capacity to absorb impact energy and noise. They have features in common with the natural cellular materials wood and bone. Overall properties depend on the particular metal or alloy and the cell topology, i.e. the size and shape of the pores and whether they are open or closed (Fig. 7.14). Again, much of the developmental work has been carried out on aluminium and its alloys for which foam densities relative to the solid state commonly range from 0.1 to 0.5, although values as low as 0.04 (i.e. 108 kg/m^3) have been recorded. Some foams have also been produced using magnesium or titanium, and a density as low as 50 kg/m^3 has been achieved with the magnesium alloy AZ91 (Mg–9Al–1Zn).

Foams can usually be made from either molten metals or powders and costs in the case of aluminium may range from less than $US 10 to several thousand dollars per kg. Stable foams cannot be formed in pure liquid metals simply by blowing in a gas because the bubbles are too buoyant and rise quickly to the surface. It is necessary therefore to increase the viscosity of the melt by adding 10 to 30% of fine particles either as fine ceramics, such as alumina, or which form because of presence of suitable alloying elements e.g. 1.5%Ca. Foaming can then be achieved in three ways: by injecting a gas (air, nitrogen or argon) into the melt, by causing in-situ gas formation through the introduction of a gas releasing agent such as 1 to 2% TiH$_2$, or by precipitating a gas that had previ-ously been dissolved in the melt under pressure. It is then possible to solidify the melt while the bubbles remain in suspension. The alternative metal powder route is generally more expensive but offers greater opportunities for near net shape forming. This method involves mixing the powdered alloy with a gas blowing agent and TiH$_2$ is again commonly used. The powder mixture is first compacted and then heated to a mushy condition to release the hydrogen.

Design rules are being developed to facilitate the use of foams in engineer-ing structures. For example, Ashby and colleagues at Cambridge University

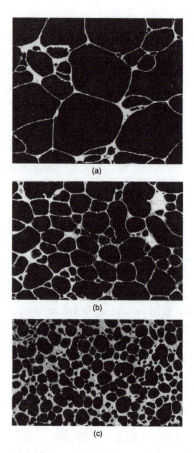

Fig. 7.14 Cross-sections of aluminium alloy foams. (courtesy M. F. Ashby)
(a) Cymat (Canada) foam formed by gas injection into the melt. Relative density 0.04 (108 kg/m³).
(b) Alporas (Japan) foam formed from the melt using a Ti H_2 blowing agent. Relative density 0.09 (240 kg/m³).
(c) Alulight (Austria) foam produced from a powder compact. Relative density 0.25 (435 Kg/m³).

have shown that strength and elastic modulus are functions of relative density and can be described by:

$$\text{(i)} \quad \frac{\sigma_{pl}}{\sigma_{ys}} = C_1 \left(\frac{\rho_f}{\rho_s} \right)^{3/2} \qquad \text{(ii)} \quad \frac{E_f}{E_s} = C_2 \left(\frac{\rho_f}{\rho_s} \right)^2$$

where σ_{pl} = plastic-collapse of a foam block under compressive load, σ_{ys} is the uniaxial yield strength of foam struts, ρ_f and ρ_s are the densities of the foam block and the solid metal or alloy from which the foam was made, E_f and E_s are the respective elastic moduli, and C_1 and C_2 are constants of about 0.3 and 0.1 respectively. Relationships have also been established for other properties such as toughness.

Some metallic foams are potentially inexpensive, particularly when cost is measured in volumetric terms and commercial products are now appearing. One example is for the core of sandwich panels described in the previous section and Karmann GmbH in Germany has a concept design for a small, low-weight automobile in which 20% of the structure could be made from foam panels with an estimated weight saving of 60 kg. The Cymat Corporation in Canada has entered into an agreement with a supplier of automotive components to develop a new bumper bar system based on aluminium foam that will improve crash resistance. Foams may be attached to steel and concrete structures and, in Japan, they have been used experimentally as baffles on the underside of a highway bridge to absorb traffic noise. Other examples of potential applications are structural panels in aircraft and trains, and firewalls, in a range of integrally moulded components.

A foam core may be encased by a solid cast skin if spacers are used to position the core within a suitable mould. The molten alloy is then introduced into the mould so that it solidifies around the core creating a mechanical bond with the rough foam surface. Either low pressure die casting or gravity casting must be used to avoid damage to the relatively fragile core. If metallurgical bonding is required, the foam surface must first be coated with a suitable flux which dissolves the oxide film during casting. Structural members can also be produced by filling extruded aluminium tubes and sections with molten foam as shown in Fig. 7.15. Continuous foam-filled aluminium alloy panels 1.5 m wide and 20–150 mm thick have been produced by Alcan at a rate of 900 kg per hour by bubbling air through a melt and casting the foam between sheets produced on a belt caster (Fig. 3.5).

Much less attention has been given to the production of magnesium foams. Some success has been achieved using a high pressure die casting machine if the molten metal is first injected through a separate chamber containing the

Fig. 7.15 Cross-section of an extruded aluminium alloy tube filled aluminium alloy foam (courtesy of Cymat Corporation).

chemical blowing agent MgH_2 before it enters the die. Hydrogen released in the mould causes a porous structure to form in the centre of the cast part whereas a solid skin forms at the walls due to the fast solidification rate.

7.3 RAPID SOLIDIFICATION PROCESSING

During the last two decades, much attention has been given to the technique of rapid solidification processing (RSP) as a means of producing entirely new ranges of alloys having mechanical and corrosion properties superior to those obtainable by conventional ingot metallurgy practices. RSP involves cooling at extreme rates from the melt to produce a powder or splat particulate. Powders are most conveniently prepared by some form of gas atomization in which the molten alloy is sprayed through a nozzle into a stream of a high velocity gas such as nitrogen or argon. Fine particles are formed which are roughly spherical and most have diameters in the range 10–50 μm. Cooling rates are faster the smaller the particles and may be as high as 10^5 °C s^{-1}. The method of splat quenching that appears to have the best practical potential is so-called melt spinning in which the molten alloy is forced through an orifice on to an internally water cooled, rotating wheel made from a metal such as copper which has a high thermal conductivity. Essential details are shown in Fig. 7.16(a) and the stable solidification condition that is established has been termed planar flow casting (Fig. 7.16b). Thin (e.g. 20 μm) ribbons are produced at cooling rates of 10^6 °C s^{-1} or higher, which are then pulverized into flakes for subsequent consolidation. The powders or pulverized ribbons are processed by canning, degassing and hot pressing to produce solid billet for subsequent forming by forging or extrusion. The major obstacle is oxide contamination which may prevent interparticle bonding during processing with consequent deleterious effects on mechanical properties.

Conventional gas-atomization of liquid metals is well known, having been used since the 1930s to produce a wide variety of metallic powders for diverse

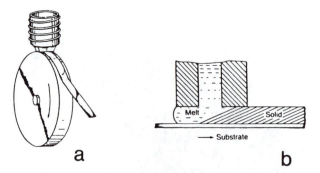

a   b

Fig. 7.16 Rapid solidification processing by melt spinning; (a) metal is induction melted in a crucible and forced on to a rotating wheel producing RSP ribbon; (b) stable solidification conditions for planar flow casting.

applications. For example, each launching of the United States Space Shuttle consumes 160 000 kg of atomized aluminium powder as part of the solid fuel propellant mixture. Scaling up of other techniques of RSP has occurred in several countries and pilot facilities for continuous melt spinning are available with capacities for making 50 to 500 kg of pulverized ribbons. Consolidated billets up to 550 mm dia., weighing 270 kg, have been processed.

7.3.1 Aluminium alloys

With the extreme rates of cooling, it is possible to extend the solid solubility of elements in aluminium which is particularly useful because, as shown in Table 2.1, relatively few elements have an equilibrium solid solubility exceeding 1 at%. Moreover, these latter elements, such as magnesium, zinc and copper, all have high diffusivities in aluminium and result in alloys that have relatively poor thermal stability. Table 7.3 gives examples of the diffusivities of some metals together with the increases in solid solubility that can be achieved by RSP. What is particularly significant are the large amounts of the sparingly soluble elements such as iron and chromium that can be retained in supersaturated solid solutions since each has a very low diffusivity in aluminium. Special interest has been centred on alloys containing one or more of these elements as they have the potential to compete with titanium alloys up to temperatures of 300 °C or higher. This compares with a maximum of only around 125 °C for existing age-hardenable aluminium alloys prepared by ingot metallurgy.

RSP also has several desirable effects on microstructure. A fine, stable grain size can be achieved (e.g. 1 μm or less) which can be retained during subsequent processing due to the presence of second-phase particles in the grain boundaries. Small metastable precipitates may also form within the grains during cooling or processing of the powders or splat particulates. Moreover, if these phases contain elements such as iron (e.g. Al_6Fe, Al_3Fe), they have high thermal stability

Table 7.3 Increased solid solubility in some binary aluminium alloys due to RSP (courtesy H. Jones)

Solute	Maximum equilibrium solubility		Reported extended solubility		Diffusion coefficient at 425 °C
	(wt%)	(at%)	(wt%)	(at%)	D_0 (m^2 s^{-1})
Cr	0.72	0.44	8–10	5–6	7.7×10^{-21}
Cu	5.65	2.40	40–42	17–18	4.3×10^{-15}
Fe	0.05	0.025	8–12	4–6	2.2×10^{-1}
Mg	17.4	18.5	34–38	36–40	1.5×10^{-14}
Mn	1.82	0.90	12–18	6–9	2.3×10^{-18}
Ni	0.04	0.023	2.4–15.4	1.2–7.7	5.2×10^{-15}

and are resistant to coarsening (Ostwald ripening) at relatively high temperatures. Finally, the scale of the microstructure is so fine that the alloys are chemically very homogeneous.

The microstructures of aluminium alloys obtained by RSP can be better understood by referring to Fig. 7.17 in which diagram (a) gives a schematic representation of the respective volume fractions of what are called microcellular (or microeutectic), combined cellular and eutectic, and primary intermetallic structures, as a function of powder diameter (i.e. cooling rate) in an RSP Al–8Fe alloy. As expected, the greater the powder diameter (i.e. the slower the cooling rate), the coarser the microstructure becomes. Figure 7.17(b) shows a section through some atomized powders which all have cellular (or so-called 'zone B') structures with a grain size around 1 μm. Another view of this cellular structure is shown at B in Fig. 7.17(c) which is a transmission electron micrograph containing a central region in which the microstructure has not been resolved ('zone A'). This is the site at which nucleation of the solid powder has occurred and where cooling rate has been highest. This so-called zone A region has the microcellular or microeutectic structure and much higher magnifications are needed to resolve the cells or grains that are revealed as having an exceedingly small average diameter of as little as 20 nm (0.02 μm) (Fig. 7.17d).

Relatively little is known about the zone A microstructures except to say that they may contain metastable phases such as the so-called O phase that has been identified in some melt spun alloys based on the Al–Fe system cooled at rates of around 10^7 °C s^{-1}. This phase has an icosahedral crystal structure and it decomposes during subsequent thermomechanical processing to produce substantial alloy strengthening.

Comparative studies show the zone A structure to be much harder than zone B. Microstructures of alloys based on the Al–Fe system can contain only large amounts of the zone A structure if they are prepared by a process such as melt spinning rather than gas atomization. In one such case for the alloy Al–12Fe–2V, an elastic modulus at room temperature of 96.5 GPa has been recorded which is some 40% higher than that for conventional aluminium alloys. Moreover, the relatively high values are retained at elevated temperatures (e.g. 78 GPa at 316 °C). Tensile strength values for the same alloy can exceed 600 MPa and 350 MPa respectively at these two temperatures.

Most of the RSP aluminium alloys that have been studied are based on the Al–Fe system. Examples are Al–8Fe–4Ce (Alcoa), Al–8Fe–2Mo (Pratt & Whitney) and Al–8.5Fe–1.3V–1.7Si (former Allied Signal). The main role of the other elements seems to be to alter the kinetics of precipitation that normally occurs in the binary Al–Fe system. The powders or pulverized ribbons are consolidated and fabricated in the manner described above. Figure 7.18 gives an example of the elevated temperature properties of several related alloys prepared by RSP in terms of their tensile strength and compares them with the ingot alloys 7075–T6 and 2219–T851. The specific strength needed to equal the widely used titanium alloy Ti–6Al–4V is shown as a dashed line.

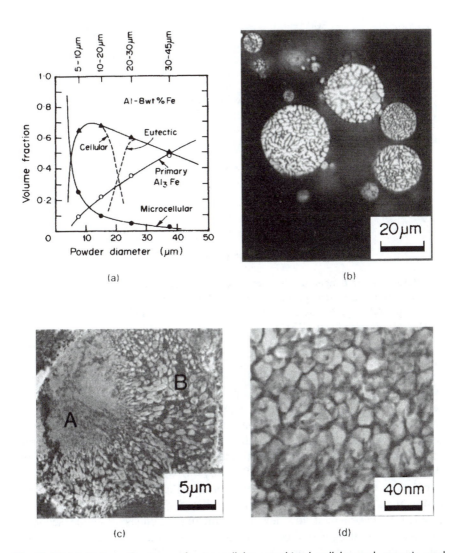

Fig. 7.17 (a) Volume fractions of microcellular, combined cellular and eutectic, and primary intermetallic structures as a function of powder diameter in RSP Al–8Fe powders. Possible separate curves for the cellular and eutectic structures are shown dotted (courtesy U.S. National Bureau of Standards). (b) Sections through gas-atomized aluminium alloy powders showing a combined cellular and eutectic (zone B) microstructure having an average grain size of 1 μm. (c) Section of an Al–8Fe powder particle showing the zone B structure and an unresolved zone A region (transmission electron micrograph courtesy U.S. National Bureau of Standards). (d) Zone A microcellular region of an RSP Al–Fe alloy showing an average cell or grain size of 0.02 μm (courtesy C. M. Adam).

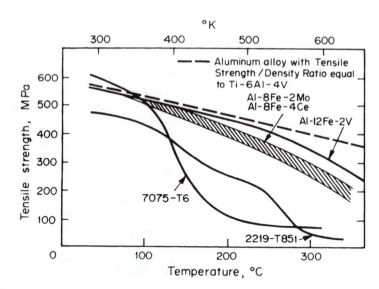

Fig. 7.18 Elevated temperature properties of RSP alloys compared with ingot alloys 7075 and 2219. The specific strength needed to equal the titanium alloy Ti–6Al–4V is shown as a dotted line (Adam, C. M. and Lewis, R. E., *Rapidly Solidified Crystalline Alloys*, Das, S. K. *et al.* (Eds), AIME, Warrendale, Pa, 157, 1985).

RSP powder compacts produced by melt spinning cost some US$100 per kg when they were first produced. These costs have been reduced but apparently remain well above an estimate of $US15 per kg that may be possible with facilities for large-scale production. Early developments were concentrated on aerospace components such as blading and vanes for compressor sections of gas turbines, and forged aircraft landing wheels. In this latter case, it was thought that the potential of the RSP alloys to withstand higher temperatures could allow the use of brakes which generate higher frictional forces. However, commercial interest in these RSP alloys has declined because of two major problems. One is their low levels of fracture toughness and the second is that they show progressive decreases in ductility when stressed at slow strain rates at elevated temperatures (e.g. 250 °C). These effects are greatest in alloys with fine microstructures and may arise from enhanced diffusion of solute atoms along grain and sub-grain boundaries.

The fact that RSP extends the solid solubility of alloying elements in aluminium has also led to a study of experimental compositions containing excess amounts of normally soluble metals such as zinc and magnesium. For example, T.S. exceeding 800 MPa with elongations of around 4% have been achieved after consolidating atomized powders made from an alloy Al–10Zn–3Mg–2Cu–1.7Mn–0.2Cr. This compares with typical values of 570 MPa and 11% respectively for the alloy 7075–T6 (Al–5.6Zn–2.5Mg–1.6Cu–0.2Cr) prepared from ingots. It should be noted, however, that the experimental RSP alloys may have the disadvantage of being more quench sensitive if heat treatment is required (Section 3.1.6). This follows because, during quenching, loss of vacancies and enhanced heterogeneous nucleation of precipitates may occur at the very many sites provided by the finely dispersed compounds and much greater number of grain boundaries in these materials.

Experimental RSP alloys have been produced that are based on the aluminium-lithium system. These alloys incorporate higher lithium contents than ingot alloys and this offers further potential for reducing density and increasing stiffness (Section 3.4.6). One example is the composition Al–3.5Li–1Cu–0.5Mg–0.5Zr for which data has been provided by the former Allied-Signal Inc. in the United States. This alloy has a density of 2.47 g cm^{-3} and an elastic modulus of 80.6 GPa. Typical mechanical properties of extrusions given a T6 heat treatment are as follows:

0.2% proof stress	455 MPa
tensile strength	595 MPa
elongation	8.8%
fracture toughness	25 MPa m$^{1/2}$

Corrosion rates, measured as weight loss over a period of 70 days in a salt fog test, are only some 15% of that recorded for both the conventional aircraft aluminium alloy 2014–T6, and the powder metallurgy alloys 7090 and 7091. In addition, the rate of growth of fatigue cracks in the RSP alloy is significantly less (Fig. 7.19).

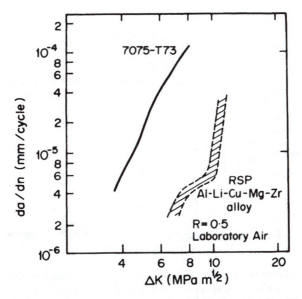

Fig. 7.19 Fatigue crack growth rates for the extruded RSP alloy Al–3.5Li–1Cu–0.5Mg–0.5Zr aged to the T6 condition compared with extruded 7075–T73 (courtesy former Allied-Signal Inc).

7.3.2 Magnesium alloys

Billets of a number of magnesium alloys have also been produced by rapid solidification processing as thin ribbons which are mechanically comminuted into powder, sealed in cans and extruded to form bars in the manner described above. One commercial alloy has been available which is known as EA55RS (Mg–5Al–5Zn–5Nd) and may develop tensile strengths exceeding 500 MPa. Some compositions have shown enhanced creep resistance but others in fact undergo accelerated deformation due to enhanced grain boundary sliding in the fine-grained microstructures. However, this behaviour can make some RSP magnesium alloys amenable to superplastic forming at temperatures as low as 150 °C. Even higher tensile strengths combined with a ductilities of 5% have been achieved with small (6 mm diameter) extruded rods prepared from helium gas atomised powders of the alloy $Mg_{97}Zn_1Y_2$. The α-magnesium phase of these rods has fine grain sizes of 100 to 150 nm, and is hardened by cubic $Mg_{24}Y_5$ particles of around 10 nm.

As with other RSP alloys, the corrosion resistance of magnesium alloys can be notably improved because microstructures are more homogeneous with respect to particulates that normally act as cathodic centres (Section 5.7). Moreover, the extended solubility of various elements may shift the electrode potentials of the alloys to more noble values. Corrosion rates of RSP magnesium alloys compared with conventionally produced compositions are shown in Fig. 7.20.

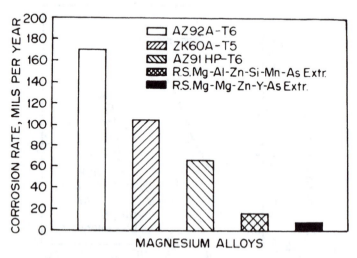

Fig. 7.20 Corrosion rates of rapidly solidified magnesium alloys tested in 3% NaCl at 21 °C, as compared with some commercial alloys (from Das, S. K. and Chang, C. F., *Rapidly Solidified Alloys*, Das, S. K., Kear, B. H., and Adam, C. M. (Eds), AIME, Warrendale, Pa, 137, 1985).

7.3.3 Titanium alloys

RSP has been successfully applied to a number of titanium alloy systems using techniques similar to those described to aluminium and magnesium alloys. Solubility limits can be extended and microstructures refined. However, the fact that relatively high temperatures are required to consolidate powders or melt spun ribbons presents difficulties because the ultrafine, rapidly solidified structures tend to coarsen leading to reduced mechanical properties.

The microstructures of rapidly quenched titanium alloys can be significantly more complicated than for aluminium or magnesium alloys because of the β/α allotropic transformation and the occurrence of martensitic and other phase transformations. Special attention has been given to using RSP as a means to improve elevated temperature performance in the following ways:

1. Whereas dispersion-hardened alloys generally display good strength and creep resistance at elevated temperatures, attempts to exploit such microstructures with titanium alloys have generally been hampered by insufficient supersaturation of alloying elements in the as-quenched condition and by rapid coarsening of precipitates. Stable oxides may be formed if rare earth elements are added to titanium and its alloys because they scavenge dissolved oxygen. However, these elements are normally ineffective because of their low solubilities under equilibrium conditions. RSP extends these solubilities and an extensive study has been made of erbium additions for which supersaturations of up to 1 at.% have been achieved. Uniform dispersions of Er_2O_3 particles in the size range 5–25 nm have been obtained by ageing at comparatively high temperatures (e.g. 700 °C). Dysprosium, gadolinium, lathanium and yttrium also form stable oxides and, in α-titanium alloys, all are reported to be resistant to coarsening at 800 °C. Coarsening is significant, however, in β-phase alloys apparently due to increased diffusivity.

2. Metalloid phases have also been formed in RSP alloys to improve creep strength. Volume fractions of the precipitate Ti_5Si_3, which is present in some creep-resistant near-α alloys (Section 6.2.3), may be significantly increased. Experimental Ti–B alloys have also been produced containing as much as 10 at.% boron which have the additional advantage of reduced density. Boron additions have also been made to alloys such as Ti–6Al–4V produced by RSP.

3. In Chapter 6, brief reference was made to alloying elements such as iron, chromium and nickel that form eutectoid systems with titanium. Such reactions are not exploited in conventionally produced alloys because they are sluggish and the microstructures are normally heavily segregated. RSP minimizes segregation and some experimental alloys such as Ti–6Al–3Ni show relatively high tensile strengths at room temperature (e.g. 1000 MPa).

However, because most eutectoid reactions occur below 900 °C, any potential applications of these alloys will be confined to intermediate temperatures or lower.
4. Titanium combines with aluminium to form several titanium aluminide intermetallic compounds. These are considered later in Section 7.9 in which reference is made to experimental studies involving RSP techniques.

7.4 QUASICRYSTALS

Solids have generally been classified as being either crystals or glasses. The essential feature of crystals is that they contain a periodic arrangement of identical unit cells, the centres of which are always equi-distant from each other. One consequence of such periodicity is the fact that it is only possible to have two-, three-, four- and six-fold rotational symmetry. In 1984, Schechman in Israel, reported that a metallic solid $Al_{86}Mn_{14}$ exhibited five-fold rotational symmetry that is forbidden with normal crystals. Whereas a crystal is said to have periodic translational order, a quasicrystal has quasiperiodic translational order. Such an arrangement allows the spacing between unit cells to be different.

Quasicrystals have since been found in a number of aluminium alloys and Fig. 7.21 shows a single grain formed in an Al–Cu–Fe alloy that was arc melted and annealed for 48 hours at 840 °C. Such quasicrystals commonly have pentagonal facets because growth is favoured along planes of atoms having five-fold rotational symmetry. They have also been found to have the symmetry of a tetrahedron, cube or prism. Since the transformation of a liquid to a quasicrystal has been shown to proceed by nucleation and growth, the undercooling that occurs during rapid solidification often leads to ultrafine grain sizes.

Fig. 7.21 Scanning electron microscope image of an icosahedral single-grain quasicrystal of the aluminium alloy $Al_{65}Cu_{20}Fe_{15}$ (from Lutz, D., Mat. Technology, 11, No. 5, 195, 1996).

Table 7.4 Mechanical properties of powder compacts produced from aluminium alloys hardened by quasicrystalline particles (from Inoue, A. and Kimura, H.M., *Mater. Sci. and Eng.*, **A286**, 1, 2000)

Type	Alloy System	Mechanical Properties
High strength	Al–Cr–Ce–M Al–Mn–Ce	Tensile strength 600–800 MPa Elongation 5–10%
High ductility	Al–Mn–Cu–M Al–Cr–Cu–M	Tensile strength 500–600 MPa Elongation 12–30%
High elevated temperature strength	Al–Fe–Cr–Ti	Tensile strength 350 MPa at 300 °C

M = elements such as Ti, Co, Ni, Mo

A series of experimental aluminium alloys with a range of interesting mechanical properties have been developed from extruded compacts of atomised powders in which nanoscale quasicrystalline particles (30–50 nm) solidfy first from the melt as the primary phase surrounded by thin films of an α-aluminium matrix. These quasicrystals are present in volume fractions of 60–70%. The alloys contain a range of transiton metals and may be divided into three types which are summarized in Table 7.4

Quasicrystals have also been observed in as cast magnesium alloys containing rare earth elements and in some rapidly solidified compounds, such as $Mg_{32}(Al,Zn)_{49}$ and Mg_4CuAl_6, that can form as precipitates in aluminium alloys. An as-cast and hot rolled Mg–Zn–Y alloy with the composition $Mg_{95}Zn_{4.3}Y_{0.7}$ has been found to have a microstructure consisting of an α-Mg matrix hardened by a high volume fraction of fine, stable quasicrystalline particles. The alloy shows good stability at temperatures up to 200 °C. A quasicrystalline titanium alloy, $Ti_{45}Zr_{38}Ni_{17}$, has shown promise for the reversible storage of hydrogen gas.

Quasicrystals themselves tend to be brittle but have some interesting mechanical, thermal and chemical properties. Some are very hard (up to 10 GPa) and offer the prospect of being used in wear-resistant coatings. They also have a low coefficient of friction comparable to Teflon and have been proposed for non-stick coatings for cooking utensils where they would have the advantage of being scratch-resistant. They have remarkably low thermal conductivities. As an example, the thermal conductivity of the quasicrystalline form of one alloy that contains as much as 70% aluminium is less than 1 percent that of aluminium metal at room temperature. Nano-sized quasicrystals of an Al–Pd alloy have been reported to be more efficient than pure palladium when used as a catalyst for cracking methanol.

7.5 AMORPHOUS ALLOYS

It is well known that the mechanical strength of alloys that can be produced with amorphous (or glassy) atomic structures by liquid quenching techniques, such as melt spinning (Fig. 7.16), may be notably greater than those obtained for the normal crystalline state. One prerequisite for this behaviour is a large negative enthalpy of mixing of constituent elements which is a feature of certain aluminium and magnesium alloys containing both rare earth and transition metal elements. Moreover, the rare earth elements have a larger atomic size than either aluminium or magnesium, whereas the atoms of transition metals such as copper and nickel are smaller, suggesting that localized strain energy will be reduced if the three types of atoms cluster together. These two factors are presumed to reduce overall atomic diffusivity during cooling from the molten state, thereby retarding nucleation of crystalline phases. Composition limits for some binary and ternary aluminium and magnesium alloys that are able to form amorphous structures in the rapidly quenched condition are shown in Fig. 7.22.

The presence of an amorphous phase in aluminium alloys was first observed in rapidly solidified binary systems with metalloid (e.g. silicon) and transition metal (e.g. copper) elements in which there was coexistence with a crystalline phase. The formation of completely amorphous, single-phase, aluminium-based alloys was first achieved in Japan in the Al–Fe–B and Al–Co–B systems. These materials proved to be extremely brittle and, more recently, the Japanese group has obtained good bending ductility combined with T.S. exceeding 1000 MPa in ternary alloys such as Al–7La–5Ni(at.%). This compares with values of 500–600 MPa for strong, conventionally produced wrought alloys (Table 3.5). Values of hardness (e.g. 300 DPN) and elastic modulus (e.g. 90 GPa) are also

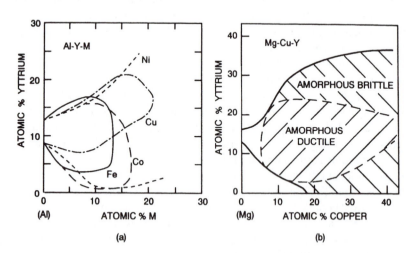

Fig. 7.22 (a) Composition loops for binary and ternary aluminium alloys that may form amorphous structures if rapidly quenched from the liquid state (from Inoue, A. and Masumoto, T., *Encyclopaedia of Materials Science & Engineering* (Second Supplementary Volume), Cahn, R. W. (Ed), Pergamon, Oxford, 1990). (b) Mg–Cu–Y phase diagram showing compositions of ductile and brittle amorphous alloys (from Kim, S. G. *et al., Mater. Trans. Japan Inst. Met.,* **31**, 929, 1990).

significantly higher for the amorphous alloy. Another advantage is that the coefficient of thermal expansion of amorphous alloys is generally lower than values for conventional crystalline alloys.

Bulk specimens of the amorphous alloys can be prepared by extruding pressed, atomized powders and some compositions have now shown some capacity to deform plastically (e.g. elongations of 1–2%). The alloys commonly undergo an amorphous to crystalline transition on heating to temperatures in the range 250–350 °C. Even higher mechanical properties have been obtained from ribbons made from aluminium alloys in which partial crystallization has been encouraged by decreasing cooling rates during melt spinning. Fine crystalline precipitates with sizes as small as 3–4 nm are formed which consist of fcc α-aluminium saturated with solute elements. One such alloy, Al–8Ni–2Y–2Mn(at.%), which contains these particles within an amorphous matrix has recorded a tensile fracture strength as high as 1470 MPa.

Amorphous structures have also been obtained with a range of magnesium alloys, most of which are also ternary compositions containing rare earth and transition metal elements (e.g. Fig. 7.22b). Melt spun ribbons of Mg–Ni–La alloys have shown values of T.S., elastic modulus and hardness in the ranges 610–850 MPa, 40–60 GPa and 190–230 DPN respectively which greatly exceed maximum values of around 300 MPa, 45 GPa and 85 DPN for the strongest conventionally cast magensium alloys. Most compositions show good bending ductility, although tensile fracture strains (including elastic strains) lie in the range 0.014–0.018 indicating little or no capacity for plastic deformation. Again, even higher values of T.S. can be obtained in partially crystallized alloys and some compositions, e.g. Mg–12Zn–3Ce (at.%), show some capacity for plastic deformation.

Amorphous magnesium alloys also show good thermal stability (i.e. crystallization temperatures >300 °C) and some compositions can retain amorphous structures at slower cooling rates than those required aluminium alloys. This has opened up the prospect of obtaining these structures in bulk castings, as well as ribbons, and the Japanese group has succeeded in retaining amorphous structures in chill cast cylinders prepared by pressure injecting molten alloys into a copper mould. Amorphous structures have been confirmed in 2 mm dia. cast bars for the alloy Mg–10Cu–10Y (at.%) and in up to 7 mm dia. bars in Mg–25Cu–10Y (at.%) (Fig. 7.23). Mechanical properties have been found to be similar to those obtained for more rapidly cooled, melt-spun ribbons even though cooling rates for casting in the copper mould are as low as 10^2 °C s^{-1}. This behaviour suggests that amorphous structures may be obtained in industrially produced, thin-walled magnesium alloy castings which would show much higher strength and wear resistance. However, practical applications of these materials have yet to be exploited.

Many experimental titanium alloys have been obtained in the amorphous condition by rapid solidification techniques. They may be classified into metal–metal and metal–metalloid systems; examples are Ti–40Be, Ti–30Co,

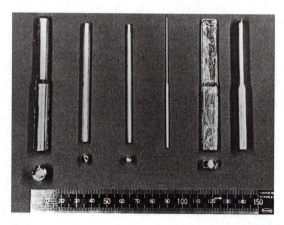

Fig. 7.23 Injection chill cast magnesium alloy products showing amorphous structures (courtesy A. Inoue).

Ti–25Al–25Ni and Ti–35Be–5Si, Ti–30Co–10B, Ti–40Ni–10P (all at.%) respectively. The metalloid systems, in particular, are characterized by very high values of hardness and strength while retaining significant bend ductility almost up to the crystallization temperature. Some properties are summarized in Table 7.5 and may be compared with values for conventional titanium alloys shown in Table 6.1.

More recently, attention has also been focused on titanium alloys that may retain their amorphous structures in bulk form. Japanese workers have studied alloys based on the eutectic composition $Ti_{50}Cu_{42.5}Ni_{7.5}$ to which were added elements having atomic radii larger or smaller than titanium. The addition of the larger element zirconium was found to promote formation of a new eutectic, and the critical diameter of a rod capable of retaining its amorphous structure was increased from 200 μm to 1.2 mm. Small amounts of hafnium serve further to enhance the stability of the rapidly solidified liquid and move the modified composition closer to the modified eutectic point. However, the largest effect was recorded when the smaller element silicon was added, and the complex alloy $Ti_{41.5}Cu_{42.4}Zr_{2.5}Hf_5Si_1$ remained amorphous when chill cast in a copper block as a 5 mm diameter rod. The alloy has a tensile strength of 2040 MPa and a compressive strength of 2080 MPa No plastic deformation was recorded in tension but some slight ductility was evident in compression. The glass transition temperature was found to be 407 °C.

One special feature of some amorphous titanium alloys is that they exhibit superconducting behaviour if cooled to sufficiently low temperatures (<10 °K). This applies to both the glassy and crystalline states. Examples are Ti–40Nb–15Si and Ti–40Nb–12Si–3B (at.%).

Table 7.5 Mechanical and thermal properties of amorphous titanium metalloid alloys (from Suryanarayana *et al., Int. Mater. Rev.*, **36**(3), 85, 1991)

Metalloid alloy (at. %)	Hardness (DPN)	Yield strength (MPa)	Density (g cm^{-3})	Crystallization Temperature (°C)
Ti–15Si	510	1960	4.1	429
Ti–20Co–10Si	570	2105	5.0	495
Ti–20Fe–10Si	580	2150	4.8	549
Ti–35Be–5Si	805	2490	3.9	462

7.6 MECHANICAL ALLOYING

Mechanical alloying was devised by Benjamin at the International Nickel Company to introduce hard particles, such as oxides and carbides, into a metallic matrix on a scale that is much finer than can be achieved by conventional powder metallurgy practices. The process involves the high-speed attrition of dry, elemental or simple alloy powders in modified, high energy ball mills. During milling, ball–powder–ball and ball–powder–container collisions occur which repeatedly deform, cold weld and fracture the powder particles (Fig. 7.24). The interaction between particle fracture and welding, combined with strain-enhancing diffusion, progressively homogenizes the powders, resulting eventually in alloy formation. Extremely fine grain sizes (<1 μm) can be achieved and solid solubilities may be extended beyond their equilibrium values, although not to the extent described above for rapid solidification processing.

For aluminium alloys, fine dispersions of Al_2O_3 are introduced from the existing oxide films on the powders whereas the carbide, Al_4C_3, is formed following the breakdown of organic surface reagents, such as stearic acid, that are added to minimize cold welding during processing. The final powder is consolidated by pressing and vacuum degassing and billets weighing as much as several hundred kilograms have been prepared. Sections or components can then be produced by hot extrusion, forging or isostatic pressing.

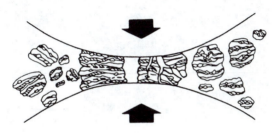

Fig. 7.24 Representation of deforming, welding and fracturing of powders by high-energy ball milling during mechanical alloying (courtesy F. H. Froes).

A special type of milling, known as cryomilling, has been used to produce very fine nanocrystalline powders. This process involves introducing liquid nitrogen during milling and was first used with aluminium and dilute aluminium alloys strengthened by fine aluminium oxy-nitride particles. The advantages of cryomilling include reduced oxygen contamination from the atmosphere, and faster heat transfer between particles and the cryogenic medium which favours particle fracturing rather than welding when ductile materials are being milled. However, if the methods used for the subsequent consolidation of powders involve elevated temperatures, it is difficult to retain the initial nanocrystalline structure because grain growth may occur.

Two aluminium alloys developed by Inco Alloys International that are available for use in structural applications are designated IN9052 (Al–4Mg–0.80–1.1C), IN9021 (Al–4Cu–1.5Mg–0.8O–1.1C). Each may have a fine grain size of 0.5 μm or less that is stabilized by the presence of oxides and carbides commonly in the size range 30–50 nm (Fig. 7.25). The first relies on solid solution and dispersion strengthening (e.g. T.S. at room temperature 550 MPa, elongation 8%) whereas IN9021 combines precipitation hardening with dispersion strengthening (e.g. T.S. 600 MPa, elongation 11%). Because of the fine microstructures, mechanical properties in the longitudinal and transverse directions are more isotropic than those found with conventionally fabricated wrought sections (Section 2.5) which is a significant advantage. Both alloys also show exceptional resistance to general corrosion, pitting and exfoliation attack, as well as to stress-corrosion cracking.

Experimental lithium-containing alloys (e.g. Al–4Mg–1.3Li–0.4O–1.1C) have also been prepared by mechanical alloying to take advantage of lower density and higher elastic modulus. One such alloy, designated AL 905XL, has been specified for undercarriage forgings for the European EH101 helicopter

Fig. 7.25 Transmission electron micrograph showing fine-grain structure of IN9052 produced by mechanical alloying (courtesy J. Weber)

(Fig. 3.41). At elevated temperatures improved performance has been obtained with mechnically alloyed Al–Ti alloys containing 6–12% titanium. Fine particles of stable Al_3Ti are formed which provide dispersion strengthening and values of elastic modulus in the range 85–100 GPa have been recorded which are comparable with those found in metal matrix composites (Section 7.1.3).

The application of mechanical alloying to the preparation of magnesium and titanium alloys is less well advanced. Magnesium alloys can present problems because they are soft and tend to adhere to the balls and container. Nevertheless the process has been used to produce "supercorroding" magnesium alloys that operate effectively as short-circuited galvanic cells with the result that they will corrode (react) rapidly with an electrolyte, such as sea water, to produce heat and hydrogen gas. Such an alloy system is suitable as a heat source for warming deep sea divers, as well as a gas generator to provide them with buoyancy. It has also been tried as a source of fuel in hydrogen engines or fuel cells. Corrosion rates are maximized because the finely dispersed microstructure formed by mechanical alloying provides very short electrolytic paths, exposes a large surface area, and makes a strong connection (weld) between cathodes and anodes. As one example, extremely fast reaction rates and high power outputs have been achieved with a range of Mg–12% to 15% Fe alloys. Several hydrides based on mechanically alloyed magnesium alloys are being evaluated for the safe storage of hydrogen because they have the potential to contain a higher volume density than is present in liquid hydrogen.

Another novel development has been an attempt to prepare titanium and certain alloys, such as Ti–6Al–4V, at ambient temperatures by the direct reduction of $TiCl_4$ with magnesium during grinding. Normally this reduction process is carried out in an inert atmosphere at temperatures around 1000 °C (Chapter 1). Some attention has been given to the synthesis of the titanium aluminides Ti_3Al and TiAl by mechanical alloying since attempts to prepare these intermetallics by ingot or powder metallurgy routes have met with limited success (Section 7.9). High-energy grinding has been found to introduce some interesting metastable amorphous phases as well as nanocrystalline regions (Section 7.8) but this work is also at an early stage.

7.7 PHYSICAL VAPOUR DEPOSITION

Physical vapour deposition (PVD) involves the high-temperature evaporation of metals and other elements and their re-deposition on to a suitable substrate. The technique has been used mainly to produce thin films and coatings, one example being the deposition of titanium nitride to improve the wear resistance of steel tools. Now the development of high-energy-rate processes involving, for example, the use of intense electron beams, is enabling PVD to be applied to produce alloys in bulk. Individual elements can be evaporated and then co-deposited to give new compositions and microstructures having extremely fine grain sizes, extended solubilities and freedom from segregation.

As one example, an experimental aluminium alloy RAE72, containing the normally insoluble elements chromium (7.5%) and iron (1.2%) has been produced in England in slab form using PVD. The alloy vapour was deposited on a 500×300 mm collector plate at a rate of 6 mm/h to a thickness of 44 mm and then warm rolled to sheet. Elevated temperature tensile results show that the alloy has a specific strength exceeding that of titanium alloys at temperatures up to 300 °C. The high strength of the alloy is attributed to a combination of solid solution strengthening by chromium, precipitation of fine particles of $AlFe_3$ and the very fine grain size of the matrix.

Another use of PVD to prepare compositions that cannot be produced by ingot metallurgy has been the alloying of titanium with the comparatively volatile metal magnesium which, in fact, boils below the melting point of titanium. Ti–Mg alloys have been found to respond to age hardening and the precipitates that form appear to be exceptionally stable. Magnesium also has the advantage of reducing the density of titanium by more than 1% for each 1 wt. % added.

7.8 NANOPHASE ALLOYS

Nanostructuresd materials are part of a rapidly expanding field referred to as nanotechnology and they have been identified as a key to the development of many new industrial processes and products. In its original context, the name applies to materials with grain sizes of 10 nm or less, the volume occupied by grain boundary structure can be 30% or more so that the density of defects in the material is abnormally high (Fig. 7.26). In effect, a nanophase material can be seen as a composite containing a mixture of crystalline and amorphous regions. As a consequence, the characteristic bulk behaviour of a material may be significantly changed leading to modified, and often enhanced, physical or mechanical properties.

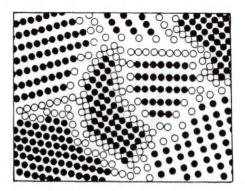

Fig. 7.26 Proposed atomic structure of a nanocrystalline material. Black circles represent atoms in normal lattice positions within grains and white circles indicate atoms that are associated with grain boundaries (courtesy H. Gleiter).

Nanophase materials were detected in samples of lunar soils but terrestrial evidence for their existence was lacking until the introduction of the new processing techniques, mechanical alloying, consolidation of amorphous powders and physical vapour deposition. So far, nanophase materials have only been produced in small quantities so that interest in them is still at an experimental stage.

Because grain size is so fine, it is to be expected that nanophase materials may be susceptible to creep at elevated temperature due to the opportunity for grain boundary sliding, and this does appear to be one of their characteristics. However, a desirable complementary feature should be the ability to undergo superplastic deformation providing their high-energy microstructures remain stable at the appropriate temperatures. Such behaviour has been observed in nano- and near-nano scale aluminium alloys produced by mechanical alloying (e.g. IN9052 (Section 7.6)) and by consolidating rapidly solidified amorphous powders (Section 7.3). These materials differ from commercial superplastic alloys such as 7475 (Section 3.6.8) in that the abnormal elongations are achieved at much higher strain rates. For example, whereas maximum superplasticity is observed in sheet made from the conventionally produced aluminium alloy 7475 (average grain size 15 μm) deformed at a slow strain rate of approximately 10^{-4}s^{-1}, the equivalent strain rate for mechanically alloyed IN9021 (average grain size 0.5 μm) may be as high as $10-10^2 \text{ s}^{-1}$. This increase in strain rate of as much as one million times opens up the interesting prospect of achieving superplasticity under impact conditions in nanophase materials.

Much attention is now being directed to obtaining ultrafine grain sizes in bulk materials and one method for achieving this is through processes involving severe plastic deformation such as equal channel angular pressing (ECAP), which is also known as equal channel angular extrusion (ECAE). This process involves pressing (extruding) a metallic billet through a die containing two channels of equal cross-section that intersect at an angle φ (Fig. 7.27a).

To date, ECAP has been carried out at ambient or a moderately elevated temperatures, and at deformation rates substantially faster than most conventional metal working processes. Under these conditions, the billet experiences simple shear and, if back pressure is applied, a hydrostatic pressure component is introduced at the intersection of the two channels. Since the dimensions of the pressed material do not change during processing, this process can be repeated several times. Progressively higher levels of strain are accumulated within the material and the total strain ε_N that is achieved in a series if pressings N and a channel intersection angle φ is given by

$$\varepsilon_N = \frac{2N}{\sqrt{3}} \cot \left(\frac{\phi}{2}\right)$$

If the pressed billet is rotated around its longitudinal axis after each pass, the microstructure and texture can be modified to achieve specific engineering purposes. During the early stages of deformation, the relatively coarse grains of

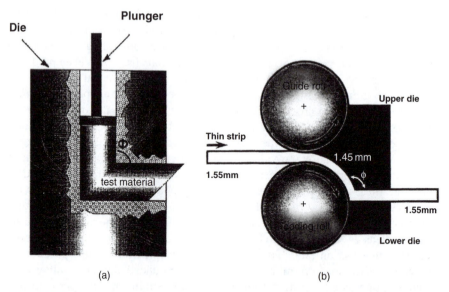

(a)   (b)

IFig. 7.27 Schematic diagrams of equal channel angular pressing (ECAP) involving (a) extrusion (Figure I from T.C. Lowe and R.Z. Valiev, *Producing Nanoscale Microstructures Through Severe Plastic Deformation, JOM,* **52** (4), 27, 2000) and (b) the concept of continuous confined strip shearing (CCSS) of strip cast sheet (from Lee, J.-C. *et al, Metall. and Mater. Trans A,* **33A**, 665, 2002).

the original billet are formed into arrays of subgrains that fragment on further straining to form much finer grains separated by what are potentially high angle boundaries. These ultrafine grains (1 μm or less) contain relatively few dislocations and therefore have the appearance of an annealed structure but with a very much finer grain size (Fig. 7.28.)

ECAP has been applied successfully on a laboratory scale to a number of commercial aluminium alloys including 3004, 5052, 5083, 6061 and 6016. One outcome has been significant increases in strength properties. For example, the solid solution hardened alloy 5052 (Al–2.5Mg–0.25Cr) typically has a 0.2% proof stress of 255 MPa and a tensile strength of 270 MPa in the H38 strain hardened condition. After processing by ECAP to an equivalent true stain of 8, these values may be as high as 395 and 420 MPa respectively, without loss of ductility. Similarly, higher strengths have been achieved in age hardened alloys such as 6061. These alloys also show promise for superplastic forming at higher rates than are possible with conventionally produced alloys (Section 3.6.8).

The industrial challenge is to convert ECAP from a batch to a more cost effective continuous process. One promising development has occurred in South Korea which has been termed "continuous confined strip shearing" (CCSS). This process involves passing continuously strip cast sheet between rolls

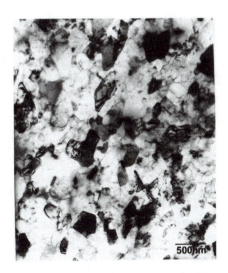

Fig. 7.28 Transmission electron micrograph showing an ultrafine grain structure in the Al–Mg–Si alloy 6016 that was solution treated at 560 °C for 1 h, furnace cooled and deformed to an equivalent true strain of ~14 (i.e. 12 passes) by ECAP (courtesy P. McKenzie). The grain size has been reduced from an original average value of ~190 μm and final value of ~0.3 μm.

that then force it at high speed through a thinner, curved ECAP channel as shown schematically in Fig. 7.27(b). Once the strip experiences shear through this channel, it expands to its original thickness and exits via an outer channel. This process offers the prospect of significantly raising the strength of strip cast alloys for which, normally, the opportunity for doing this by cold rolling is limited.

Magnesium alloys are also good candidates for ECAP because of their limited capacity for forming by conventional methods. Experiments with alloys such as AZ31 and ZK31 have confirmed that ECAP produces a fine and uniform microstructure, as well as textures in which the basal planes of the lattice tend to become aligned parallel to the extrusion direction. Because the critical resolved shear stress is low along these planes, ECAP processing results high ductilities in this direction. Tensile properties increase as the ECAP temperature is reduced. Some prototype components such as automotive knuckle arms, which have been forged from ECAP-processed magnesium alloy feedstock, have shown high resistance to fracture under impact conditions.

Small quantities of some nanophase titanium alloys including Ti–Cu and several titanium aluminides (Section 7.9) have been produced by mechanical alloying. Little information is available concerning mechanical properties although it has been reported that the nano-phase γ-TiAl intermetallic compound may have double the hardness of the equivalent as-cast ingot at room temperature. Softening occurs above 200 °C and the hardnesses of the two materials become similar at 300 °C.

7.9 TITANIUM ALUMINIDES

As discussed in Chapters 1 and 6, conventional titanium alloys cannot withstand prolonged exposure to air at temperatures above approximately 600 °C, mainly because oxidation causes surface scaling and internal embrittlement. Moreover, despite their high melting points, the best of these alloys undergo excessive creep at or above this temperature. There are, however, ordered titanium–aluminium intermetallic compounds that have the potential to be used for components such as turbine blades operating at much higher temperatures. These compounds are now seen as the probable next stage in the development of ingot titanium alloys.

The Ti–Al phase diagram is shown in Fig. 7.29 and the titanium aluminides of interest are Ti$_3$Al (known as α_2) and TiAl(γ), both of which exist over a range of compositions, together with the stoichemetric compound TiAl$_3$. Physical properties of each of these compounds are summarized in Table 7.6 and crystal structures are shown in Fig. 7.30.

Immediate advantages of these aluminides are improved levels of specific stiffness arising from desirable combinations of lower densities, which are approximately half those of the nickel-based superalloys, and higher values of elastic modulus than normal titanium alloys. Moreover, resistance to oxidation improves progressively as the aluminium content is raised. The ordered crystal structures also promote enhanced creep resistance because the strong bonding between the two different atoms in the superlattices restricts both dislocation

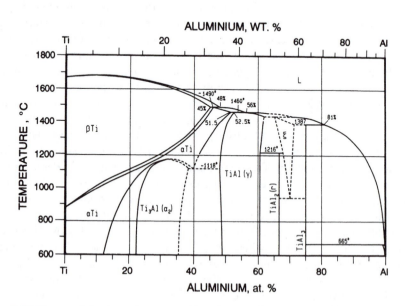

Fig. 7.29 Proposed titanium–aluminium phase diagram (from Massalski, T. B. (Ed.), *Binary Alloy Phase Diagrams*, 2nd Edn, Vol. I, ASM International, Materials Park, Ohio, 1990).

Table 7.6 Physical properties of titanium aluminides (from Froes, F. H. *et al.*, *J. Mater. Sci.*, **27**, 5113, 1992)

Compound	Crystal structure	Lattice parameters (nm)	Melting point (°C)	Density (g cm^{-3})	Elastic modulus (GPa)
Ti$_3$Al	DO$_{19}$ ordered hexagonal	$a = 0.5782$ $c = 0.4629$	1600	4.3	145
TiAl	L1$_0$ ordered f.c. tetragonal	$a = 0.4005$ $c = 0.4070$	1460	3.9	175
TiAl$_3$	DO$_{22}$ ordered tetragonal	$a = 0.3840$ $c = 0.8596$	1340	3.4	200

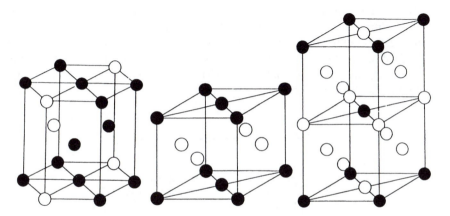

Fig. 7.30 Crystal structures of titanium aluminides. (a) Ti$_3$Al, (b) TiAl, (c) TiAl$_3$. Black balls Ti atoms, white balls Al atoms.

motion and atomic diffusion at elevated temperatures. However, these particular features of intermetallic compounds, together with a limited capacity to undergo slip, result in low values of ambient temperature ductility. This situation for titanium aluminides is exacerbated by the presence of interstitial impurity elements such as oxygen and hydrogen.

Titanium aluminides also present formidable problems when preparing castings. Firstly, the molten alloys are difficult to contain because they react chemically with virtually all ceramic refractories. It has therefore been necessary to use cold-wall (internally water cooled) furnaces which limits the ability to superheat to about 60 °C, and presents difficulties with metal flow when pouring into moulds. Furthermore, because considerable vaporisation of aluminium occurs if melting is carried out in vacuum, it is necessary to introduce an inert gas such as argon which can be entrapped in the castings as they tend to solidify rapidly. The reactive nature of the titanium aluminides also presents difficulties in developing suitable materials for moulds when producing cast components. For wrought products, formability at elevated temperatures is restricted and much attention has been directed at improvements through alloying and control of microstructure. Each of the above problems have delayed the wider introduction of the titanium aluminides.

7.9.1 Ti₃Al(α₂)

Compositions based on the compound Ti_3Al were the first titanium aluminides to be studied in detail. As shown in Fig. 7.31, they may display outstanding strength:weight ratios when compared with conventional titanium alloys, although improvements in creep strength have been relatively small (Fig. 6.2) and the upper limit for operation in air has been little changed.

Ti_3Al undergoes an order/disorder transition within the composition range 22–39% aluminium to form an ordered DO_{19} hexagonal structure. Low ductility at room temperature is attributed to a coplanar mode of slip (Fig. 2.22a) and the lack of sufficient slip systems parallel, or inclined, to the hexagonal direction of the unit cell. Moreover, in contrast to most other hexagonal metals and alloys, Ti_3Al does not undergo deformation by twinning.

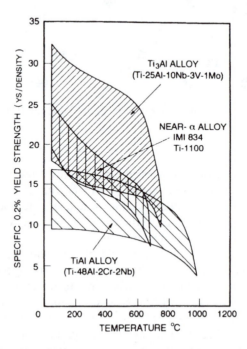

Fig. 7.31 Specific yield strength values of titanium aluminides and the conventional near-α alloys IMI 834 and Ti1100 (from Kumpfert, J. and Ward, C. H., in *Advanced Aerospace Materials*, Buhl, H. (Ed.), Springer Verlag, Berlin, 1992).

Both the microstructures and mechanical properties of Ti_3Al tend to follow the behaviour of more conventional titanium alloys. Thus thermomechanical processing can occur in the β- or $α_2$-β-phase regions (Fig. 7.27) as was described for the α/β-alloys in Section 6.3. Similarly, Ti_3Al alloys quenched from the β-phase field undergo a variety of martensitic transformations. Additions of β-stabilizing elements such as niobium, molybdenum and vanadium to Ti_3Al also promote formation of a ductile ordered body-centred cubic phase (B2) in which ω and ω-related phases may form on cooling to room temperature. In summary, transformed β microstructures that produce a basket weave configuration of secondary Widmanstätten plates of $α_2$ (similar to Fig. 6.4d) together with the phase B2 are considered to provide the best overall combination of mechanical properties.

Niobium has become the most widely used and important addition to Ti_3Al. Niobium atoms substitute for titanium atoms in the crystal lattice which has the effect of improving low-temperature ductility by increasing the number of active slip systems. Molybdenum, in smaller quantities, behaves in the same way and provides the additional advantages of solid solution strengthening and improved creep strength. Vanadium has been used as an alternative to niobium because of its lower density although this element has the disadvantage of reducing oxidation resistance. Examples of alloys that were developed are Ti–24Al–11Nb and Ti–25Al–10Nb–3V–2Mo. Each has a microstructure containing the two phases $α_2$ and B2.

During the 1990s it was realised that the $α_2$-Ti_3Al alloys suffered severe environmentally-induced embrittlement at temperatures as low as 550 °C and development was discontinued in favour of compositions based on the orthorhombic compound Ti_2AlNb (O phase). A balance needed to be obtained between the contents of aluminium, which decreases density and improves oxidation resistance, and niobium which favours formation of the O phase but increases density and reduces oxidation resistance. Compositions close to Ti–22Al–25Nb appear to provide the best compromise. This alloy is ductile at room temperature ductility, displays good formability, high creep and fatigue strength, and has moderate oxidation resistance. It also has a coefficient of thermal expansion that is less than that for conventional titanium alloys and γ TiAl.

7.9.2 TiAl(γ)

The development of alloys based on TiAl is of more recent origin because this compound, intrinsically, has even less room temperature ductility than Ti_3Al. Tensile properties are also significantly lower (e.g. Fig. 7.31). Nevertheless γ alloys are now attracting special attention because of their lower densities combined with superior values of elastic modulus (Table 7.5), oxidation resistance and thermal stability (Fig. 6.2)

TiAl has an Ll_0 ordered face-centred tetragonal structure in which titanium and aluminium atoms form as successive layers on (002) planes. The composition may extend from 48.5 to 66 at.% aluminium although alloys of possible practical interest lie at the lower end of this range. Tetragonality (c/a ratio) is close to unity, varying from 1.01 to 1.03 for the two extremes of aluminium content, and the compound remains ordered up to its melting point of 1460 °C. Deformation at low temperatures involves slip although dislocation mobility is again severely restricted. Twinning occurs at higher temperatures and is considered to account for the increased plasticity that is observed. Fracture is predominantly by cleavage at both low and high temperatures.

γ alloys of special interest lie in the composition range Ti–46 to 52Al–1to10M (at %) where M represents at least one element from a list that includes the transition metals V, Cr, Mn, Nb, Mn, Mo and W. Both single-phase γ and two-phase (γ + α_2) alloys are possible for this range of compositions. Generally, two-phase alloys are again favoured and these usually have aluminium contents within the range 46–49 at% together with 1–4% of the individual additional elements. These latter elements can be divided into three groups:

1. Cr, V, Mn and Si, which improve ductility but reduce oxidation resistance.
2. Nb, Ta, Mo and W, which enhance oxidation resistance.
3. Si, C and N which, in relatively smaller amounts, improve creep resistance.

Examples of compositions of two-phase alloys that have been of interest are Ti–48Al–2Nb–2Cr (at.%), Ti–48Al–2V and Ti–47Al–2.5Nb–2(Cr + V). Densities lie in the range 3.9–4.1 g cm^{-3}.

TiAl alloys are particularly susceptible to segregation during solidification because of what has been described as the double cascading effect of the two peritectic reactions in this region of the phase diagram (Fig. 7.29). Accordingly special attention must be given to high-temperature homogenization treatments that are usually carried out within a comparatively narrow temperature range in the single-phase α-field prior to subsequent processing.

In addition to alloying, microstructural control is exercised by heat treatment and thermomechanical processing as with other titanium alloys. Lamellar microstructures are common (e.g. Fig. 7.32) but processing can be adjusted so that equi-axed, lamellar or duplex morphologies are obtained. The equi-axed microstructure consists entirely of γ grains in single-phase alloys, whereas, in two-phase alloys, this structure is predominantly γ grains with small amounts of grain boundary α_2 particles. A fully lamellar microstructure consists of colonies (i.e. grains) of γ plates or, in two-phase alloys, alternating plates of γ + α_2. In this regard ductility is generally improved up to 10 vol. % α_2. Duplex microstructures contain mixtures of equi-axed grains and lamellar colonies. Different mechanical properties are favoured by one of these three types of microstructures, but a two-phase, lamellar morphology is generally considered to provide the best balance so long as the grain (colony) size can be kept fine.

γ-TiAl alloys can show good workability, some tensile plasticity (1–3%) at room temperature and fracture toughness values of 10 to 35 MPam$^{-1/2}$. Current

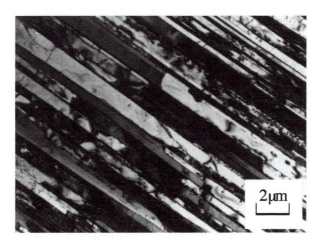

Fig. 7.32 Lamellar microstructure in a γ-TiAl alloy (courtesy C. Suryanarayana).

creep resistance limits them to an operating temperature of around 700 °C. Improvements in strength and creep resistance can be achieved through precipitation hardening from oxides, nitrides, and silicides and carbides. Special attention has been paid to the alloys containing small additions of carbon and holding at 1250 °C, quenching and ageing at 750 °C leads to precipitation of a high density of particles of the phase Ti_3AlC. The formation of precipitates tends to embrittle the γ-TiAl alloys, although this effect can be reduced by ensuring that the alloys are processed, e.g. by extrusion, to refine the microstructure before the alloys are aged.

7.9.3 TiAl₃

This compound, which has an ordered DO_{22} tetragonal crystal structure, has the highest specific stiffness and oxidation resistance of all the titanium aluminides (Table 7.6). It exhibits some compressive ductility above 620 °C but is brittle at lower temperatures at which deformation occurs solely by $(111)[11\bar{2}]$ twinning that does not disturb the DO_{22} symmetry.

TiAl₃ has received least attention of the titanium aluminides and the main strategy being followed in attempts to improve ductility is to make ternary additions of transition metals that encourage formation of the structurally related, but more symmetric, cubic Ll_2 structure. One example is Ti–65Al–10Ni(at.%) in which the aluminium and nickel atoms occupy the face-centred sites with titanium atoms at the cube corners. Ternary additions of copper, manganese, zinc, iron and chromium have also been made. However, although these ternary compounds do satisfy the von Mises criterion for plastic deformation by deforming at room temperature by slip on five independent systems of the type $\langle 110 \rangle$ $\{111\}$ at low temperatures, they are still brittle and exhibit cleavage fracture.

7.9.4 Processing methods

As mentioned earlier, melting and casting of titanium aluminides presents special difficulties. Moreover, the subsequent fabrication of ingots by forging or other methods requires the use of higher temperatures and more stages (smaller reductions) as compared with the working of conventional titanium alloys. Thus the production of near-net shape components by casting, or by powder metallurgy techniques, are potentially attractive since the working operations are avoided.

Other more innovative practices have been studied including the production of titanium aluminide powders by rapid solidification processing, although the usual cost penalties are imposed on components made in this way (Section 7.3). Rapid solidification processing offers the potential to improve the ductility through disordering of the crystal structures, grain refinement, and deoxidation of the matrix. The opportunity is also available to enhance elevated temperature strength through dispersion strengthening by the introduction of fine, thermodynamically stable, particles that was discussed in Section 7.3. One example of this latter effect is the formation of a rare earth dispersion of Er_2O_3 alloys in Ti_3Al alloys. As compared with cast ingots, improved homogeneity has been achieved in TiAl alloys which, when combined with refined microstructures, does appear to offer significantly higher ductilities than are present in the other material.

Some progress has been made in attempts to incorporate continuous alumina and silicon carbide fibres into hot-pressed foils or powders of the α_2- and γ-titanium aluminides as a means of improving fracture resistance and creep strength. The fibres must be pre-coated to prevent interfacial reactions at high temperatures and are usually prepared as woven mats. Tensile strengths as high as 1100 MPa have been achieved at 750 °C. However, cost have been as high as US$10 000 per kg and must be substantially reduced before these materials could find practical applications.

7.9.5 Applications of titanium aluminides

As mentioned earlier, potential applications of these materials reside mainly in the aerospace industry. Substantial weight savings would be possible if monolithic or composite titanium aluminides could be used to replace some nickel-based superalloys for gas turbine engine components such as discs, blades, vanes and spacer rings. In fact, it has been stated that titanium aluminide composites are the key to the achievement of the goals of a critical engine programme in the United States. Cast turbine blades and other cast components such as compressor casings made from γ-TiAl alloys have been successfully tested in gas turbine engines. Potential applications for rolled sheet include exhaust nozzles and internal engine flaps. The titanium aluminides are also seen as contenders for use in the structure of possible hypersonic transatmospheric aircraft of the future which would suffer severe aerodynamic heating during the ascent and descent stages of each flight. In the meantime, applications are likely to be confined to non-critical aerospace applications and to parts

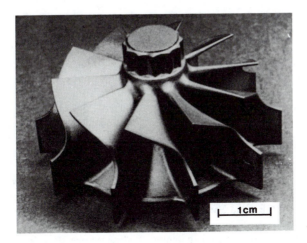

Fig. 7.33 Cast γ-TiAl turbocharger rotor (courtesy Y. Nishiyama).

for motor car engines such as the cast turbocharger rotor shown in Fig. 7.33. Providing cost reductions can be achieved, they also have the potential to be used for forged automotive engine valves as their lower density could lead to reductions in fuel consumption of 3 to 5%.

FURTHER READING

Evans, A.G., Lightweight materials and structures, *MRS Bulletin*, **26**, No. 10, 790, 2001

Vogelesang, L.B., Schijve, J., and Fredell, R., Fibre-metal laminates ; damage tolerant aerospace materials, in *Case Studies in Manufacturing with Advanced Materials*, Vol. 2 Eds, Demaid, A., and de Wit, J.H.W., Elsevier Science B.V., Holland, 1995.

Vlot, A., *Glare: history of the development of a new aircraft material*, Kluwer Academic Publications, Dordrecht, The Netherlands, 2001

Wu, G. and Yang, J.-M., The mechanical behaviour of GLARE laminates for aircraft structures, *JOM*, **57**, No. 1, 72, 2005

Allison, J.E. and Cole, G.S., Metal-matrix composites in the automotive industry : opportunities and challenges, *J. Metals*, **45**, (1), 19, 1993

Shercliff, H.R. and Ashby, M.F., Design with metal matrix composites, *Mater. Sci. and Tech.*, **10**, 443, 1994

Lloyd, D.J., Particle reinforced aluminium and magnesium metal matrix composites, *Int. Mater. Rev.*, **39**, 1, 1994

Ashby, M.F., Evans, A.G., Fleck, N.A., Gibson, L.J., Hutchinson, J.W. and Wadley, H.N.G., *Metal Foams: A Design Guide*, Butterworth Heinemann, Oxford, 2000

Banhart, J., Manufacture, characterization and application of cellular materials and metal foams, *Progress Mater. Sci.*, **46**, 559, 2001

Lavernia, E.J., Ayers, J.D., and Srivatsan, T.S., Rapid solidification processing with specific application to aluminium alloys, *Inter. Mater. Rev.,* **37**, No. 1, 1, 1992

Inoue, A., Amorphous, nanoquasicrystalline and nanocrystalline alloys in Al-based systems, *Progress Mater. Sci.*, **43**, 365, 1998

Suryanarayana, C., Froes, F.H., and Rowe, R.G., Rapid solidification processing of titanium alloys, *Inter. Mater, Rev.*, **36** (3), 85, 1991

Lieberman, H.H. Ed., *Rapidly Solidified Alloys*, Marcel Dekker, New York, 1993

Suryanarayana C., Mechanical alloying and milling, *Prog. Mater. Sci.*, **46**, 1, 2001

Inoue, A., High-strength aluminium alloys containing quasicrystalline particles, *Mater. Sci, and Eng.*, **A286**, 1, 2000

Gleiter, H., Nanocrystalline materials, *Progress Mater. Sci.* **33**, 223, 1989

Gleiter, H., Nanostructured materials : basic concepts and microstructure, *Acta mater.*, **48**, 1, 2000

Valiev, R.Z., Islamgaliev, R.K. and Alexandrov, I.V., Bulk nanostructured materials from severe plastic deformation, *Prog. Mater. Sci.*, **45**, No. 2, 103, 2000

Froes, F.H., Suryananayana, C. and Eliezer, D., Review : Synthesis, properties and applications of titanium aluminides, *J. Mater. Sci.*, **27**, 5113, 1992

Kim, Y.-W., Wagner, R., and Yamaguchi, M., (Eds.) *Gamma Titanium Aluminides*, TMS, Warrendale, Pa, USA, 1995

Camm, G. and Koçak., M., Progress in joining of advanced materials, *Inter. Mater. Rev.,* **43**, No.1, 1, 1998

APPENDIX

Table A.1 Unit conversion factors

Property	To convert B to A multiply by	SI units (A)	Non-SI units (B)	To convert A to B multiply by
Mass	0.4536	Kilogram (kg)	Pound (lb)	2.204
	0.4536×10^{-3}	Tonne	lb	2204
	1.0163	Tonne	U.K. ton	0.9839
Stress	6.894×10^{-3}	Megapascal (MPa) (Meganewtons per square metre)	Pounds force per square inch (psi)	145.04
	15.444	MPa	U.K. tons force per square inch (tsi)	6.475×10^{-2}
	9.8065	MPa	Kilograms per square millimetre ($kg\ mm^{-2}$)	0.10197
Fracture toughness	1.0989	Megapascal (metre)$^{\frac{1}{2}}$ (MPa m$^{\frac{1}{2}}$)	Kilopounds force per square inch (inch)$^{\frac{1}{2}}$ (ksi in$^{\frac{1}{2}}$)	0.91004
Thermal conductivity	4.1868×10^2	Watts per metre per kelvin	Calories per centimetre per second per degree Celsius	2.3885×10^{-3}
Specific heat capacity	4.1868×10^3	Joules per kilogram per kelvin ($J\ kg^{-1}\ K^{-1}$)	Calories per gram per degree Celsius	2.3885×10^{-4}

INDEX